AF531534

# PLANT BIOCHEMISTRY

# PLANT BIOCHEMISTRY

*By*

**Dr. Shubhrata R. Mishra**
*Dept. of Botany*
*Vikram University*
*Ujjain (M.P.)*
*(India)*

**DISCOVERY PUBLISHING HOUSE PVT. LTD.**
**NEW DELHI-110 002**

*Published by:*
**Namit Wasan**
**DISCOVERY PUBLISHING HOUSE PVT. LTD.**
4383/4B, Ansari Road, Darya Ganj
New Delhi-110 002 (India)
*Phone* : +91-11-23279245; 23253475; 43596065
*E-mail* : discoverybooksindia@gmail.com
discoverypublishinghouse@gmail.com
namitwasan9@gmail.com
*web* : www.discoverypublishinggroup.com

*Edition:* 2020

**ISBN: 978-81-8356-629-2**

**Plant Biochemistry**



*Printed at:*
Infinity Imaging Systems
Delhi

# Preface

Biochemistry today has made spectacular progress in unraveling the mysteries of animate nature. This progress has allowed us to gain deeper insight into the principles of vital activity and has to a very significant extent stimulated the development of applied disciplines, especially medicine. The present title *"Plant Biochemistry"* is intended for those who wish to understand living organisms, especially man. Biochemistry is essential for this purpose, but it would be almost impossible for a student to survey on his own the massive body of existing knowledge, constantly augmented by a remarkable torrent of brilliat discoveries. The purpose of the book, then is to organize our knowledge into something that can be comprehended in a relatively short time and still convey a reasonable complete picture of the chemical structure and function of man. Readability without sacrifice of coverage has been a prime goal. An important device in gaining that goal is to keep attention constantly focused on function, with repeated use of rationalization to show that the chemical facts are not isolated, but part of a whole.

Biochemistry has two major goals as a fundamental science, it treats the vital functions from the standpoint of physical chemistry, and as an applied discipline, it points out practical applications for the wealth of scientific knowledge it has acquired. The dual purpose has been, as far as possible, take into account in the writing of this book. We have also tried to summarize our pedagogical experience in teaching biochemistry to students specializing in medicine and pharmacy. The material of this book has been organized according to the principle of functionally in order to trace the close relationship between the functions of the living organism and the structures of its constituents molecules as well as the chemical and physico-chemical process in which they are involved.

The aim of this book is to present a core of biochemical knowledge that is desirable for undergraduate and postgraduate students and also those involved in the field of medical, microbiology, biotechnology and pharmaceutical. Every attempt has been made to keep abrest of the advances in the subject and at the same time to include the fundamentals.

To make the work more comprehensive and informative, the author has consulted many authoritative books, research journals, abstracts, monographs etc. He is grateful to all those great scholars whose work are cited or substantially reproduced.

There can be no claim to originality except in the manner of treatment and much of the information has been obtained from the books and scientific journals available in the different libraries.

The author expresses his thanks to his friends and colleagues whose continue inspirations have initiated him to bring out this book.

The author expresses his gratitude to Mr. Wasan and Staff of M/s Discovery Publishing House Pvt. Ltd., for their whole hearted co-operation in the publication of this book.

In the mean time, the author will remain sincerely responsible for any shortcomings of the book and be grateful to the readers for their suggestions and constructive criticism for the continuous betterment of the book. He takes this opportunity to appeal to the readers to send their suggestions straightway to his publisher.

**Author**

# CONTENTS

# CHAPTER 1

# INTRODUCTION

Plants have played a significant role in maintaining the health and improving the quality of human life for thousands of years. The majority of the earth's inhabitants rely on traditional medicine for their primary health-care needs, and a major part of this therapy involves the use of plants, plant extracts, or their active principles. For centuries, American Indians have utilized a number of native herbs for medicinal purposes. During the past decade, many more Americans have turned to herbal remedies for the treatment of a variety of medical conditions including coughs and colds, insomnia, digestive problems, headache, premenstrual syndrome, prostate problems, anxiety, and depression. The increasing use of herbs in the United States has been fueled by the high cost of drugs, the fear of side effects experienced with conventional drugs, and the desire to take more personal responsibility for one' s health in a way that is perceived as more natural. Some of the more popular herbs in use today include echinacea, ginkgo, garlic, ginseng, goldenseal, saw palmetto, St. John's Wort, chamomile, cranberry, aloe vera, kava, valerian, milk thistle, and feverfew.

Recently, research has validated the usefulness of echinacea for stimulating the immune function; saw palmetto berries and stinging nettle for the treatment of benign prostate hypertrophy; hops, passionflower, and valerian for the treatment of insomnia; St. John's Wort for anxiety and depression; chamomile for its anti-inflammatory effects; cranberry for urinary tract infections; aloe vera for healing burns and wounds; kava for anxiety disorders; milk thistle to protect and restore liver function; and feverfew for the relief of migraine headaches

## UNIQUE HERBAL FLAVOURS AND USES

Herbs and spices can be defined as fragrant, aromatic, or pungent edible plant substances (bark, buds, bulbs, flowers, leaves, fruit, seeds, rhizomes, and roots) that contribute flavour to food and beverages. Herbal teas have become very popular as an alternative to caffeinated beverages. Dried herbs usually smell and taste somewhat differently from their fresh counterparts. The drying process usually causes a loss or change in the volatile oil of the plant material. Hence, the nonvolatile components become concentrated, resulting in the domination of bitter elements. Furthermore, the flavour of most dried herbs diminish with time. Culinary herbs and species have been used since antiquity to flavour and preserve food.

The ability of herbs and spices to delay food spoilage is due in large part to their rich content of antioxidants. The unique flavours associated with different herbs are provided by the aromatic ingredients of their essential oils and oleoresins (mixtures of terpenes such as thymol, menthol, carvone, cineole, etc.), whereas their pungency is due to their alkaloids (such as piperine in black pepper and capsaicin in red pepper). In addition, some herbs such as saffron, paprika, and turmeric are used to add colour to food. With the current emphasis on eating more healthful diets that are low in fat and salt, people are turning to various herbs and spices to flavour their food. This trends in line with the recommendations of various government health agencies and professional health organizations.

The culinary herbal seasonings that can be safely used to enhance the flavour of vegetables, soups, stews, and pasta dishes include basil, caraway, cilantro, coriander, cumin, dill, fennel, ginger, marjoram, oregano, parsley, pepper, rosemary, sage, and thyme. Sometimes these herbs are blended together, as in Italian seasoning, to produce a richer, more pleasing aroma and flavour to the food. During the past two centuries, immigrants to the United States brought with them their own culture and ethnic dishes resulting in the highlighting of new flavours. The present popularity of Italian, Mexican, Indian, and other ethnic dishes has resulted in an increased use of herbal seasonings. For example, oregano is essential in the preparation of Italian and Spanish food and is commonly used to flavour pizza.

Chili pepper is commonly used in many Italian, Mexican, and Indian dishes. Italians use sweet basil for flavouring beans and many of their tomato dishes. Thyme has a prominent place in French cuisine, and rosemary is a common ingredient in Italian and French dishes. For centuries, garlic has been used in Mediterranean, Indian, and Oriental dishes. Many Americans are now using garlic in a variety of dips, vegetable dishes, soups, and some baked goods. Similarly, onions or dehydrated onion can be used to enhance the flavour of most vegetables, salads, soups, gravies, and many entrees. A number of the culinary herbs contain physiologically active compounds that make them useful in treating certain disorders, or generally to promote health.

In this fashion, the herb can provide a typical drug like action. For example, fennel and various mints have been successfully used to treat coughs and colds; ginger is effective as an antiemetic for preventing nausea and vomiting associated with motion sickness as well as for morning sickness; licorice root can help heal stomach and duodenal ulcers; and peppermint is used in the treatment of irritable bowel syndrome. These physiological effects of herbs are parallel to those observed with some conventional medicines, but without the undesirable side effects. Many commonly consumed plant foods, such as broccoli, carrots, citrus, grapes, soy, and tomatoes, contain physiologically active phytochemicals or phytonutrients that promote health and protect against chronic diseases. This has given rise to the term "functional foods" to designate those foods that provide health benefits beyond basic nutrition.

However, the distinction between a basic nutrient and a phytochemical is increasingly blurred. This situation is highlighted when substances such as soluble fiber and linolenic acid provide a variety of health-promoting properties (see Table 1 ), in addition to their traditional roles in normal digestion and metabolism. This has led some people to look on food not only as a source of nutrients but also as medicine.

## HERBS FOR CARDIOVASCULAR PROBLEMS

A plant-based diet, rich in whole grains, fruits, vegetables, and legumes, modest in nuts, and low in saturated fat, along with a regular aerobic exercise program, is recommended for anyone with an elevated risk of cardiovascular disease. In addition, there are herbs that appear to provide help for persons with hyperlipidemia, an abnormal tendency to form blood clots, impaired blood flow, or other cardiovascular problems. Garlic *(Allium sativum* L.) has been used therapeutically for many centuries.

The compound producing much of the activity of garlic is allicin, which is released when intact cells of a clove are cut or crushed. Allicin inhibits the growth of a wide variety of bacteria, molds, yeasts (including *Candida),* and viruses. The regular use of garlic can also be useful in lowering the risk of heart attacks and strokes, as it lowers both total and low-density lipoprotein (LDL) cholesterol levels and triglyceride levels, without affecting high-density lipoprotein (HDL) cholesterol levels. On average, consuming one-half to one clove of garlic per day for 3-6 mo reduces hypercholesterolemia by about 10% of its initial value. Garlic also increases fibrinolytic activity and inhibits platelet aggregation due in part to the presence of ajoenes, allyl methyl trisulfide, vinyldithins, and other sulfur compounds produced from the breakdown of allicin.

Different forms of garlic exhibit various levels of activity. The odor-modified garlic extract (Kyolic) has been found to be just as effective as fresh garlic for lowering blood cholesterol levels. The use of enteric-coated garlic pills, which dissolve in the intestinal tract, are another way to cut down on odor problems while still enjoying the benefits of garlic. However, a steam-distilled garlic oil preparation was recently found to be inactive when fed for 3 mo to hypercholesterolemic patients. The results of a meta-analysis suggest that garlic may also be useful for patients with mild hypertension. Onions *(Allium cepa* L.) contain many compounds that are identical or similar to those found in garlic.

**Table 1.1. Commonly Used Health-Promoting Herbs, Their Functions and Phytochemical Content**

| *Physiological function* | *Herb with desired function* | *Active phytochemicals* |
|---|---|---|
| Lowers total and LDL cholesterol | Garlic | Disulfides |
| | Psyllium | Soluble fiber |
| | Flaxseed | Soluble fiber, phytosterols |
| | Fenugreek | Soluble fiber, saponins |
| | Gugulipid | Guggulsterones |
| | Lemon grass oil | Terpenoids |
| Inhibits LDL oxidation | Green tea | Catechines |
| | Black tea | Theaflavins |
| | Licorice root | Glabridin |
| | Grape seed | Proanthocyanidins |
| | Pine bark | Pycnogenols |
| Inhibits blood clots | Garlic | Ajoenes, Vinyldithiins, diallyl trisulfide |
| | Onions | Alpha-sulfinyl disulfides |
| | Flaxseed | Linolenic acid |
| | Ginger | Diterpene dialdehydes |
| | Asian ginseng | |
| | Hawthorn | Proanthocyanidins |
| | Evening primrose oil | Linolenic acid |

| *Physiological function* | *Herb with desired function* | *Active phytochemicals* |
|---|---|---|
| Improves circulation or vascular function/ arterial compliance | Ginkgo biloba | Flavonoids, terpenoids |
| | Hawthorn | Flavonoids |
| | Horse chestnut | Escin |
| | Red clover | Isoflavones |
| Improves glycemic control or insulin utilization | Bitter melon | Steroids, polypeptide |
| | Fenugreek | Soluble fiber |
| | Gurmar | |
| | Flaxseed | Soluble fiber |
| | Psyllium | Soluble fiber |
| | Asian ginseng | |
| | Cinnamon | |
| Enhances immune function | Echinacea | Isobutylamides, flavonoids |
| | Garlic | |
| | Cat's claw | Flavonoids, terpenoids |
| | Licorice root | Glycyrrhizin, chalcones |
| | Astragalus | Saponins |
| | Black current seed | |
| Cancer chemopı ,ventive activity | Garlic | Sulfides, disulfides, and trisulfides |
| | Onions, chives, leeks | Disulfides and trisulfides, |
| | Labiatae herbs (mint family) | Terpenoids, flavonoids, ursolic acid |
| | Umbelliferous herbs (carrot family) | Coumarins, phthalides, terpenoids, polyacetylenes |
| | Licorice root | Glycyrrhizin, chalcones |
| | Green tea | Catechins |
| | Flax | Lignans |
| | Ginger | Curcuminoids, gingerols, diarylheptanoids |
| | Turmeric | Curcuminoids |
| | Asian ginseng | Ginsenosides |

However, garlic is considered more potent because it contains about three times the level of sulfur compounds found in onions. Onions are considered anticlotting agents because they possess substances with fibrinolytic activity and can suppress platelet aggregation. A whole family of sulfinyl disulfides isolated from onions have been shown to strongly inhibit platelet aggregation. However, onion consumption has not been linked to modifications of the plasma lipid profile. Diets rich in soluble fiber are also known to influence the lipid profile. Psyllium *(Plantago psyllium)*, the major component of Metamucil, is a rich source of soluble fiber in the diet and hypercholesterolemic patients have experienced decreases in total and LDL cholesterol levels of about 10-15 mg/dL (0.26-0.4 mmol/L), respectively.

In a meta-analysis of 12 studies hypercholesterolemic adults experienced decreases in total cholesterol and LDL cholesterol levels of 0.31 and 0.35 mmol/L (5% and 9%), respectively, without changes in HDL cholesterol levels after consumption of a psyllium-enriched cereal for an average of 42 *d*. The regular use of ground flaxseed *(Linum usitatissimum)* can also lower both total cholesterol and LDL cholesterol levels about 10% as well as producing a substantial decrease in platelet aggregation, without altering HDL cholesterol and triglyceride levels.

Flax has a very low saturated fat content, and a high content of polyunsaturated fat and phytosterols, in addition to its soluble fiber content. Other herbal products have also been shown to exert effects on cardiovascular risk factors. Blood cholesterol levels may be reduced by fenugreek *(Trigonella foenum-graecum)*, gugulipid, the oils of lemon grass *(Cymbopogon citratus)* and evening primrose (*Oenothera biennis*) and by the terpenoids found in *Labiatae* (mint family) and *Umbelliferae* (parsley family) herbs.

Asian ginseng *(Panax ginseng)* is reported to inhibit platelet aggregation and increase blood clotting time. Many herbs also contain a variety of antioxidant phenolic compounds such as caffeic acid, ferulic acid, and ellagic acid which can inhibit atherosclerosis. Flavonoids are plant pigments responsible for the colours of flowers, fruits, and some levels. Flavonoids have extensive biological properties that promote human health and help reduce the risk of disease.

Suggested activities include the extension of the activity of vitamin C, acting as antioxidants, protection of LDL cholesterol from oxidation, inhibition of platelet aggregation, as well as anti-inflammatory and antitumor activity. In the Zutphen study, it was shown that flavonoid intake from fruit, vegetables, and herbs was inversely associated with heart disease mortality, and incidence of heart attack and stroke over a 5-yr and 1 5-yr period, respectively. Those in the highest tertile of flavonoid intake had a 68% lower risk of mortality from heart disease (after adjustment for potential confounders) compared with those in the lowest tertile of flavonoid intake.

Similarly, those in the highest quartile of flavonoid intake had a 73% lower risk of stroke compared with those in the lowest quartile of flavonoid consumption. In a prospective study of postmenopausal women in Iowa, total flavonoid intake was associated with a 38% decreased risk of coronary heart disease (CHD) mortality (for the highest pentile of flavonoid intake vs the lowest) but no association was observed between flavonoid intake and stroke mortality. Among the commonly consumed herbs, there are a number that contain substantial levels of flavonoids. These include chamomile, dandelion, ginkgo, green and black tea, hawthorn, licorice, passionflower, milk thistle, onions, rosemary, sage, and thyme. Anthocyanins are the water-soluble pigments responsible for the red, pink, mauve, purple, blue, and violet colour of most flowers and fruits.

They are useful for the treatment of vascular disorders and symptoms associated with capillary and venous fragility. Grape seeds *(Vitis vinifera* L.) are a good source of proanthocyanidins, polyphenolics that provide protection against LDL oxidation and show good promise for the treatment of vascular disorders such as inadequate circulation. Similar compounds are found in the bark of the French maritime pine, *Pinus maritima,* and are marketed as pycnogenol. An extract of horse chestnut seeds contains escin, a triterpene glycoside, that is also useful for chronic venous insufficiency. *Ginkgo biloba* and hawthorn appear to exert their cardiovascular effects by acting as vasodilators.

*Ginkgo biloba* leaf extract appears to be somewhat effective, especially in geriatric patients, against conditions such as memory loss, dizziness, depression, confusion, and other ailments.

These conditions often respond to the vasodilation and improved cerebral blood flow induced by the Ginkgo extract which contains flavone glycosides and diterpenoids (ginkgolides). The leaves, fruits, and flowers of hawthorn *(Crataegus* spp.) can improve blood flow and have been suggested to improve the pumping capacity of the heart. Hawthorn probably causes dilation of the smooth muscles of the coronary vessels, thereby increasing blood flow and reducing the tendency for angina, Proanthocyanidins, the active principles in the flower heads of hawthorn *(Crataegus oxyacantha)*, can inhibit platelet aggregation. LDL oxidation is believed play a key role in atherosclerosis.

Plants contain a variety of antioxidants that can offer defense against LDL oxidation. LDL cholesterol isolated from 10 normolipidemic subjects, who consumed licorice root extract for 2 wk, was more resistant to oxidation than LDL isolated before the licorice was consumed. When a licorice extract (free of glycyrrhizinic acid) or glabridin, a flavonoid found in licorice, was fed to apo-E-deficient mice, they also experienced a reduced susceptibility of their LDL cholesterol to oxidation and a reduction in the extent of atherosclerotic lesions compared with the control mice. The use of licorice had no effect on either the total or LDL cholesterol levels, and blood coagulation was unaffected.

Because glabridin was found to be less active than the whole licorice extract, it is thought that licorice contains a number of other antioxidants, including licochalcones and other polyphenols. LDL oxidation is also inhibited by tea flavonoids—the catechins from green tea or the theaflavins (catechin dimers) from black tea. Of the catechins, epigallocatechin gallate appears to provide the most protection, whereas the theaflavins exert stronger inhibitory effects than the catechins. Several epidemiological studies have suggested that drinking either green or black tea may lower blood cholesterol and blood pressure levels and provide a degree of protection against cardiovascular disease.

The protective effect of tea may result from the ability of the catechins and similar compounds to reduce intestinal cholesterol absorption, as well as to lower blood coagulability and inhibit proliferation of human aortic smooth muscle cells.

## BLOOD SUGAR MODIFICATION

Diabetes is a disease characterized by elevated blood sugar levels. The unregulated blood sugar may result from either a lack of insulin or a reduction in its effectiveness. Careful dietary habits and regular exercise are essential components in the management of type II diabetes (non-insulin-dependent diabetes mellitus). In addition, there are a few herbs that may be therapeutically useful. These herbs may lower blood glucose levels or improve the body's ability to release and use insulin. Bitter melon or balsam pear *(Momordica charantia)* is a green, cucumber-shaped tropical fruit with a bitter taste and gourdlike bumps that is eaten unripe like a vegetabl.

It is used traditionally throughout India, Sri Lanka, Africa, and the West Indies as a diabetic remedy and is available in the United States in Asian food stores. Clinical trials have established that the use of the bitter melon extract can effectively lower blood sugar levels and improve glucose tolerance in persons with type II diabetes. Bitter melon contains a mixture of steroidal glycosides that have a potent hypoglycemic effect and a polypeptide that mimics insulin activity.

The consumption of fenugreek seeds *(Trigonella foenum-graecuim)* can also lower blood sugar levels in diabetics.

Research in India found that glucose tolerance improved, urinary glucose excretion decreased 70%, and insulin responses were reduced in diabetics after defatted fenugreek was used for 10 *d.* Total serum cholesterol, LDL cholesterol, and triglyceride levels, but not HDL cholesterol, all significantly decreased by about 20% when fenugreek was added to the diet. These changes in blood lipids are advantageous for a diabetic who has hyperlipidemia. Fenugreek, which belongs to the legume family, contains a high level of viscous gum (soluble fiber). A daily use of 25-100 mg of fenugreek seeds could serve as an effective supportive therapy in the management of diabetes. Gurmar *(Gymnema sylvestre),* a native plant of the forests of Iridia, has been effectively used in the management of diabetes mellitus (types I and II).

The leaves of this climbing vine contain certain components that block the sensation of sweetness when applied to the tongue. An extract of the leaves of *Gymnema* reduces insulin requirements (or oral hypoglycemic drug dosage), improves fasting blood glucose levels, and improves blood glucose control by enhancing the action of insulin and possibly by rejuvenating the dysfunctional beta cells of the pancreas. It may also produce lower levels of blood cholesterol and triglycerides. These effects are seen in diabetics only, and not in healthy volunteers. Flaxseed, ginseng, psyllium, cinnamon, and other herbs may also provide benefits to diabetics.

Because flaxseed *(Linum usitatissimum)* is very rich in soluble fiber it is also a candidate for the management of abnormal glucose levels. Subjects consuming bread containing 25% flax seed meal showed an almost 30% improvement in a glucose tolerance test compared with those who ate plain bread. For centuries, ginseng has been used to treat diabetes by practitioners of traditional Chinese medicine. In a double-blind, placebo-controlled study, patients with type II diabetes who took 200 mg of ginseng daily for 8 wk experienced improved fasting blood glucose levels and improved glycated hemoglobin levels. Men with type II diabetes who took 5 g of psyllium *(Plantago psylliutm)* twice a day for 8 wk experienced an 11 % drop in daily blood glucose levels in addition to a 13 % drop in LDL cholesterol levels. An extract from cinnamon has been found to potentiate insulin activity.

It has been suggested that compounds in cinnamon may find use in adult-onset diabetes. Preliminary studies have reported antihyperglycemic activity or improved glucose tolerance from a number of other herbs including garlic, onions, bay leaves, cloves, coriander, cumin, cloves, juniper berries, prickly pear cactus, turmeric, and ivy gourd leaves. Further research is needed to validate these findings and discover if there is any clinical significance to the hypoglycemic effects of these herbs.

## HELP FOR THE IMMUNE SYSTEM

*Echinacea,* licorice, cat's claw, and garlic are some of the herbal products that may help enhance the immune system. In the early 1900s, *Echinacea* (purple coneflower) was the major plant-based antimicrobial medicine in use. With the development of sulfa drugs, the use of *Echinacea* rapidly declined. *Echinacea* is known to promote the activity of lymphocytes, increase phagocytosis, and induce interferon production. *Echinacea* appears to be useful in moderating the symptoms of the common cold, flu, and sore throat. However, not all studies have shown that *Echinacea* extracts significantly decrease the incidence, duration, or severity of colds and respiratory infections.

The immuno-enhancing activity of *Echinacea* is believed to be provided by certain polysaccharides and isobutylamides. Glycyrrhizin, a sweet tasting triterpenoid saponin in licorice root *(Glycyrrhiza glabra* L.), and its aglycone (glycyrrhetinic acid) have been reported to augment interferon activity and natural killer cell activity. The chalcones in licorice also possess antiviral activity against the human immunodeficiency virus (HIV), and glycyrrhizin possesses noticeable anti-inflammatory and antiallergic properties. For over 2000 year the Peruvian Indians have used for medicinal purposes two species of cat's claw, *Uncaria guianesis* and *U. tomentosa.*

Today, preparations of the bark from the root and stalk of these plants are attracting much attention in the West because of their immunostimulant properties and their potential to help fight AIDS and leukemi. Extracts of cat's claw are reported to stimulate T-cells, macrophages, and other components of the immune system. Extracts are also reported to have antimutagenic and anti-inflammatory properties. Recently, blackcurrant seed oil, an oil rich in linolenic acid, has a moderate immune-enhancing effect because of its effect on prostaglandin metabolism. Garlic preparations have also been reported to stimulate the immune system of patients with AIDS.

Garlic can increase the number of helper cells and killer cell activity as well as improving AIDS-related conditions such as diarrhea and fungal and viral infections. Validation of the immunostimulant activities of garlic may be of particular importance for the treatment of AIDS in developing countries, where a lack of hard currency may limit access to Western drugs and medicine.

## HERBS WITH CANCER CHEMOPREVENTIVE ACTIVITY

A number of commonly used herbs have been identified as possessing cancer-protective properties. These include members of the *Allium* sp. (garlic, onions, chives, leeks), members of the mint family (basil, mints, oregano, rosemary, sage, sweet savory, thyme), turmeric, ginger, licorice root, green tea, flax, and members of the *Umbelliferae* (carrot) family, such as anise, caraway, celery, chervil, cilantro, coriander, cumin, dill, fennel, and parsley. Unique cancer chemoprotective phytochemicals have been identified in all of these herbs.

In addition, many herbs contain a variety of phytosterols, terpenoids, flavonoids, saponins, and carotenoids, which also have cancer chemoprotective activity. These beneficial substances act as antioxidants, immune system stimulants, inhibit the formation of DNA adducts from carcinogens, inhibit hormonal actions and metabolic pathways associated with the development of cancer, or induce protective phase I or II detoxification enzymes such as glutathione-*S*-transferase. Examples of phytochemicals that stimulate glutathione-S-transferase activity include the phthalides in the umbelliferous herbs, the sulfides in garlic and onions, curcumin in turmeric and ginger, and terpenoids such as limonene, geraniol, cineole, alpha-pinene, and carvone found in commonly used culinary herbs.

Rosemary and sage contain the antioxidant diterpenoids (rosmanol, carnosol, rosmarinic acid, carnosic acid, epirosmanol, and isorosmanol) and ursolic acid, a triterpenoid with antitumor activity. Garlic *(Allium sativum)* has been shown to reduce the development of bladder, skin, stomach, and colon cancer. In a review of case-control and cohort studies of all types of cancer, 27 out of 34 studies revealed an inverse association between cancer and the consumption of allium vegetables. A prospective study in Iowa revealed that risk of colon cancer was 32% less in those in the highest quartile of garlic consumption compared with those in the lowest quartile.

Garlic can inhibit the formation of nitrosamines, which are potent carcinogens, and also inhibit the formation of DNA adducts. The rich content of sulfides, disulfides and trisulfides in garlic also helps to explain its cancer chemopreventive properties. In China, those in the highest quartile of intake of garlic, onions, and other allium herbs have a risk of stomach cancer that is 40% less than those in the lowest quartile. Case-control studies in Greece have also shown a high consumption of garlic, onion, and other allium herbs to be protective against stomach cancer.

Finally, a Dutch study revealed that cancer in the noncardia section of the stomach for those consuming the highest level of onions (at least half an onion a day) was about 50% lower than that in persons consuming no onions. Flaxseed, turmeric, and ginger have all been suggested to prevent cancer proliferation. Flaxseed *(Linum usitatissimum)* contains a rich supply of lignans. Metabolites of these lignans act as phytoestrogens by binding to estrogen receptors and inhibiting the growth of estrogen-stimulated breast cancer.

Turmeric *(Curcuma longa)* contains phenolic compounds that inhibit cancer development as well as having antimutagenic activity. Turmeric has been shown to suppress the development of stomach, breast, lung and skin tumors. Its activity is largely a result of its content of the antioxidant curcumin. Ginger also contains curcumin in addition to a dozen powerful antioxidant phenolic compounds, known as gingerols and diarylheptanoids. Carotenoid pigments are also effective antioxidants that quench free radicals, provide protection against oxidative damage to cells, and stimulate immune function. Persons with high levels of serum carotenoids typically have a reduced risk of cancer.

Carotenoids are the pigments found in rose hips, paprika, and the green, leafy herbs. However, studies using β-carotene supplements have not provided the expected health benefits, and in some cases have produced adverse effects such as an increased risk of lung cancer and overall mortality in smokers. These findings suggest that although the consumption of carotenoid-rich fruits and vegetables promote health, caution should be exercised in the use of β-carotene supplements. Polyphenolics in green tea *(Camellia sinensis)* are known to possess antimutagenic and anticancer activity.

Some evidence suggests a protective effect of tea against cancer of the stomach and colon. Tumor incidence and average tumor yield in rats with chemically induced colon carcinogenesis was significantly reduced when the rats received (-)-epigallocatechin gallate, a major polyphenolic constituent of green tea. Extracts of both black and green tea have significantly inhibited leukemia and liver tumor cells. Extracts of gotu kolu *(Centella asiatica)* were recently shown to be very effective in killing cultured tumor cells. Centella appears to have selective toxicity against tumor cells, as it lacked toxicity towards human lymphocytes. In follow-up studies, Centella extract more than doubled the life span of mice with tumors, and showed a remarkable lack of toxicity even at high doses.

Centella is also therapeutic for the treatment of ulcers, wounds, and eczema. Recent Korean studies suggest that ginseng *(Panax ginseng)* may also lower the risk of cancer in humans. Ginseng extract and powder has been found to be more effective than fresh sliced ginseng, the juice, or a ginseng tea, in reducing the risk of cancer. In a case-control study, the incidence of human cancer was seen to steadily decrease with duration of ginseng use and total lifetime use of ginseng. Those who had taken ginseng for 1 yr had 36% less risk of cancer than nonusers, whereas those who used ginseng for 5 yr or more had 69% less risk.

Ginseng seems to be most protective against cancer of the ovaries, larynx, pancreas, esophagus, and stomach but less effective against breast, cervical, bladder, and thyroid cancer.

The protective properties of ginseng root are believed to be partly due to its content of ginsenosides, a family of triterpene saponins.

## CONCERNS WITH HERBAL USAGE

Health practitioners, when taking clinical histories, often fail to ask patients, about their use of herbal products. In addition, patients may not volunteer this information because they do not realize its importance. Furthermore, patients may not wish to frustrate the health practitioner by telling them they are self-medicating with botanical materials or herbal supplements. Because herbs can often enhance or negate the effect of a conventional drug, it is important for physicians to know which herbs their patients take, how much, and how often. There are a number of possible drug-herb interactions to be considered.

For example, nonsteroidal anti-inflammatory drugs may negate the usefulness of feverfew in the treatment of migraine headaches; ginkgo, garlic, ginger, and ginseng should not be used concomitantly with warfarin, as they may alter bleeding time; immunostimulants such as *Echinacea* should not be given with immunosuppressants such as corticosteroids; evening primrose oil should not be used with an anticonvulsant, as it may lower the seizure threshold; excessive sedation may result when kava and barbiturates are used together; and licorice may offset the effect of spironolactone. Another problem with the use of herbs or herbal extracts is the fact that the contents of many are not standardized so that the dose of a particular active ingredient is unknown.

The phytochemical content of an herb can vary from plant to plant based on where the herb was grown, the light conditions, and the maturity of the plant when harvested. The level of phytochemicals can vary greatly even between cultivars of the same species. Furthermore, different parts of an herb usually contain different amounts of the active ingredient. Methods of preparation of the herbal extract also influence the activity of the final product for the consumer. The level of ginsenosides has been found to vary as much as 60-fold among 10 common brands of ginseng on the market. Occasionally, there are also some reports of adulteration of herbal products. For example, *Echinacea* is sometimes adulterated with the inactive root of wild quinine, *Parthenium integrifolium,* for reasons of cost.

## CONCLUDING REMARKS

A variety of commonly used herbs containing different phytochemicals have the potential for use in the treatment of chronic diseases. Some of these herbs provide assistance in the treatment of hypercholesterolemia, some provide protection against cancer, some help with blood sugar control, and others are known to stimulate the immune system. Furthermore, a diet in which culinary herbs are generously used to flavour the food will provide a variety of active phytochemicals that promote health and protect against chronic diseases. Herbs have been described as both a friend of physicians and the praise of cooks. Although the discriminate use of some herbal products is safe and some therapeutic benefits may be derived from their proper usage, the indiscriminate or excessive use of herbs can be unsafe.

# CHAPTER 2

# POLYSACCHARIDES OF PLANTS

A number of review articles and book chapters have appeared during the year under report. A section is devoted to plant and algal polysaccharides in the chapter on polysaccharides in a series entitled 'Comprehensive Organic Chemistry. The dielectric properties and d.c. conduction of cellulose are reviewed in a volume devoted to applied fibre science as well as the inter- and intra-crystalline reactions in cellulose fibres. Another book is devoted to some aspects of the properties of biological macromolecules and polyelectrolytes in solution. Information of relevance to this chapter can be found in a volume on complex carbohydrates, part of the series on 'Methods in Enzymology. The bibliography of the crystal structure of polysaccharides has been updated to the end of 1975 and the hemicelluloses of grasses and cereals have been reviewed with respect to isolation, structural features, and occurrence within the plantstuff.

## STARCHES

The $^{13}C$ n.m.r. spectra of the previously reported comb-like derivatives prepared by orthoester- and Helferich-type condensations of D-glucose on to amylose have been recorded. It was known from monomeric studies that the orthoester method favoured the β-D-glycosidic bond, whereas the Helferich method favoured the α-D-glycosidic bond. The spectra confirmed that 4, 6-disubstitution occurred on the D-glucose residues of amylose and that a degree of polymer linearity existed but the spectra also indicated that extensive, if not exclusive, β-D-glucosylation occurred with both condensation methods. Differential thermal analysis has been applied to an examination of the thermal behaviour of starch and foodstuffs.

Potato starch showed an endotherm at *ca.* 65 °C which was not influenced by either the rate of heating, or the concentration and the initial amount of starch. Similar analyses, applied to intact potato, Indian lotus, taro, and sweet potato, were distinct from each other but were similar to the starch from each plant except that the isotherm was shifted by 3-7°C to the higher temperature region. The absorption and circular dihcroism spectra of the blue colour produced by tri-iodide ions in amylose have been characterized. Four fundamental Raman lines were observed for $I_3^-$ bound to amyloses which were not dependent on the DP in the range 20-100 or the excitation wavelength.

Concanavalin A does not react with amylose but does react with the amylose-iodine complex although the complex can be dispersed by shaking. The complex is non-specific as it cannot be displaced by concanavalin A specific sugars. A simplified determination of amylose in milled rice has involved an initial dispersion in dilute alkali. The amylose-iodine complex at pH 7.5 was estimated from the absorption at 630 nm. Ultrafiltration was considered to be a suitable method for a hydrodynamic study of the amylose-iodine complex formation. No partially complexed amylose molecules were present in the permeate and so it was considered that complex formation took place instantaneously in a whole molecule and not successively in different zones of the molecule?

The conformation of amylose in aqueous solution is entirely that of loose and extended helices. The formation of the iodine complex is visualized as a trapping of iodine atoms by the contraction of the loose helical macromolecule into tight and stiff helices. The specific volumes of the crystalline and amorphous portions of starch have been determined. All the values were corrected by extrapolation to zero moisture content. The amorphous portion gave values very similar to those for cellulose, being *ca.* 0.673 for potato, waxy corn, and waxy rice starch and *ca.* 0.684 for sweet potato, tapioca, corn, and wheat starch. Digital oscillator densimeter techniques have been used to determine the density of dispersions of starch, glycogen, dextrin, amylopectin, and amylose in water at 40 °C but at different concentrations. The conformation of amylose in aqueous solution was determined by light-scattering sedimentation-equilibrium measurements and was found to be dependent on the molecular weight. When the molecular weight was outside the 'dissolving gap' ($6.5 \times 10^3 < Mr < 1.6 \times 10^5$) the molecule behaved as a random coil while inside the range the molecules aggregated to a rigid coil. The conformation of the rigid coil was that of a double helix and resembled β-type (retrograded) amylose. When potato starch granules were heated with a limited amount of water (water : starch < 1.5) two endothermic transitions were observed on differential scanning calorimetry.

The lower endothermic transition always occurred at 66°C and was the only transition observed when excess water was present. The temperature of the higher transition was inversely proportional to the water content and decreased in intensity with increasing water content. The appearance of the higher endothermic transition allowed the stoicheiometry for full hydration of 14 water molecules per hexose unit; to be determined. Synthetic amyloses have been prepared from maltopentaose and D-glucose 1-phosphate by the action of potato phosphorylase.

The products were characterized by the analysis of the products of β-amylolysis and were used as standards in the determination of the molecular weight of starches by gel chromatography. To elucidate the most probable types of chelation that exist in the chains of amylose and cellulose, studies have been undertaken on molecular models which represent, in part, the maltose and cellobiose entities in the polymers. These models were some intra- and extra-cyclic acetals of the 2-cyclohexyloxytetra-hydropyran type, which contained primary or secondary hydroxy-groups whose covalent disposition, relative to the acetal function, were analogous to those of OH-2, OH-3′, OH-6, and OH-6′ of amylose and cellulose. Most of the diastereoisomers, containing at least two asymmetrical carbon atoms were separated by gel permeation chromatography and their structures were confirmed by $^1$H and $^{13}$C n.m.r. spectroscopy. It was possible, by *i.r.*, spectroscopy, to observe absorption bands corresponding to chelation in rings containing 5, 6, 7 or 8 members.

In maltose, methyl β-maltoside, and amylose chelation was observed between OH(2) and OH(3′) and possibly between OH(2) and O(1). A Raman spectroscopic study of the amylose-monoglyceride complex, using monostearin as the model compound, indicated that the hydrocarbon conformation inside the amylose helix was ordered as if in a crystalline state.

Three helical turns existed for each hydrocarbon chain. The average chain length of corn starch has been shown to be reduced by γ-irradiation as well as by acidic hydrolysis, the extent being dependent on the dose rate. There was no evidence for the formation of dialdehydes by ring cleavage of the D-glucose residues. The degrees of branching of commercial D-glucose syrups and D-glucose syrups prepared by reverse osmosis have been investigated. The results of methylation and enzymatic degradation analyses showed no significant differences. The acidic and thermal stabilities of unmodified pea starch are destroyed on preparing the acetate or monophosphate derivatives. These properties are similar to those of corn starch.

The thermal stability was not affected on acetylation of a cross-linked pea starch. In this paper, a unique purification step was outlined that removed damaged starch granules along with pea hull fragments, A higher concentration of oval or irregularly shaped granules were found in the starch from *Vicia faba* than corn starch. The granules also had a higher amylose content, gelatinization temperature range, and water-binding capacity, but a lower swelling power and solubility than either wheat or corn starch. Starches have been prepared from finger millet *(Eleusine coracana)* and Foxtail millet *(Setaria italica)* and were characterized by their low solubility. The mature starches were susceptible to α-amylase action.

The amylose contents varied between 15.5 and 17.5%, but successive leaching with water and alkali gave amylose components with different molecular weights and β-amylolysis limits. The results suggested that the amylose fraction was heterogeneous. Recent studies on oat starch showed that the iodine affinity, solubility, and $\lambda_{max}$ of the iodine complex were all lower than for other cereal starches. The concentration of 'fractions between the size of amylose and amylopectin was higher than in wheat and the β-amylolysis limit was lower, indicating that the starch was more highly branched. Flours, prepared from the five legumes, navy, pinto, faba, mung bean, and lentil all had similar starch and moisture contents but the starch content was lower than that of wheat flour.

The legume starch granules are oblong and do not have such a wide variation in size as wheat flour granules. The range of amylose contents was lower than that for wheat starches. The acetylated wheat starch derivative had a higher molecular weight than the legume starches. A method for the separation of the amylose and amylopectin components of cotton leaves, but which should be applicable to other plants, has been reported. After extraction with aqueous dimethylsulphoxide, the amylose component was separated from the amylopectin and cell-wall particles by chromatography on macroporous agarose (Sepharose 2B), while the amylopectin was purified by its interaction with concanavalin A. It is now generally accepted that the digestion of starch by many mammalian species involves, firstly, the action of pancreatic α-amylase to give oligosaccharide end-products which are hydrolysed further to D-glucose by the enzymes of the small intestine brush border.

The enzymology of these reactions has been reviewed with respect to both nutritional and clinical implications. The controlled hydrolysis of corn starch with resin-bound α-amylase has yielded a modified starch of high molecular weight in yields of *ca.* 93% α. The amylose component was hydrolysed preferentially and due to the *exo* nature of the enzyme, oligosaccharides of DP 1-8 were isolated. The modified starch had a lower tendency to retrograde. The hydrolysis of gelatinized modified and unmodified starch with porcine pancreatic α-amylase has been studied α. Modification of the starch with hydroxypropyl and acetate groups reduced the extent of digestion, acetylation being better than hydroxypropylation.

Cross-linking of the unmodified starch with phosphate groups also lowered the extent of digestion. The ability of native and dioxan-extracted starch from potato and maize to bind $Ca^{2+}$ ions influenced their digestibility *in vitro* with pancreatic α-amylase. With maize starch, the digestibility was improved on extraction but lowered by $Ca^{2+}$ ions, whereas the digestibility of potato starch was improved with both native and extracted starch. Scanning electron microscopy showed that numerous pin holes could be observed on the surface of starch granules when attacked by various amylases. It was also noticed that these pores penetrated into the inner layers of the granules during enzymic action and in some granules, a terraced or step-shaped appearance in their inner portions was observed. These internal characteristics are probably indicative of a layered internal structure in the granules.

The other observations from scanning electron microscopy were the presence of striated structures on the surface of starch granules from banana, lily, and lotus which had been attacked by pancreatin. Rapid heating to 200 °*C* and rapid cooling of starch with a 20% moisture content resulted in a product which could be gelatinized at room temperature. The thermally modified starch was a better substrate than the native starch for saccharification with D-glucoamylase. The action of Q-enzyme on a model amylodextrin which contained a single-branch linkage has been studied. It was concluded that the enzymatic process for synthesizing the branch linkage in amylopectin is a random action of the Q-enzyme on a complex, possibly a double helix, formed between two (1→ 4)-linked α-D-glucan chains.

The action pattern predicts a novel arrangement of the unit chains in amylopectin. The intestinal surface membrane from rats contains both (l → 4)-α- and (1 → 6)-α-D-glucosidase (-glucanase) activities and these enzymes have been given the trivial names of 'sucrase' and 'isomaltase', respectively. The concurrent action of the sucrase and isomaltase active sites of the hybrid sucraseisomaltase enzyme complex has been studied on the hexasaccharide (1). Hydrolysis occurred by sequential removal of D-glucose residues from the non-reducing terminal end of the molecule through the pentasaccharide (2), tetrasaccharide (3), maltotriose, maltose, and finally D-glucose. The free isomaltase monomer could hydrolyse each step outlined, but displayed maximal specificity for the hydrolysis of (3) to maltotriose. The free sucrase was active against all 1,4-linked residues but could not hydrolyse (3) to maltotriose. Instead of acting synergystically, the two α-D-glucosidase sites appear to cleave α-limit dextrins by their complementary specificity.

α-D-Gl*cp*-(1 → 4)α-D-Gl*cp*-(1 → 4)α-D-Glcp

1

↓ →

6

α-D-G1*cp*-(1 → 4)-α-D-Gl*cp*-(1 → 4)-D-Gl*cp*

(1)

α-D-Gl*cp*-(1 → 4)-α-D-Gl*cp*

1

↓

6

α-D-Gl*cp*-(1 → 4)-α-D-Gl*cp*-(1 → 4)-D-Gl*cp*

(2)

α-D-Glc*p*
1
↓
6
α-D-Glc*p*-(1 → 4)-α-D-Glc*p*-(1 → 4)-D-Glc*p* → Maltotriose → Maltose → D-Glucose

(3)

The retrogradation of amylose in solutions of enzymatically liquified starch depended on the degree of starch hydrolysis and on the manner of preparing the solution. Even at DE values of 20-30%, retrogradation occurred in solutions of < 5 % and > 50% concentration. The subcellular localization of the starch biosynthetic and degradative enzymes of spinach leaves was carried out by measuring the distribution of the enzymes in a crude chloroplast pellet and in separated components of a protoplast lysate.

The enzymes which were involved in the syntheses of starch were detected in the chloroplasts, whereas some of the enzymes involved in the degradation of starch were mainly in the soluble protein fraction but were also found in the chloroplast. The digestion pattern of the amylase on amylopectin as substrate indicated that the enzyme was an α-amylase from its *endo-iytic* activity, but displayed properties unlike the typical α-amylase isolated from endosperm tissue. A time-sequence analysis of the starch digestion pattern in germinating rice showed that activity, in the initial stage, was localized in the epithelium septum between the scutellum and the endosperm. The breakdown of starch in endosperm tissue began thereafter. Amylase activity could only be detected in the aleurone layer after two days. α-D-Glucopyranose 1, 2-cyclic phosphate is a potent inhibitor of phosphorylasecatalysed starch elongation in potato. The affinity of the enzyme for the cyclic ester is *ca*. Thirty times that of D-glucose 1-phosphate. Under conditions where elongation has been retarded by a factor of three, UDP-D-glucose, ADP-D-glucose, D-glucose 6-phosphate, and D-glucose 2-phosphate cause a rate reduction of less than 10%.

## FRUCTANS

Methylation analysis of the D -fructose polymers from cocksfoot *(Dactylis glomerata)* indicated that both (2 → 1)-β-D-(inulin) and (2 → 6)-β-D-(levan) types of linkages were present. Chemical analysis failed to detect any low molecular weight oligosaccharides during fructan synthesis in the leaves of cocksfoot (*D. glomerata)*. These results suggest that direct transfer of D-fructose to existing fructan polymers takes place. As the percentage dry weight of individual onion bulbs increased, the D-fructan content also increased, with a higher proportion of molecules with higher molecular weight being present. The percentage dry matter and fructan content increased from outer to inner layer of the bulbs.

## CYCLOAMYLOSES

Cycloamyloses have been separated by h.p.l.c. on a μ-Bondapak-carbohydrate column using acetonitrile- water mixtures as eluant. The molecular dynamics of the inclusion complexes formed between cyclohexa-amylose and some aromatic amino-acids and dipeptides have been studied by $^{13}C$ n.m.r. spectroscopy. The forces binding the complexes were found to be weak.

The c.d. spectra of cyclohepta-amylose which had been complexed with 2-substituted naphthalenes were measured at various concentrations of cyclohepta-amylase and temperatures between 10-70 °C. The complex with 2-naphthoxyacetic acid showed 1 : 1 stoicheiometry. The molar ellipticity and thermodynamic parameters were determined and enthalpy and entropy ranges calculated.

The correlation was explained by a cyclohepta-amylose guest molecule interaction where the guest molecule was highly solvated. The induced c.d. spectra of cyclohepta amylose complexes with substituted benzenes confirmed that an axial inclusion was favoured, whereby the long axis of the substituted benzenes is parallel to the axis of the cyclohepta-amylose cavity. The binding forces contributing to the association of cyclohexa- and cyclohepta-amyloses with alcohols in aqueous solution have been determined from a spectrophotometric examination of the inhibitory effect of the various alcohols on the association of the cycloamyloses with Azo dyes.

Hydrophobic and van der Waals interactions were found to be of primary importance in complex formation. As the bulkiness of the alcohol increased, the stability of the cyclohexa-amylose adduct decreased, whereas the stability of the cyclohepta-amylose adduct was enhanced owing to the larger cavity in the cyclohepta-amylose molecule allowing better contact of the alcohol within the cavity. Inclusion complexes of poly-(cyclohepta-amylose) with two separate fluorescent probes have been used as models to measure the pressure effects upon ligand-protein complexes. The data permitted the calculation of compressibility curves, which were in broad agreement with those of liquid aliphatic and aromatic hydrocarbons in the low-pressure range, but indicated a reduced compressibility at higher pressures. Metastable crystals of cyclohepta-amylose complexes with propan-1-ol have been prepared and their structure has been discussed in terms of the membrane diffusion model proposed previously by these authors.

## CELLULOSE

Single crystals of cellulose II have been grown from dilute aqueous solutions of cellulose acetate by deacetylation and consequent precipitation. The crystals had a ribbon-like lamellar appearance and the best results were obtained using methylamine, at 90 °C, as the deacetylating agent. It has been suggested that the transformation from cellulose I to cellulose II during mercerization is the result of a progressive shift of the sheets of cellulose chains within the crystallites of a microfibril from the quarter-staggered relationship in cellulose I to the complete correspondence found in cellulose II. This idea agrees with the observed changes in lateral disorder, cell dimensions, swelling, and *X-ray* diffraction reflections of cellulose fibres.

Such a shift may take place in the transformation of native celluloses with antiparallel structures as well as those with parallel polarity. The dielectric constant and dielectric loss for moist cellulose fibres was measured over a frequency band, a temperature range, and relative humidity range. The dielectric constant increased with increasing frequency and temperature owing to an increase in the rotation and polarization of the flexible part of the fibre. As the relative humidity was increased, the dielectric constant and dielectric loss both increased. This may have been due to the presence of the polar water molecules; the freeing of polar groups; or the freeing of ions in the fibre molecule. The $^{13}C$ n.m.r. spectra of peracetylated cello-oligosaccharides have been recorded and compared with that of cellulose acetate.

All of the signals from cellulose acetate were only observed in the spectra of cellotetraose and cellopentaose peracetates. Some signals were not observed in the spectra of callotetaose and cellopentaose peracetates. Some signals were not observed in the spectra cellobiose and cellotriose peracetates. A bioengineering approach has been used for the separation of cellodextrins. Cellulose was hydrolysed with HCl and a combination of columns of Sephedex G15 and Dowex AG5OW × 4 allowed a one step desalting and separation of the hydrolysate using water as the eluant. The column was stable and did not require to be regenerated after every run. Up to 3 g of cellodextrins could be produced per day. A method for the gel permeation chromatographic analysis of the molecular weight distribution of wood pulp holocellulose as the carbanilate derivative has been applied to red maple *(Acer rubrum)* and loblolly pine *(Pinus taeda)*.

Either the chlorine-ethanolamine or acid-chlorite method could be used to prepare the holocellulose and the derivative was obtained by heating at 80 °C with phenylisocyanate in pyridine. Higher temperatures caused depolymerizatiom. Scanning electron microscopy of cellulose xerogels indicated that a membraneous phase existed with individual pores being present which were separated by thin walls. These pores had a short diameter of 50-90 μm and a long diameter of 80-300 *μm*. The cellulose xerogels were distinct from agarose xerogels but were similar to chitin xerogels which had been chemically prepared from chitosan by *N*-acetylation. Dye fixation on modified celluloses depended on the pH, temperature, time of thermofixation, type and concentration of the dye used. Diffusion data have been obtained for helium, nitrogen, oxygen, and carbon dioxide gases through membranes of the acetate derivative of cellulose.

The thermodynamic data have been evaluated and the mechanism of diffusion was discussed. To assist with the statistics of cross-linking, the relationship between relative reactivity and accessibility of the hydroxy groups in cellulose have been determined. When epichlorhydrin was used as the cross-linking agent, the reactivity was in the order OH-3 > OH-6>OH-2. Considerable interest has been shown in the use of anthraquinone in wood pulping operations. The optimal yield of pulp and pulp viscosity at a given Kappa number was obtained *with 0.25 M-NaOH*. The rate of delignification was increased by the addition of anthraquinone (38 mg $1^{-1}$).

Reducing end groups of hydrocelluloses were stabilized more effectively against degradation during alkaline cooking by the presence of anthraquinone 2-sulphate than by anthraquinone itself. This is probably due both to the better solubility and oxidation potential of the sulphate. Neither anthraquinone or naphthaquinone had any effect on the cooking of wood meal with sodium hydrogen carbonate in the presence of oxygen, but the presence of amines did retard the rate.

Pretreatment of wood meal in the presence of oxygen with anthraquinone gave an increased yield of pulp on NaOH cooking. The rates of delignification and depolymerization were increased as well as the stabilization of the carbohydrate towards an endwise β-alkoxy elimination reaction. The presence of magnesium hydroxide retarded the rate of depolymerization and delignification. During the pulping of the tracheid cell wall from the sapwood of *Pinus sylvestris* by alkali in the presence of anthraquinone, the rate of delignification of the secondary wall and middle lamella was increased, with the middle lamella being degraded more rapidly than the secondary wall. Low molecular weight products from the alkaline degradation of cellulose have been identified by g.c.-m.s. of their trifluoroacetate derivatives.

Both the residual aqueous phase and the floating oily product were examined and found to contain essentially the same products. These included unsaturated, aliphatic and alicyclic hydrocarbons, aldehydes, ketones, alcohols, and furans. The synthesis of molecules with more than six carbon atoms also occurred. Reductic acid, separated as its lead salt, has been isolatecd from the pyrolysis products of cellulose by reaction at 450 °C in the presence of oxygen. Amylose, maltose, and D-glucose also gave reductones. A mass spectrometer with a data system was required to study the products of the direct pyrolysis of cellulose and cellulose which had been treated with a fire-retardant.

The amount of products evolved at any point during, or throughout, the entire pyrolysis could be analysed. A philosophical discussion of cellulose as a vast and renewable resource and potential source of food has been published. The properties of cellulose and the uses to which it can be put, as well as a discussion on celluloses, which can convert the potential value of the cellulose into reality are presented. An immunochemical approach has been used to study the cellulolytic complex of *Trichoderma reesei* QM91414. *Endo-(*1 → 4)-β-D-glucanase (CM-cellulase) and cellobiohydrolase (Avicelase) activities were separated on diethylaminoethyl-Sepharose CL 6B. The *endo*-(1 → 4)-β-D-glucanase activity was located in a single peak, but the cellobiohydrolase activity was associated with two peaks. Each peak showed several commponents on isoelectric focusing. The digestion of cellulose by anaerobic micro-organisms from a variety of habitats has been investigated. Of the humen organisms, *Ruminococcus albus* was the most effective, being able to digest 10 g cellulose per litre producing 1.4 g dry cells, 2.8 g acetic acid, and 2.6 g ethanol.

A strain from a compost, similar to *Clostridium thermocellum,* digested 4 g cellulose per day giving 0.8 g dry cells, 0.6 g acetic acid, and 0.9 g ethanol. In studies on the biosynthesis of cell-wall polysaccharides in cultured carrot cells, D-glucose was only incorporated into the neutral sugar residues, particularly cellulose, whereas *myo*-inositol appeared in the uronic acid and pentose residues. An increase in cellulose content was observed during the thickening of the cell walls of parenchymous tissue of *Discorea dumetorum* tubers during storage. The $^{18}O$ : $^{16}O$ ratio of the cellulose from two sets of wheat plants, grown under conditions similar in all respects except for a large difference in the $^{18}O$ : $^{16}O$ ratio of the carbon dioxide supplied to them, only differed by a small amount.

The difference in ratio of the cellulose was similar to the $^{18}O$ : $^{16}O$ ratio of the water present in the plants. These results indicate that the oxygen of the carbon dioxide undergoes complete exchange with the oxygen of the water in the plant during cellulose synthesis and that the $^{18}O$ : $^{16}O$ ratio of the water within the plant is the primary influence on the $^{18}O$ : $^{16}O$ ratio of cellulose in terrestrial plants.

## HEMICELLULOSES

Two different L-arabinans have been isolated from azuki beans *(Phaseolus radiatus)* by alkaline extraction and separated by gel chromatography. No other sugar residues were present and, from structural analysis, the polymers were composed of chains of 1,5-linked L-arabinofuranose residues with branch points at O-2 and/or O-3. These branches were composed of non-reducing -r. terminal L-arabinofuranose residues and there were more (1 → 2)-linkages than (1 → 3)-linkages.

The polysaccharides differed in their molecular weights and solubilities in aqueous ethanol. An endo-L-arabinanase and two α-L-arabinofuranosidases are produced when *Bacillus subtilis* is grown in a medium containing sugar beet arabinan. The *endo*-L-arabinanase was homogeneous by sodium dodecyl sulphate-polyacryl-amide gel electrophoresis, had a molecular weight of $3.2 \times 10^4$, and a pH optimum of 6.0. It is able to release residues of L-arabinose from suspension-cultured sycamore cell walls. α-L-Arabinofuranosidase activity was determined in 306 species of *Actinomycetales*. *Streptomyces* species produced these enzymes at a faster rate than *Ascomycetes* or *Basidiomycetes* species.

Activity appeared in the *Streptomyces* species after 2-4 days and, although the pH of the medium was *ca.* 8.5, the pH optimum of the enzyme was *ca.* 6.0. An L-arabino-D-galactan, in 2% yield, has been isolated from the fruits of *Dillenia indica*. From the results of partial acid hydrolysis and methylation analyses, when 2,6-di-O-methyl-D-galactose, 2,3,6-tri-O-methyl-D-galactose, 2,3,4,6-tetra-O-methyl-D-galactose, and 2,3,5-tri-O-methyl-L-arabinose residues in the molar ratio 7:3:1:8 were identified, it was concluded that the poly-saccharide was a (1 → 4)- β D-galactan that contained L-arabinofuranose residues linked at O-3 of some of the D-galactose residues in the main chain. The occurrence, isolation, chemistry, and physico-chemistry of the plant L-arabino-D-galactans and their protein conjugates have been reviewed.

The structural relationship between those found in woods, gum exudates, and plant callus cells, and whole tissue is discussed and the nature of the glycoproteins compared with L-arabinose- and D-galactose-containing cell-wall glycoproteins. The possible biological role of these materials is also considered. Structural studies have been carried out on the L-arabino-D-galactan-protein which is the major component of the mucilage of *Gladiolus* style. It was isolated by affinity chromatography on the D-galactose-binding lectin from the clam coupled to macroporous agarose. The preparation contained 3% protein and a D-galactose: L-arabinose molar ratio of 6:1. Carbohydrate and protein were co-eluted even on chromatography in urea buffers.

Ultracentrifugation showed the material to be polydisperse in the molecular weight range 1.5-4.0 x $10^5$. From the results of methylation, partial acid and enzymatic hydrolysis analyses, the structure was confirmed as that of a (1 → 3)-D-galactan with (1 → 6)-D-galactose side chains, some of which contain terminal non-reducing α-L-arabinofuranosyl residues. Removal of these residues reduced the solubility of the complex. A new water-soluble polysaccharide has been isolated from the seeds of *Cassia multifuga,* and the polysaccharide, on methylation analysis, gave residues of 2,3-di-O-methyl-D-galactose, 2,3,6-tri-*O*-methyl-D-galactose, 2,3,4,6-tetra-*O*-methyl-D-galactose, 2,3-di-*O*-methyl-D-mannose, 2-*0*-methyl-D-xylose, 2,3-di-*O*-methyl-D-xylose, and 2,3,4-tri-*O*-methyl-D-xylose in the molar ratio 2:4:4:2:1:2:1, respectively.

Partial acid hydrolysis gave the oligosaccharides (4)-(7) and periodate oxidation studies indicated that 32.4% of the monosaccharide residues were end groups. The polysaccharide was therefore considered to have a highly branched structure. The alkali-extracted gum from oats contained a β-D-glucan as the major component. Some endogenous D-glucanases survived the alkaline treatment so deactivation of these enzymes increased the yield of the β-D-glucan with a corresponding increase in the viscosity of the extract. The high viscosity gums had a high

α-D-Galp-(1 → 6)-D-Gal

(4)

α-D-Galp-(1 →6)-D-Man

(5)

β-D-Galp-(1 → 6)-D-Xyl

(6)

β-D-Xylp-(1 → 3)-D-Xyl

(7)

limiting viscosity number (17-18 $dLg^{-1}$) in both dimethylsulphoxide and 7 M urea. Low concentrations of the dyes Calcofluor White M2R, Tinopol ia CBS-X, or Congo Red caused preferential precipitation of the β-D-glucan from an aqueous solution of oat gum in the presence of protein, pentosan, and starch.

The presence of salt suppressed the precipitation while higher temperatures (>25 °C) dissolved the precipitate. The precipitation seemed to be specific for (1 → 4)-β-D-glucose residues. A(1 → 3 )-β-D-glucan occurs in cotton fibres. Its solubility varies with fibre age, a proportion being soluble in cold water, while some cannot even be extracted with alkali. Enzymatic digestion studies indicated that all of the linkages are of the β-configuration, whereas methylation analysis revealed that the predominant linkage is *(1→* 3) although a small proportion of (1 → 6)-linkages are also present.

The timing of the deposition of the (1 → 3)-β-D-glucan during fibre development coincides closely with the onset of secondary wall-cellulose synthesis. In the hemicellulosic fraction from rice endosperm cell walls, methylation and Smith degradation analyses as well as cellulose fragmentation studies have confirmed the presence of a mixed (1 → 3), (1 → 4)-β-D-glucan in which blocks of (1 → 3)-β-D-glucose residues occur. D-Glucose-containing polysaccharides have been isolated from *Colletotrichum trifoli, C. destructivwn,* and *C. lindemuthianum* , which elicit browning and phytoalexin production in *Phaseolus vulgaris.*

*A* high molecular weight D-glucan was isolated from each organism and this polysaccharide intiated the browning symptoms at a concentration of $10^{-7}$g D-glucose equivalents per hypocotyl. The molecular weight distribution of sinistrin, a polysaccharide isolated from the bulbs of the red squill and which has been used for renal clearance, covered the range 8.0 x $10^2$ to 1.6 x $10^4$. Viscosity measurements indicated a highly branched molecule in which residues of D-glucose were present at the reducing end of most molecules. The fine structures of curdlan and paramylon, the (1 → 3)-β-D-glucans from *Alcaligenes faecalis* and *Euglena gracilis,* respectively, have been determined, and it was found that they resembled the native-regenerated cellulose system with curdlan being the regenerated polysaccharide and paramylon the native one.

From the negative birefringence of annealed fibres and positive birefringence of paramylon granules, a tangential disposition of these chains in the granules may be concluded. A new α-D-glucan, elaborated by *Elsinoe leucospila,* has a structure based on the repeating sequence (8).

...[ → 3)-α-D-Glc*p*-(1 → 4)-α-D-Glc*p*-(1 → 3)-α-D-Glc*p*-(1 → $]_n$

(8)

The D-glucan produced by *Botrytis cinerea,* when grown on grape berries, has a structure based on a backbone of (1 → 3)-β-D-glucopyranose residues with (1 → 6)-β-D-glucopyranose side chains. The D-glucan is not produced from cellulose, but only from simple hexoses.

Two novel tetrasaccharides, (9) and (10), have been isolated and characterized from cotton seed.

β-D-Fru*f*
2
↓
1
α-D-Gal*p*-(1 → 6)-α-D-Glc*p*-(1 ↔ 2)-β-D-Fru*f*

(9)

α-D-Gal*p*
1
↓
6
α-D-Gal*p*-(1 → 4)-α-D-Glc*p*-(1 ↔ 2)-β-D-Fru*f*

(10)

β-D-Glc*p*-(1 → 4)-D-Glc*p*
6
↑
1
α-D-Xyl*p*

(11)

β-D-Glc*p*-(1 → 4)-β-D-Glc*p*-(1 → 4)-D-Glc*p*
6 6
↑ ↑
1 1
α-D-Xyl*p* α-D-Xyl*p*

(12)

The oligosaccharides (11) and (12) have been isolated from the degradation products obtained from cellulose treatment of the L-arabino-D-xylo-D-glucan from tobacco leaves. Their structures were determined from the results of methylation analysis both before and after reduction with borohydride ion as well as by $^{1}H$ and $^{13}C$ n.m.r. spectroscopies. Present in the hemicellulose fraction from the endosperm cell walls of rice was a D-galacto-D-xylo-D-glucan.

This polysaccharide was based on a (1 → 4)-β-D-glucan backbone to which were attached residues of D-xylose and D-galactose at O-6. The modes of action of β-D-glucanases of *Penicillium notatum* on galactoxylo-D-glucans from various sources have provided a means of establishing their structure. The main chains are based on (1 → 4)-β-D-glucans although an occasional (1 → 3)----linkage may also be present. The side chains are nearly equimolar amounts of β-D-xylopyranosyl and O-β-D-galactopyranosyl-(l → 2)-β-D-xylopyranosyl units linked at O-6 of some of the main chain D-glucose residues. The regions with disaccharide substitution were inaccessible to the enzyme, supposedly confirming the irregularity of this substitution. Auxin only induced growth in segments of rice coleoptile when the cell wall was rich in non-cellulosic D-glucose residues. The effect decreased with age and with the total amount of non-cellulosic polysaccharide present.

A decrease in the relative amount of non-cellulosic D-glucose and an increase in the content of D-xylose residues were found in the growth cycle of cell walls of rice coleoptiles irrespective of the culture conditions. D-Galacto-D-mannans, which differed in their average molecular weight, polydispersity, and D-mannose: D-galactose molar ratio, have been isolated from seven legumes. The Mark-Houwink relationship between intrinsic viscosity and average molecular weights held for these D-galacto-D-mannans despite the variation in the D-mannose: D-galactose molar ratio. Estimates were made of the polydispersity of the polysaccharides from both the sedimentation coefficient distribution and also from the diffusion coefficient ratios, $D_M/D_A$, and the two methods gave similar results. The apparent viscosities of aqueous solutions of guar and locust bean D-galacto-D-mannans were measured over a range of shear rates both alone and in the presence of D-glucose, sucrose, or D-glucose syrups.

For all of the solutions, the variation of the relative viscosity with shear rate fitted a power law equation. The addition of the sugar solutions had no effect on the non-Newtonian behaviour of the gum solutions. The presence of casein had no effect on the relationship between relative viscosity and shear rate for the two D-galacto-D-mannans but its presence had a marked effect on K-carrageenan. The location of inter-residue cyclic acetals formed following the oxidation of guar D-galacto-D-mannan and other heteroglycans by the periodate ion, has been determined by methylation analysis of the oxidized glycans.

The formation of cyclic hemiacetals led to the protection of hydroxy-groups during methylation in dimethylsulphoxide and their positions were located by the analysis of the resulting mono- and di-methyl ethers. These derivatives were not observed in either the native glycans or the glycans treated sequentially with periodate and borohydride ions. Inter-residue hemiacetals were identified in all glycans and were formed between aldehydo-groups at C-2 or C-3 of the oxidized residues and the hydroxy-groups at C-3 or C-2 of the adjacent unoxidized residues. The selective removal of six 0-substituents from the oxidized residues resulted in a decreased ability of the latter to form inter-residue hemiacetals. An analysis of the types and proportions of the methyl ethers resulting from inter-residue hemiacetal formation could yield further structural information on the glycan. A new D-galacto-D-mannan, isolated from the seeds of *Ipomoea fistulosa* possessed a D-galactose : - D-mannose molar ratio of 3 :10.

The results of methylation and periodate-ion oxidation analyses confirmed that the polysaccharide was based on a main chain of (1 → 4)-D-mannose residues to which were attached single residues of D-galactose at O-6. Five oligosaccharides were found in the hydrolysis products of the polysaccharide from the seeds of *Caisas-Grandis*. There were three disaccharides, a trisaccharide, and a tetrasaccharide, whose structure was confirmed by methylation, periodate ion oxidation, and Smith degradation analyses. The tri- and tetrasaccharides had the unusual structures (13) and (14).

β-D-Galp-(l → 4)-β-D-Galp-(1 → 4)-D-Man

(13)

β-D-Manp-(l → 4)-β-D-Galp-(l → 4)-β-D-Galp-(l → 4)-D-Man

(14)

A number of enzymic complexes containing β-D-mannanase activity were purified by affinity chromatography on D-gluco-D-mannan-derivatized AH-Sepharose. Isoelectric focusing gave multiple protein bands all of which showed β-D-mannanase activity, while a single band was obtained on sodium dodecyl sulphate-polyacrylamide gel electrophoresis. The purified enzymes differed in their physical properties and had wide variations in their ability to hydrolyse highly substituted D-galacto-D-mannans. A series of papers has appeared on the gelation of Konjac D-mannan. On reaction with either sodium hydroxide or sodium carbonate, an induction period occurred before gelation due to the requirement for deacetylation.

The rate of the induced reaction was of the form $V = 4.8[OH^-]^{0.61}$ where *V.* is the reaction rate ($min^{-1}$). The power value of 0.61 suggested that the rate of deacetylation was proportional to the concentration of hydroxide ion and that gelation of the deacetylated molecules was inhibited by the hydroxide ion. The degree of deactylation, calculated from the Arrhenius equation, increased with a decrease in the concentration of D-mannan and an increase in the concentration of hydroxide ion, but was not influenced by temperature or the nature of the cation ($Na^+$ or $K^+$).

Deacetylation was an induced reaction which was essential for gelation and occurred after the equilibrium of the deacetylation and peptization reactions by hydroxide ion. To clarify the macromolecular and chemical properties of Konjac D-mannan, light scattering and viscosity measurements have been carried out on partially methylated derivatives. The weight average molecular weight was found to be 1.0-1.2 × $10^6$ and it was suggested that a random coil configuration applied.

The intrinsic viscosity was related to the molecular weight by a simple mathematical relationship, which meant that the molecular weight could easily be determined from viscosity measurements. The $^{13}C$ n.m.r. spectra of D-gluco-D-manno-oligosaccharides have been recorded. Considerable structural heterogeneity was shown by this method, mainly with respect to the sequence of linkages in the oligosaccharides. Four partially acetylated D-gluco-D-mannans have been isolated from the leaves of *Aloe* vera. The polysaccharides differed in the ratio of D-glucose to D-mannose as well as in their acetyl content. The results of methylation and periodate-ion oxidation analyses showed that the polymers were all linear with only (1 → 4) linkages present. The hemicellulosic fraction of the seed endosperm of groundnuts *(Arachis hypogea)* contains a (1 → 4)-β-D-gluco-D-mannan.

The homogeneity of the polymer was confirmed by electrophoresis, sedimentation, and sugar analysis. Hydrolysis of the methylated polysaccharide yielded 2,3,4,6-tetra-O-methyl-D-glucose and -D-mannose, 2,3,6-tri-O-methyl-D-glucose, and 2,3,6-tri-O-methyl-D-mannose in the molar ratio 1.6 :21.2 : 5.6, respectively. High-resolution electron microscopy of fractions from potassium hydroxide extracts of sprucewood holocellulose showed the presence of D-galacto-D-gluco-D-mannan and 4-0-methyl-D-glucurono-D-xylan in the form of microfibrils with different diameters and flexibility.

The supramolecular elements of the D-mannans from the 5% and 24% KOH extracts were thin fibrils. The longer D-mannan fibril obtained from the 24% KOH extract had a higher molecular weight-Lignin was sometimes present in the fractions. Methyl glycosides of D-xylo-oligosaccharides (linear and branched) and 4-O-methyl-D-glucuronic acid-containing oligosaccharides, closely reflecting the main structural features of native D-xylans, were studied by thermal analysis and pyrolysis-g.c. The number of monomeric units in the oligomers markedly affected the course of active thermal decomposition.

Although the thermal stability increased with increasing numbers of monomeric residues, the ratio of 2-furaldehyde to 2-hydroxy-2-penteno-l,5-lactone produced remained constant. Considerable differences between reducing sugars and glycosides were observed during the course of thermal decomposition and other major differences were also observed when an ionic dehydrating catalyst was added to the pyrolysis sample. The results suggested that the 4-O-methyl-D-glucopyranosyl uronic acid linkage was the most thermally stable in the acetylated 4-O-methyl-D-glucurono-D-xylan, and acetyl groups do not significantly accelerate, the thermal decomposition of the polysaccharide the kinetics of the hydrolysis of D-xylo-oligosaccharides, ranging from di- to penta-saccharide, in dilute sulphuric acid have been studied.

The rate of hydrolysis of the glycosidic bond in 4-*O*-β-D-xylopyranosyl-D-xylose was similar to that of the same bond at the reducing end of the other oligosaccharides and was 1.8 times the rate of each internal bond. The yield of D-xylose from each oligosaccharide could be calculated from this ratio and from the empirical equation which describes the rate constant for the above xylobioser. The hemicellulose component of the endosperm cell walls of rice is mainly composed of L-arabino-D-xylan. This L-arabino-D-xylan is based on the usual (1 → 4)-β-D-xylan backbone to which are attached L-arabinose residues at O-3 of 78% of the D-xylose residues. Alkaline extraction of the husks *of Ispaghula* has yielded a highly branched acidic L-arabino-D-xylan.

The polysaccharide is unusual in containing both (l → 3)- and (1 → 4)-D-xylose residues in the main chain with the majority of residues further substituted at O-2 or O-3 with residues of L-arabinose, D-xylose, or D-galactopyranosyluronic acid-(l → 2)-L-rhamnose. Another acidic polysaccharide, based on a (1 → 4)-D-xylan main chain has been isolated from the seeds of *Ocimum basilicum.* The polysaccharide contained D-xylose, L-arabinose, L-rhamnose, and D-galacturonic acid in the molar ratio 15:9:7:12 and the branch points were present at both O-2 and O-3 of the D-xylose residues in the main chain. Two oligosaccharide sequences, (15) and (16), were isolated from the side chains.

D-Gal*p*A-1 → L-Rha*p*-(1 → 4 or 5)-L-Ara-1 → D-Xyl*p*-(1 → 3)-L-Rha

(15)

α-D-Gal*p*A
1
↓
3 or 2
D-Gal*p*A-(1 → 4)-D-Xyl*p*-(l → 2 or 3)-L-Ara

(16)

Exhaustive extraction of the milled-wood lignin fraction from beechwood was unable to remove small amounts of carbohydrate. N.m.r. spectroscopy of model compounds as well as the acetylated milled-wood lignin confirmed that the carbohydrate was a xylan and was covalently bonded to the lignin. Both milled-wood lignin and lignin obtained after exhaustive cellulase treatment of Douglas fir and red alder sapwood contained covalently bound lignin. The method of pulping, duration of pulping, type of polysaccharide, and method of isolation all affected the lignin-carbohydrate bond. During studies on the etherification reactions of vanillyl and veratryl alcohols, it was observed that lignin-carbo-hydrate bonds could be formed, subsequent to lignification, through esterification of the p-hydroxybenzyl alcohol groups in the lignin polymer. The carbohydrates present in groundnuts have been reviewed. As well as starch and cellulose, the more abundant polysaccharides were D-gluco-D-mannans and highly-branched D-xylans. Higher levels of non-starchy polysaccharides were present in legume flours than in wheat flour.

The distribution of mono saccharide residues and consequently the polysaccharides present varied considerably between the different types of legume although D-xylans were present in each species examined except field bean.

## PECTINS

The effect of the degree of polymerization of D-galacturonic acid oligosaccharides (DP 2-5) and D-galacturonans on their c.d. spectra has been examined in aqueous solution. The chiroptic properties were correlated with the degree of polymerization in terms of the optical superposition of monomeric unit increments. The interpretation of the results led to the conclusion that the conformation of D-galacturonans in solution is close to that of a helical structure. Above the critical concentration of pectin (0.02 g $1^{-1}$ ) in 0.1 m-NaCl, the addition of an increased amount of $Ca^{2+}$ ion resulted in an apparent increase in molecular size owing to the formation of intermolecular cross-linkages.

Below the critical concentration, any increase in the apparent molecular weight was slight and due to a stretching of the polymer chains with expansion to a rod-like shape although the amount of cross-linking did increase with time. When the concentration was as low as 0.004 g $1^{-1}$, no increase in molecular weight was observed and the chains contracted owing to intramolecular cross-linking. Grapefruit pectic polysaccharides have been shown to contain D-galacturonic acid residues (78%), and amongst the neutral sugar residues found were D-galactose, D-glucose, D-xylose, L-arabinose, and D-mannose of which D-galactose accounted For about half.

Oxidation of the complex with D-galactose oxidase followed by reduction with potassium borotritide gave a radio-labelled product in which about 90% of the incorporated tritium was located in D-galactose residues. D-Galacturonic acid oligosaccharides, with degrees of polymerization from 2 to 9 have been isolated in 54.8% yield by ion-exchange chromatography from an enzymic digest of D-galacturonan. The trisaccharide was the major product. Improved purification could be gained by t.l.c. on cellulose-coated plates. The recovery and degradation of sugars from potato pectin fractions by hydrolysis with 2 M trifluoroacetic acid has indicated that between 61 and 96% of each sugar residue was recovered, depending on the monosaccharide residue.

Lower recoveries were obtained from intact onion cell walls. Various Commercial pectins and pectinesterase-treated pectins have been fractionated according to their degree of esterification in the range 6-70%. These fractions were tested for their ability to destabilize the 'cloud' in orange juice. The juice was clarified by fractions with degrees of esterification between 6 and 14%, but not at all by those with degrees of esterification greater than 21%. A new dialdose dianhydride derivative has been obtained from an acid hydrolysate of the water-soluble pectic fraction of wobaku wood by successive treatment with methanolic hydrogen chloride and acetic anhydride-pyridine.

The compound was the 1,2:l',2' dianhydride of 3,4-di-*O*-acetyl-β-L-rhamno-pyranose and methyl 3,4-di-*O*-acetyl-α-D-galactopyranosyluronate. This is the first report of a dianhydride containing two different sugar residues. The role of secondary alcoholic groups in the degradation of D-galacturonan by the endo-D-galacturonanase of *A spergillus niger* has been investigated using 0-acetyl derivatives of pectic acid. The initial rate depended on the numbers of free hydroxy-groups on C-2 and C-3, with substitution giving an enhanced value for the Michaelis constant.

The rate of degradation decreased in the presence of acetyl groups according to the relationship for competitive inhibition, and it was concluded that the lowered rate of reaction and extent of degradation was due to the lowered affinity of the enzyme for the substrate caused by modifying the groups which are required for enzyme-substrate binding. From the rates of hydrolysis of D-galacturonic acid oligosaccharides from pure heptamer to trimer, as well as mixed higher oligomers, it was suggested that the active site of the endo-D-galacturonanase from *Aspergillus niger* could accommodate at least seven units.

The release of end units from larger oligomers also suggested that the interaction was with four sugar units from the non-reducing side of the bond to be hydrolysed and three sugar units from the reducing side. The content of D-galacturonic acid, D-galactose, and L-arabinose residues of tomato cell walls declined during ripening. A water-soluble D-galacturonan found in isolated walls probably arises from the action of D-galacturonanase *in vivo*. This polysaccharide and that from the walls of mature green fruit, after incubation with galacturonanase, contained fewer neutral sugar residues than intact walls.

The loss of D-galactose and L-arabinose residues on ripening was not related to the loss of D-galacturonan.The predominant change that took place in the cell walls of apples and pears on ripening was a dissolution of the middle lamella. This change could be brought about in unripe pears by application of solutions of D-galacturonanase and cellulase, while the change in unripe apples could be initiated by D-galacturonanase alone. Ripe strawberries, artificially inoculated with various fungi, contained endo-D-galacturonanase and pectinesterase activities in their culture filtrates, whereas uninfected strawberries had exo-D-galacturonanase and pectinesterase activities.

No other pectin-degrading enzymes were detected in either sound or infected berries. Since the disintegration could be substantially reduced by inhibition of the D-galacturonanase activity, it was concluded that the D-galacturonanases secreted by the fungi were primarily responsible for the breakdown in sulphited strawberries. Sodium pectate can be cross-linked with epichlorhydrin in an alkaline medium.

Different degrees of cross-linking can be obtained by varying the concentration of epichlorhydrin. These materials could be used for the separation of the two pectinesterase isoenzymes from oranges. Both components had a 7-8 fold increase in activity and could saponify 93% of the pectin. The D-galacturonanase activity secreted by an aggressive strain of *Colletotrichum lagenarium* can be separated into two proteins.These proteins differ in their molecular weight and in their mode of action on the hydrolysis of D-galacturonans. The anti-inflammatory drugs benzydarnine and ketoprofen formed co-precipitates with pectin.

A Langmuir-type of binding isotherm was obtained on equilibrium dialysis and this slow dissolution of the precipitate suggested that binding may help to reduce the adverse effects of these drugs to the stomach on oral administration.

## ACIDIC GUMS

The acidic polysaccharide from peach gum and macromolecular products obtained from it by mild acidic hydrolysis have been characterized by their molecular weight and composition.

The single-ion activity coefficient of $Ca^{2+}$ ions bound to the polysaccharide and to the corresponding aldobiouronic acids was measured and the c.d. spectra of both $K^+$ and $Ca^{2+}$ salts were recorded. $Ca^{2+}$ ions were bound to the carboxy-groups by electrostatic bonds in disperse solutions. There is a close arrangement of uronic acid residues which does not under-go alteration with a stepwise degradation of the macromolecules.

It was concluded that the main chain of the polysaccharide is substituted in long segments with uronic acid residues at least on each second D-galactose unit. The enzymic hydrolysis of peach gum polysaccharide has yielded the oligosaccharides (17)-(20), whose structures were confirmed by g.c.-m.s. of their partially methylated alditol acetates.

β-D-Galp-(1 → 6)-D-Gal

(17)

α-D-Manp-(1 → 3)-D-Gal

(18)

β-D-Galp-(1 → 3)-β-D-Galp-(1 → 6)-D-Gal

(19)

β-D-Galp-(1 → 6)-β-D-Galp-(1 → 3)-β-D-Galp-(1 → 6)-D-Gal

(20)

The Hofman degradation has been applied to mangle gum, and a variety of fragments were obtained on borohydride ion reduction and acidic hydrolysis of the product. The same fragments were obtained on the Barry degradation of the Hofman-degraded product.

The results supported the earlier conclusions about the structure of the gum. An analytical study of eight *Acacia* gums (four each from the *Gummiferae* and *Vulgares* series) has confirmed the main chemotaxonomic differences between the species that have been previously reported. During the methylation analysis of *Combretum nigricans* gum by the Hakomori procedure, losses of D-galacturonic acid and L-rhamnose residues were observed, which did not occur in the Haworth or a modified-Haworth procedure. However, during the Haworth procedure, *C. nigricans* gum was degraded to a greater extent than the gum from *A cacia seyal.* A new acidic polysaccharide, isolated from the bark of *Pterospermum aceri-foli,* contains D-galacturonic acid, D-galactose, and L-rhamnose in the molar ratio 5:3:3.

The oligosaccharide (21) was isolated from a partial acid hydrolysate and its structure was confirmed by methylation and periodate oxidation analyses. The polysaccharide of Junsai mucilage *(Brasenia schreberi)* contained D-galactose, D-glucuronic acid, L-fucose, D-mannose, L-rhamnose, D-xylose, and L-arabinose in decreasing order of magnitude. The actual composition varied between different samples. From the results of methylation analysis, the

α-D-Gal*p*A-(1 → 2)-L-Rha*p*-(1 → 4)-D-Gal

(21)

precise structure was difficult to establish but it was found to be highly branchedr. Partial acid hydrolysis of paniculatan, the mucous polysaccharide from the inner bark of *Hydrangea paniculata,* led to the isolation of five acidic oligosaccharides.

The structure of each oligosaccharide was established by methylation analysis although each has previously been isolated from other sources. A possible structure of the polysaccharide was proposed. The polysaccharide had a molecular weight of 5.46 X $10^5$. There were *0*-acetyl groups present on 40% of the L-rhamnose residues. Complete deacetylation of the polysaccharide in 0.1 M alkali led to a reduced molecular weight of 4.2 ×$10^4$. A mucilage, Abelmoschus-mucilage M, isolated from the roots of *Abelmoschus maniliot,* appeared homogeneous.

The polysaccharide contained L-rhamnose, D-galacturonic acid, and D-glucuronic acid in the molar ratio 1.2 :1.0:1.0 and had a molecular weight of 2.53 × $10^4$. Reduction of the carboxy-groups followed by methylation analysis suggested that the main chain comprised (1 → 2)-L-rhamnose and (1 → 4)-D-galacturonic acid residues. These latter residues were substituted at *O*-3 by D-glucuronic acid residues. These results were confirmed by the isolation, *inter alia,* of the oligosaccharide (22) from a partial acid hydrolysate of the polysaccharide.

α-D-Gal*p*A-(1 → 2)-α-L-Rha*p*-(1 → 4)-α-D-Gal*p*A-(1 → 2)-L-Rha

3 ↑ 1 β-D-Glc*p*A (on first Gal*p*A)

3 ↑ 1 β-D-Glc*p*A (on second Gal*p*A)

(22)

The mucous polysaccharide from *Lilium longiflorum* was homogeneous, contained D-mannose and D-glucose residues in the molar ratio 5:2 and had a molecular weight of 2.63 × $10^5$. From the results of methylation and periodate ion oxidation analyses and the products of partial acid hydrolysis studies, it was concluded that the main chain was composed of a (1 → 4)-β-D-glycan with ten residues per non-reducing terminal group.

The branch points were located on O-2 or O-3 of the D-mannose residues. The polysaccharide isolated from the mucous of *L. maculatum* was qualitatively similar to that from *L. longiflorum.* water-soluble extract from *Opuntia ficus-indica* contained a number of polysaccharides. Four acidic components were present, each of which contained residues of D-galacturonic acid, L-arabinose, L-rhamnose, D-galactose, and D-xylose in a variety of proportions. Other studies on the same mucilage suggested that the core contained D-galacturonic acid, L-rhamnose, and D-galactose residues with the pentose sugars being present on the periphery. The mucilages in callus cultures of higher plants have been investigated particularly with regard to their monosaccharide composition. In studies on the effect of polysaccharides as ice-cream stabilizers, the gum from *Khaya grandifoliola* was most effective. The viscosity effect was low but the polysaccharide had a wide range of desirable effects.

## ALGAL POLYSACCHARIDES

The vacuum-u.v. c.d. spectra of solid films of agarose have been measured and show a positive band near 180 nm and a larger negative band about 152 nm.The positive band remains accessible in aqueous solution and has been used to characterize the changes in molecular conformation and interactions during sol-gel transitions.

The temperature profile of the vacuum-u.v. c.d. spectra shows sharp discontinuous changes around melting and setting points of the gel. These changes were interpreted in terms of co-operative intermolecular associations through double helices and also show pronounced hysteresis, which was considerd as evidence of helix-helix aggregation. rates of tumbling and exchange of free nitroxides in aqueous solutions are unaffected by high concentrations of agarose gels. Low-level covalent attachment of this spin-label to the agarose gels by means of a stable acetamido-ether linkage (23) caused considerable diminution of their rate of reorientation. Dissolution of this material to form a gel, and melting and setting of the gel, caused further changes in the e.s.r. spectrum.

$$\begin{matrix}—OH \\ —OH\end{matrix} + ClCH_2CONH—\langle N^{\bullet}—O \rangle \xrightarrow{OH'} \begin{matrix}—OCH_2CONH—\langle N^{\bullet}—O \rangle \\ —OH\end{matrix}$$

(23)

Studies on the interaction between agarose and the *endo-D-galacturonanase* from *Aspergillus niger* has shown it to be of an ion-exchange nature which did not involve biospecific interactions. The sequence and composition of uronate residues in intact alginate samples have been determined by high resolution $^1H$ n.m.r. spectroscopy

Problems encountered owing to the viscosity of the samples were overcome by controlled and slight depolymerization before n.m.r. at 90 °C. The D-mannuronic acid : L-guluronic acid molar ratio was obtained from the intensities of the signals from the anomeric protons. A sequence-dependent deshielding of H-5 of the L- guluronic acid residues made it possible to determine the fractions of the four possible doublets of nearest neighbours along the chain.

The results were consistent with $^{13}C$ n.m.r. spectroscopic data. Significant deviation from results obtained by chemical analysis appeared only in samples containing a large fraction of mixed doublets. *An* enzyme preparation from the marine brown alga *Pelvetia canaliculata* contained both alginate 5-epimerase and alginate lyase activities. $Ca^{2+}$ ions activated both enzymes, but $Mn^{2+}$ ions only inhibited lyase activity. Increasing concentrations of substrate D-mannuronanate removed the inhibition of lyase activity by $Mn^{2+}$ions. The storage D-glucans of the algae *Chlorella* and *Prototheca* species are identical in that they consist of a linear polymer, akin to amylose, and a branched amylopectin component.

The branched D-glucans from the above two species are markedly different from that formed by the alga *Cyanidium caldarium* which exists in hot springs. The amylopectin component of the hot springs alga was more highly branched. The osmotic coefficients of $Na^+$, $K^+$, and $Ca^{2+}$ counter-ions have been determined in aqueous solutions of *K-*, *l-* and $\lambda$-carrageenans and the results were correlated with those calculated from the limiting law of Manning. An ordered secondary structure exists in *K-* and t-carrageenans, whose stability can be discussed as a function of temperature, ionic strength, and nature of the counterion. The viscosity and other physical properties of two carrageenan-water gels have been compared with those of an agar gel.

The carregeenan gels had a smaller viscosity gradient and a higher viscosity at a given concentration. The structure of the gels were also different as revealed by microscopic analysis; those from the carrageenans were coarse and spongy, while that from agar contained closely intertwined fibrous molecules. Carrageenans from male and female gametophytic plants of *Rhodoglossum californicum, Chondrus crispus, Gigartina pistillata, Irideae cordata,* and another unidentified *Gigartina* species have been fractionated into KCl-soluble and -insoluble components.

An anti-(K-carrageenan) antibody was used to analyse the components immunochemically. The KCl-insoluble fractions were highly reactive K-types whereas the soluble fractions were less reactive. These latter fractions showed anti-(λ-reactivity). These data suggest that the KCl-soluble carrageenans contain either λ- or μ-carrageenan as no increase in immuno-logical activity to anti-(K-carrageenan) antibody was observed on treatment with alkali. A χ carrageenanase from *Pseudomonas carrageenovora* has been isolated and purified and its identity has been confirmed by an inspection of the products obtained by its enzymic action.

The limit digest was neo-carrabiose 4-sulphate. Addition of carrageenan to flours from soft wheats gave a product that was more akin to the flour from hard wheats in its bread-making quality. The non-Newtonian behaviour of the viscosity of K-carrageenan solutions at various shear rates increased on addition of D-glucose or sucrose. The results were explained in terms of changes in the solute-solute interactions. The structure of the laminaran from *Eisenia bicyclis* has been determined by methylation and periodate ion oxidation analyses as well as by $^{1}H$ and $^{13}C$ n.m.r. spectroscopies.

The polysaccharide possessed a branched structure with both (1 → 3) and (1 → 6)-β-D-glucopyranose residues being present in the ratio of 3 : 2. (1 → 3)-β-D-Glucans were specific activators of the prophenoloxidase from crayfish serum. Laminaran chains, terminated at the reducing end with a D-glucose residue (G-type chains) were more potent activators than the laminaran chains terminated at the reducing end with a D-mannitol residue (M-type chains). The extracellular polysaccharide of *Porphyridium cruentum* contained D-xylose, D-glucose, D- and L-galactose, 3-0-methyl-D-xylose, 3(or 4)-*O*-methyl-D-galactose, D-glucuronic acid, 2-*O*-methyl-D-glucuronic acid, and a 2-0-methyl hexose in the molar ratio 3 : 1 : 2.5 : 0.1 3 : 0.1 3 : 0.8 : 0.2 : 0.1 3, whereas that from *P. aerugineum* contained the first six sugar residues in the molar ratio 1.7 : 1.0 : 1.1 : 0.3:0.6:0.5, respectively, as well as 2,4-di-0-methyl-D-galactose.

Each polysaccharide contained, in addition, 10% half-ester sulphate and *ca.* 0.5% protein. Structural studies showed, *inter alia,* that the D-glucuronic acid residues were linked to *O*-3 of the D-galactose residues and that 2-*O*-methyl-D-glucuronic acid residues were linked to *O*-4 of the L-galactose residues. The polysaccharide from *P. cruentum* had a molecular weight of $4 \times 10^{6}$, and that from *P. aerugineum* had a molecular weight of $5 \times 10^{6}$. The sulphated heteropolysaccharide from the brown alga, *Dyctyota dichotoma,* contains residues of D-glucuronic acid, D-galactose, D-mannose, D-xylose, and L-fucose.

Evidence, from chemical analyses, was provided that the polysaccharide is partially sulphated and that the sugar residues are(1 → 4)-linked except L-fucose which is (1 → 2)-linked. The carbohydrates of the brown seaweed *Padina pavonia* contained a sulphated heteropolysaccharide that contained residues of D-glucuronic acid, L-fucose, D-glucose, D-mannose, and D-xylose as well as a D-mannan-protein complex. The red alga *Laurencia spectabilis* contains a glycoproteineous species which, on purification, was found to contain 92% carbohydrate and 8% protein. The carbohydrate portion mainly consisted of D-galactose and uronic acid residues and there was no ester sulphate present. The polysaccharides from the heterocyst and spore envelopes of the blue-green alga *Anabaena cylindrica* have a backbone, which consists of the repeating units (24) to which are attached residues of D-glucose, D-xylose, D-galactose, and D-mannose as side chains.

The main chain contained between 128 and 150 sugar residues. After treatment of the polymer with various glycoside hydrolases, the degraded product gave, on treatment with *endo*-(1 → 3)-β-D-glucanase, the trisaccharide (25) and the pentasaccharide (26).

–β-D-Man*p*-(1 → 3)-β-D-Glc*p*-(1 → 3)-β-D-Glc*p*-(1 → 3)-β-D-Glc*p*-(1 → 3)–

(24)

```
β-D-Manp-(1 → 3)-β-D-Glc
                      4
                      ↑
                      1
                    D-Glcp
```

(25)

```
β-D-Glcp-(1 → 3)-β-D-Manp-(1 → 3)-β-D-Glc
    4                                  4
    ↑                                  ↑
    1                                  1
  D-Xylp                             D-Glcp
```

(26)

The use of $^{13}C$ n.m.r. spectroscopy, and methylation and Smith degradation analyses have confirmed the structure of the (1 → 2)-β-D-mannopyranan from *Crithidia deanei.* It was assumed that the polymer arises from the protozoon and not a bacterial endosymbiont, since a related organism *C. fasciculata* also contains a D-mannan. The protozoon *Herpetomonas samuelpessoai* also contains a (1 → 2)-β-D-mannopyranan as well as a branched chain D-glucurono-D-xylan.

Since injection of cells or flagellar suspensions from *H. samuelpessoai* cause protection against infection with *Trypanosoma cruzi,* the immunological protection may be due to these polysaccharides or glycoproteins. The surface coat of *T. congolense* has been characterized as a glycoprotein. At high N:P ratios in the growth medium, the marine diatom *Chactoceros affinis* produces large amounts of extracellular polysaccharide, whereas *Skeletonema costatum,* grown under similar conditions, produces only very small amounts of extracellular polysaccharide.

CHAPTER

3

# PHOTOSYNTHESIS

*In photosynthesis light energy is converted into chemical energy.*

Heterotrophic organisms, including animals, fungi, and most types of bacteria, live by degrading complex molecules provided by other organisms. Life on earth obviously could not continue indefinitely in this manner without a mechanism for synthesizing complex molecules from simple ones. The energy that sustains this synthesis comes almost entirely from the sun, and is captured in the process of photosynthesis. Plants and photosynthetic bacteria annually convert, or "fix," about 10" tons of carbon from $CO_2$ into organic compounds. In spite of the impressive magnitude of this conversion, the total amount of fixed carbon on earth is decreasing as a result of consumption.

As our reserves of energy diminish, it becomes increasingly important that we understand how photosynthesis works and how our activities affect it.

An overall equation for $CO_2$ fixation as it occurs in plants is

$$6\,CO_2 + 6\,H_2O + \text{light} \rightarrow C_6H_{12}O_6 + 6\,O_2 \qquad \text{...(3.1)}$$

Electrons and protons are removed from $H_2O$, $O_2$ is evolved, and $CO_2$ is reduced to the level of a carbohydrate. One group of photosynthetic bacteria, the cyanobacteria, carry out identical reactions. Other types of photosynthetic bacteria carry out similar overall processes but do not evolve $O_2$; these species use materials other than $H_2O$ as a source of electrons. If we leave out the light, the equilibrium for the synthesis of glucose from $CO_2$ and $H_2O$ lies vanishingly far to the left. The equilibrium constant at 27°C is $10^{-496}$! In this chapter we explore how photosynthetic organisms use light to drive the reaction in the direction of carbohydrates against this enormous thermodynamic gradient.

## The Photochemical Reactions of Photosynthesis

Although important differences exist between photosynthesis in bacteria and plants, the photochemical reactions that capture the energy of light are basically the same. In plants, the reactions of photosynthesis take place in specialized organelles, the chloroplasts. An electron micrograph of a chloroplast from a lettuce leaf. Chloroplasts are bounded by an envelope of two membranes, and they have a third internal membrane called the *thylakoid membrane*. In electron micrographs of thin-sectioned chloroplasts, the thylakoid membrane has the appearance of a large number of separate sheets of flattened vesicles. It actually is a single highly folded membrane that encloses a distinct compartment, the *thylakoid lumen*.

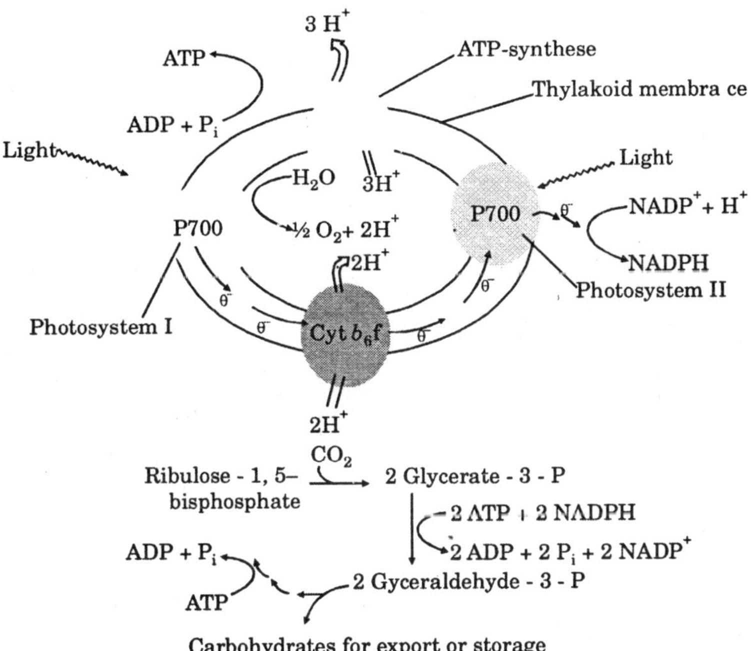


**Fig. 3.1. Photosynthesis harnesses the energy of sunlight for biosynthesis. In plants, the reactions of photosynthesis occur in chloroplasts. Pigment-protein antenna complexes in the thylakoid membrane absorb light and pass energy to the reaction centers of two distinct photosystems, where excited chlorophyll molecules (P680 in photosystem II, P700 in photosystem I) transfer electrons to a series of electron acceptors. Photosystem I generates strong reductants that ultimately reduce $NADP^+$; photosystem II generates a strong oxidant that splits $H_2O$, releasing $O_2$. Electrons flow from photosystem II to photosystem I through the cytochrome $b_6f$ complex. The electron-transfer reactions release protons in the thylakoid lumen and take up protons from the stroma, setting up a transmembrane pH gradient. Flow of protons back across the membrane through an ATP-synthase drives the formation of ATP. NADPH and ATP provided by the electron-transfer reactions are used to convert $CO_2$ to carbohydrates. $CO_2$ combines with ribulose-1, 5-bisphosphate to form two molecules of glycerate-3-phosphate. ATP and NADPH are used to reduce the glycerate-3-phosphate to glyceraldehyde-3-phosphate and to recycle some of the glyceraldehyde-3-phosphate to ribulose-1, 5-bisphosphate. The remaining glyceraldehyde-3-phosphate is exported from the chloroplast or converted to other carbohydrates. Many of the photosynthetic reactions resemble reactions seen in earlier chapters: The formation of hexoses and pentoses from trioses resembles reactions of gluconeogenesis and the pentose phosphate pathway; the cytochrome $b_6f$ complex is similar to the mitochondrial complex III, and the proton-conducting ATP-synthase is similar to the mitochondrial ATP-synthase. The major difference between the operation of mitochondria and chloroplasts is the direction of electron flow. In mitochondria, electrons flow from reduced organic compounds to $O_2$; photosynthetic organisms use the energy of light to push electrons in the opposite direction.**

In places, the folded membrane stacks into disklike structures called *grana*. The chlorophyll found in chloroplasts is bound to proteins in the thylakoid membrane, and it is here that the initial conversion of light into chemical energy occurs. The thylakoid membrane also contains a collection of electron carriers and an ATP-synthase that are similar to those in the mitochondria. The enzymes responsible for the actual fixation of $CO_2$ and the synthesis of carbohydrates reside in the *stroma* that surrounds the thylakoid membrane. Algae are members of the plant kingdom and contain chloroplasts similar to those of higher plants, but prokaryotic photosynthetic organisms do not have chloroplasts.

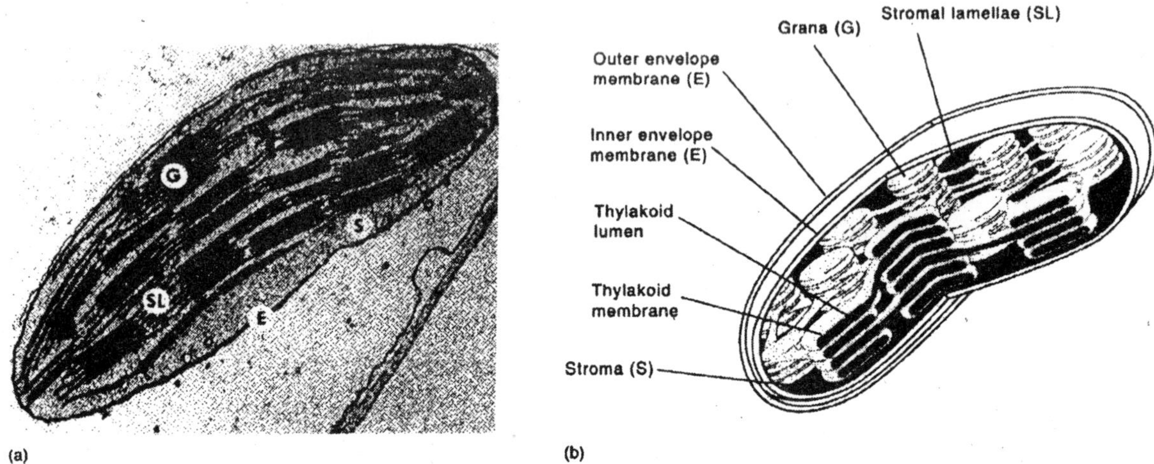


**Fig. 3.2. (*a*) Electron micrograph of a chloroplast in a lettuce leaf. The organelle is shaped like a flattened sausage with a width of about 10 μM and a thickness of about 3 μM. It is surrounded by a double outer membrane or envelope (E). The stroma (S) contains DNA, ribosomes, and the soluble enzymes of $CO_2$ fixation. Extending throughout the stroma is the thylakoid membrane, which is differentiated into stacked regions or grana (G) and unstacked stromal lamellae (SL). (*b*) A schematic drawing of the chloroplast membrane systems. The highly folded thylakoid membrane separates the thylakoid lumen from the stroma.**

In prokaryotes, the photochemical reactions occur in the plasma membrane, which has extensive invaginations resembling the cristae of the mitochondrial inner membrane. Table 3.1. summarizes the main distinction between the various types of photosynthetic organisms.

## Photochemical Reactivity of Chlorophyll

With the exception of certain archaebacteria, all known *photosynthetic organisms take advantage of the photochemical reactivity of one or another type of chlorophyll.* Chlorophylls resemble hemes but differ from them in four major respects:

1. The central metal atom is magnesium rather than iron.

**Table 3.1. Properties of Photosynthetic Organisms.**

| *Group* | *Evolve $O_2$* | *Contain Chloroplasts* | *Type of Chlorophyll* | *Number of Photosystems* |
|---|---|---|---|---|
| Algae and higher plants | yes | yes | chlorophyll *a* and *b* | 2 |
| Cyanobacteria | yes | no | chlorophyll *a* and *b* | 2 |
| Purple bacteria | no | no | bacteriochlorophyll *a* or *b* | 1 |

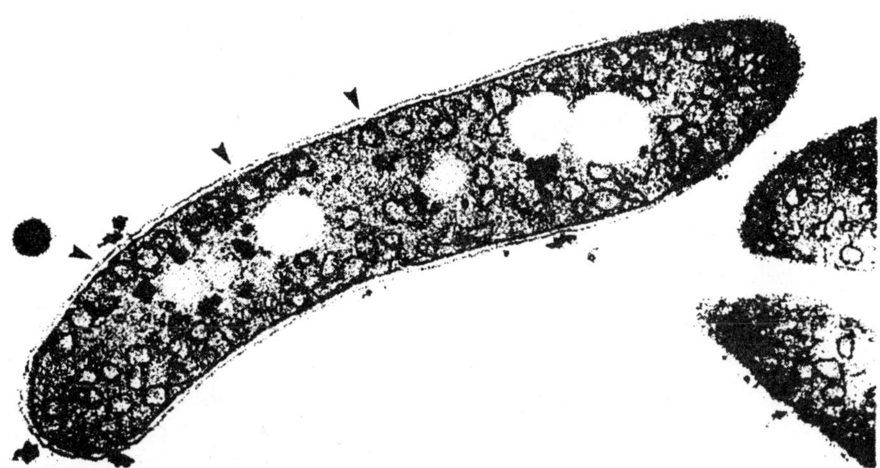

**Fig. 3.3. Cross-sectional view of a cell of *Rhodospirillum rubrum*, a purple photosynthetic bacterium. A double-membrane system surrounds the cell. The inner membrane is extensively invaginated into tubules (arrows). These look circular when they are cut in cross section.**

2. Chlorophylls have an additional ring (ring V).
3. In the case of chlorophyll *a* and chlorophyll *b*, the two major chlorophylls in plants and cyanobacteria, one of the pyrrole rings (ring IV) is reduced by the addition of two hydrogens. In bacteriochlorophylls *a* and *b*, which occur in the purple and green bacteria, two of the rings are reduced (rings II and IV).
4. The propionyl side chain of ring IV is esterified with a long-chain isoprenoid alcohol.

Chlorophylls *a* and *b* contain the alcohol *phytol*; bacteriochlorophylls *a* and *b* have either phytol or geranylgeraniol, depending on the species of bacteria. Photosynthetic organisms also contain small amounts of *pheophytins* or *bacteriopheophytins*, which are the same as the corresponding chlorophylls or bacteriochlorophylls except that two hydrogens replace the $Mg^{2+}$. We will see that pheophytins and bacteriopheophytins play special roles as electron carriers in photosynthesis. The reduction of ring IV in chlorophylls *a* or *b* changes the optical absorption spectrum of the molecule dramatically.

Whereas the long-wavelength absorption band of a cytochrome is relatively weak chlorophyll *a* has an intense absorption band at 676 nm. Chlorophyll *b* has a similar band at 642 nm. Bacteriochlorophylls *a* and *b* have strong absorption bands in the region of 770 nm. The chlorophylls thus absorb red or near-infrared light very well.

## Light Is Composed of Photons

Before we discuss how chlorophylls can transform light into chemical energy, let's review some of the basic properties of light. *Light is an oscillating electromagnetic field. It interacts with matter in packets, or quanta, called photons, each of which contains a definite amount of energy.* The relationship between the energy $\epsilon$ of a photon and the frequency $\upsilon$ of the oscillating electromagnetic field is given by the equation

$$\epsilon = h\upsilon$$

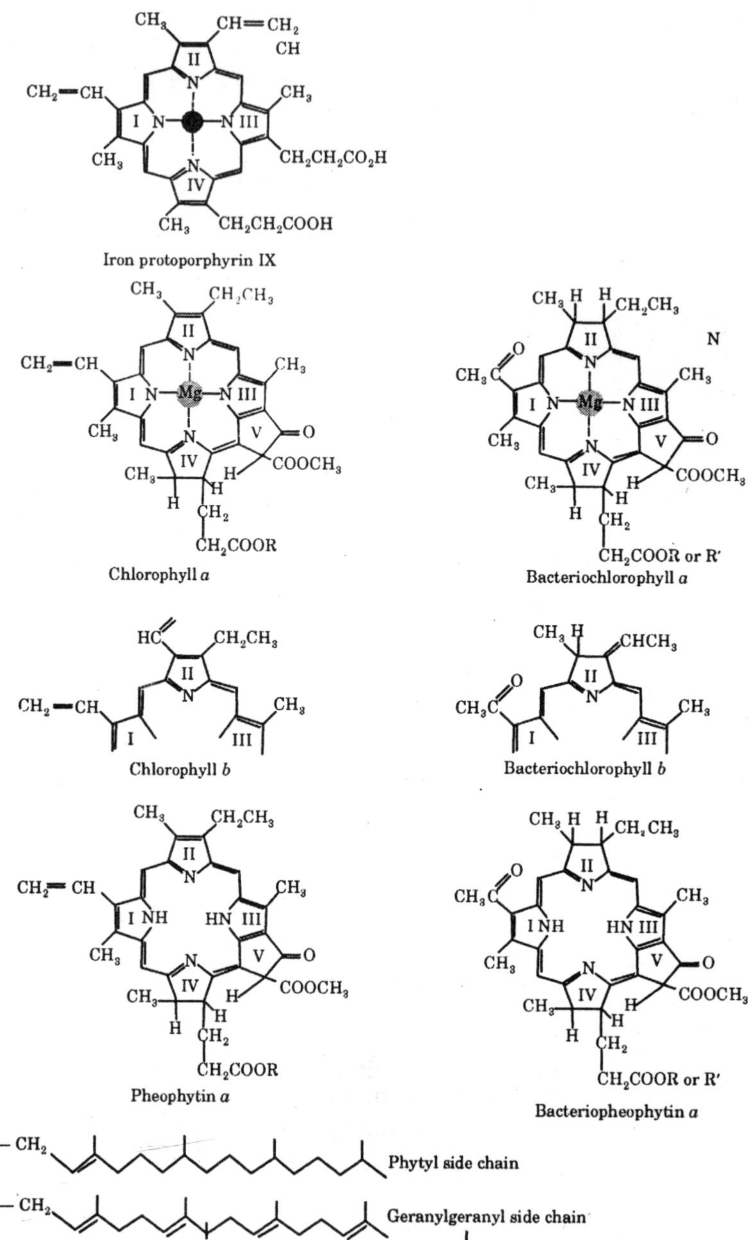


**Fig. 3.4. Structures of protoporphyrin IX(the prosthetic group of hemoglobin, myoglobin, and the c-type cytochromes), and several types of chlorophyll and bacteriochlorophyll. Chlorophyll *a* and chlorophyll *b* are the main types of chlorophyll in plants and the cynobacteria. Purple photosynthetic bacteria contain either bacteriochlorophyll *a* or *b*, depending on the bacterial species. (The green photosynthetic bacteria contain still another form of bacteriochlorophyll). Chlorophyll *b* and bacteriochlorophyll *b* are the same as chlorophyll *a* and bacteriochlorophyll *a* except for the substituents on ring II. Pheophytins and bacteriopheophytins are the same as the corresponding chlorophylls or bacteriochlorophylls, except that two hydrogen atoms replace mg.**

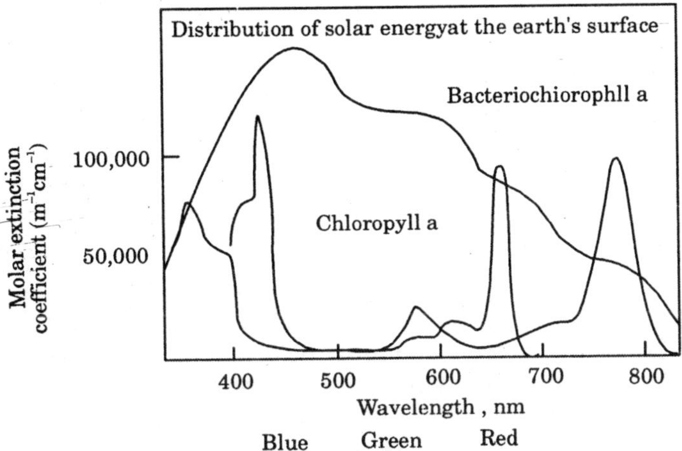


**Fig. 3.5. Absorption spectra of chlorophyll *a* and bacteriochlorophyll *a* in ether. Note that the long-wavelength absorption bands are much stronger than the α band of reduced cytochrome c. In the chlorophyll-protein complexes found in photosynthetic organisms, the long-wavelength absorption band generally is shifted to even longer wavelengths. This probably reflects interactions between neighbouring chlorophylls, which are bound to the proteins as dimers or larger groups.**

where $h$ is Planck's constant ($4.12 \times 10^{-15}$ eV · s). A mole of photons is called an *einstein*, and the energy per einstein is

$$\mathscr{E} = N\epsilon = Nh\upsilon \qquad ...(3.2)$$

where $N$ is Avogadro's number ($6.02 \times 10^{23}$ photons/einstein). The frequency $\upsilon$ is the number of oscillations per second at a given point in space. The wavelength λ of the oscillations (the distance between successive peaks in the amplitude of the field) depends on both $\upsilon$ and the velocity $c$ at which the peaks move through space

$$\lambda = \frac{c}{\upsilon} \qquad ...(3.3)$$

In a vacuum, light travels with a velocity $c = 3 \times 10^{10}$ cm/s.

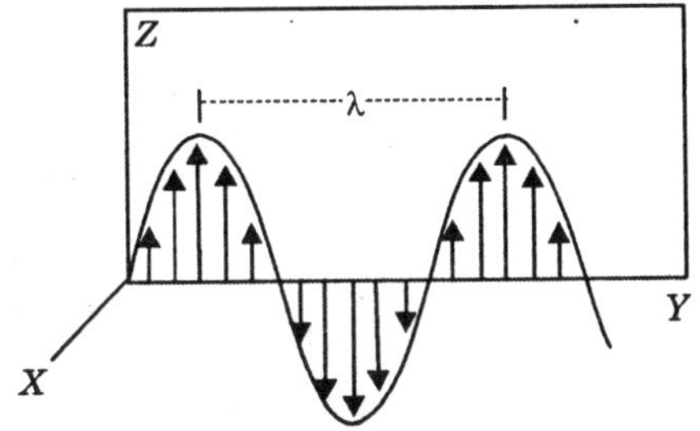


**Fig. 3.6. Light is an oscillating electromagnetic field. The lengths of the arrows in this diagram represent the strength of the electric field at a particular time, as a function of position in a ray of light proceeding along the *y* axis.**

Photon energies often are stated in units of electron volts (eV). One eV is the energy required to move an electron against a potential difference of 1 V and is equivalent to 23,060 kcal/mol (96,480 kJ/mol). Blue light with a wavelength of 450 nm ($\upsilon = 6.7 \times 10^{14}$ $s^{-1}$) has an energy of 2.75 eV, or 64 kcal/einstein, and far red light (700 nm) has an energy of 1.77 eV (41 kcal/einstein).

## Photons of Light Interact with Electrons in Molecules

The electrons in a molecule reside in a set of molecular orbitals, each of which is characterized by a particular energy. In an isolated molecule, the orbital energies depend mainly on the interactions of the electrons with each other and with the nuclei. A molecule can have a variety of energies, depending on how its electrons are distributed among the available orbitals. For an organic molecule with Zn electrons, the lowest overall energy usually is obtained when two electrons with antiparallel spins occur in each of the first n orbitals, leaving all the orbitals with higher energies empty. This is the ground state of the molecule. When light interacts with a molecule, and a photon is absorbed, *an electron moves from one of the occupied molecular orbitals to an unoccupied orbital with a higher energy.*

Two requirements must be met for this to occur. First, *the difference between the energies of the two orbitals must be the same as the photon's energy, hv.* This is why a given type of molecule absorbs light of some wavelengths and not of others. Second, the two orbitals must have different geometrical symmetries and must be oriented appropriately with respect to the oscillating electrical field of the light. This requirement explains why some absorption bands are stronger than others. If chlorophyll (or any other molecule) absorbs a photon, it is excited to a state that lies above the ground state in energy.

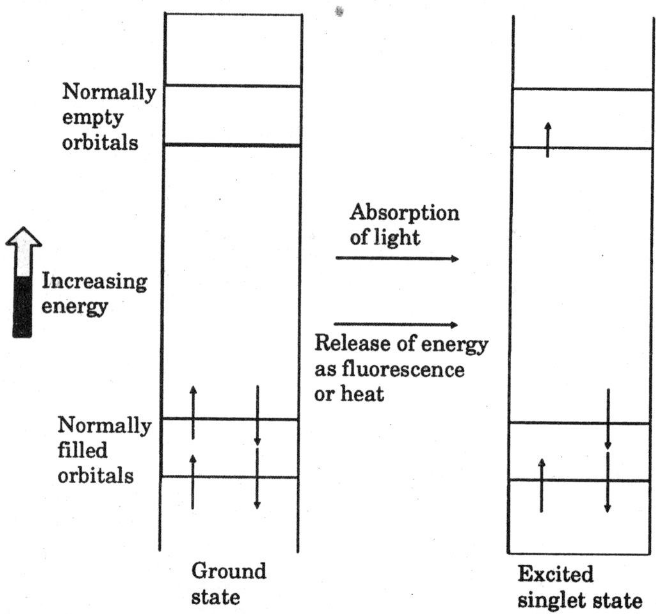


**Fig. 3.7. When a molecule absorbs light, an electron is excited to a molecular orbital with higher energy. The horizontal bars in this diagram represent molecular orbitals for electrons. Each orbital can hold two electrons with antiparallcl spins (arrows pointing upward or downward). Only the top few of the filled orbitals are shown here. Absorption of light raises an electron from one of these orbitals to an orbital that is normally unoccupied. For this to occur, the energy of the photon must match the difference between the energies of the two orbitals. The choice of an upward or downward arrow is arbitrary, but no change of spin occurs during the excitation. As long as the spins of the two unpaired electrons remain antiparallel, so that the molecule has no net electronic spin, the molecule is said to be in an excited "singlet" state. An excited molecule can return directly to the ground state by giving off energy as fluorescence or heat, or by transferring energy to another nearby molecule, or it can transfer an electron to another molecule.**

An excited molecule can return to the ground state by releasing energy in several different ways. One possibility is to emit a photon; this is fluorescence. Another decay mechanism is to transfer the energy to a neighbouring molecule. This phenomenon, *resonance energy transfer*, plays an important role in photosynthesis, as we will see. An excited molecule also can transfer an electron to a neighbouring molecule. *Electron transfer can be a favourable path for the decay of an excited molecule, because an electron in the upper, normally unoccupied orbital is bound less tightly than one in a lower, normally filled orbital.* If the absorption of a photon increases the energy of the molecule by $\in$ electron volts, where $\in \approx hv$, it makes the standard redox potential for removing an electron from the excited molecule ($E°$) more negative by approximately $\in$ volts, compared with the $E°$ for the molecule in the ground state.

In the case of chlrophyll *a*, $E°$ for oxidation in the ground state is about +0.5 V, and $hv$ for the long-wavelength absorption band is 1.7eV. $E°$ of the excited molecule is therefore about –1.2V. This means that in the excited state chlorophyll *a* is an extremely strong reductant. For comparison, recall that $NAD^+$/NADH has an $E°'$ of only –0.32 V. The basic principle underlying the photochemical process is that *excitation causes chlorophyll to release an electron.* The chlorophyll is oxidized by the electron loss while the molecule receiving the electron becomes reduced. The oxidized chlorophyll returns to its ground state by extracting an electron from a third molecule. The idea that light drives the formation of oxidants and reductants was first advanced by C. B. van Niel in the 1920s.

It was strengthened through experiments done by Robin Hill in 1939. Hill discovered that isolated chloroplasts evolved $O_2$ if illuminated in the presence of an added electron acceptor such as ferricyanide, $Fe(CN)_6^{3-}$. The electron acceptor became reduced in the process. Because no fixation of $CO_2$ occurred under these conditions, this experiment demonstrated that the photochemical reactions of photosynthesis can be separated from the reactions that involve $CO_2$ fixation.

## Photooxidation of Chlorophyll

The photooxidation of chlorophyll can be detected by changes in the optical absorption spectrum of the molecule. Oxidation results in the loss of the chlorophyll's characteristic absorption bands. Measurements of this sort were made first in the 1950s by Bessel Kok and Louis Duysens. Kok found that illumination of chloroplasts caused an absorbance decrease at 700 nm. He suggested that this reflected the photooxidation of a reactive chlorophyll complex, which he called P700. (P stood for "pigment," and 700 for the wavelength at which the unoxidized complex had its main absorption band.) Duysens made similar observations on purple photosynthetic bacteria. Here the reactive bacteriochlorophyll complex absorbed at 870 nm, and Duysens called it P870. A second type of reactive complex in chloroplasts, P680, was discovered by H. Witt and his colleagues.

Subsequent investigations showed that P700 and P680 were parts of two distinct photochemical systems: *photosystem* I and photosystem II. Photooxidation of P870, P700, or P680 generates a cationic radical ($P870^{\dot{+}}$, $P700^{\dot{+}}$ or $P680^{\dot{+}}$) in which the positive charge and the unpaired electron are delocalized over the conjugated ring system of the chlorophyll. In fact, the electron spin resonance (ESR) and electron nuclear double resonance (ENDOR) spectra of P870$^{\ddagger}$ indicate that the unpaired electron is delocalized over two bacteriochlorophyll *a* molecules, which form a closely interacting pair.

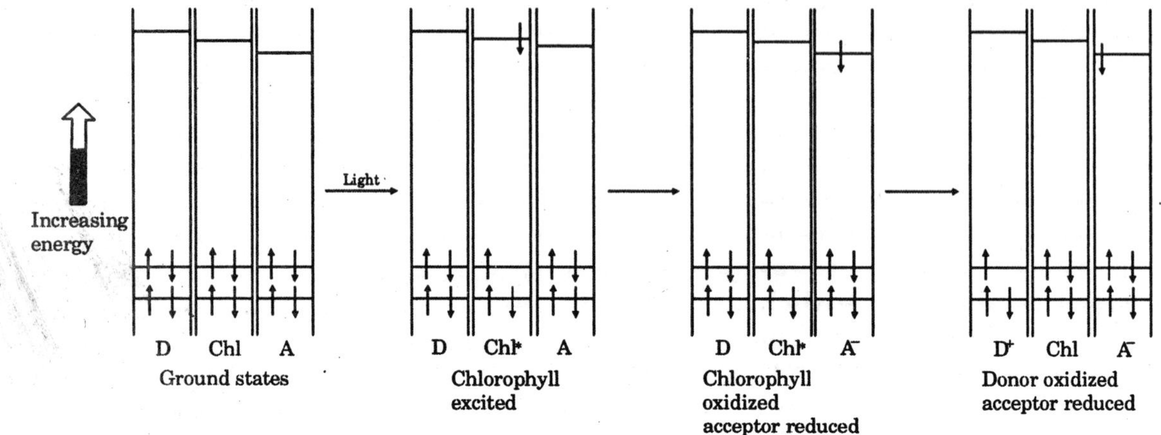


**Fig. 3.8. The photochemical process that initiates photosynthesis is an electron-transfer reaction. The horizontal bars and vertical arrows represent molecular orbitals and electrons. Absorption of light increases the free energy of a chlorophyll complex (Chl) by *hν*, making the transfer of an electron to an acceptor (A) thermodynamically favourable. The oxidized chlorophyll complex (Chl$^{+}$) extracts an electron from a donor (D).**

The strong interaction between the two bacteriochlorophylls explains why the absorption band of the complex is at 870 nm, whereas the long-wavelength band of monomeric bacteriochlorophyll *a* in solution is at 770 nm. P700 and P680 probably consist of similar dimers of chlorophyll *a,* although the arrangements of the chlorophylls in these dimers may be somewhat different.

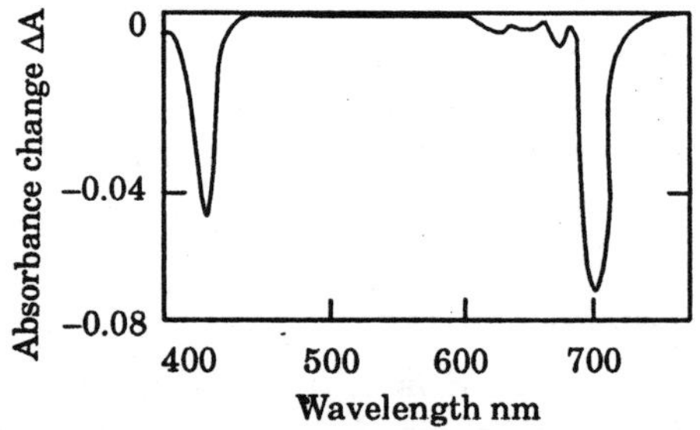


**Fig. 3.9. When a suspension of chloroplasts is illuminated, its optical absorbance decreases in the regions around 430 and 700 nm. The absorbance changes reflect the oxidation of a special chlorophyll complex (P700). Addition of a chemical oxidant such as potassium ferricyanide causes similar absorbance changes. Partial purification and determination of oxidation reduction potential of the photosynthetic chlorophyll complex absorbing at 700 mμ.**

## The Reactive Chlorophyll

The chlorophyll or bacteriochlorophyll that undergoes photoxidation is bound to a protein in a complex called a *reaction center.* Reaction centers have been purified by disrupting chloroplasts or bacterial membranes with detergents. This was first achieved with purple photosynthetic bacteria. When they are excited with light, purified bacterial reaction centers can carry out the initial photochemical transfer of an electron from P870 to a series of electron acceptors. Reaction centers of purple bacteria typically contain three polypeptides, four molecules of bacteriochlorophyll, two bacteriopheophytins, two quinones, and one nonheme iron atom.

In some bacterial species, both quinones are ubiquinone. In others, one of the quinones is menaquinone (vitamin $K_2$), a naphthoquinone that resembles ubiquinone in having a long side chain. Reaction centers of some species, such as *Rhodopseudomonas viridis,* also have a cytochrome subunit with four *c*-type hemes. The crystal structure of reaction centers from *R. viridis* was determined by Hartmut Michel, Johann Deisenhofer, Robert Huber, and their colleagues in 1984. This was the first high-resolution crystal structure to be obtained for an integral membranc protein. Reaction centers from another species, *Rhodobacter sphaeroides,* subsequently proved to have a similar structure.

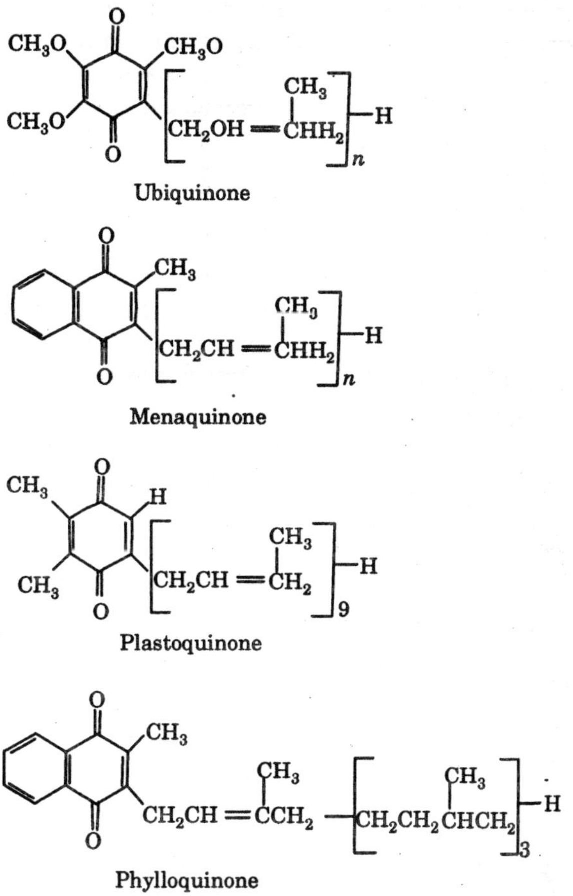

Ubiquinone

Menaquinone

Plastoquinone

Phylloquinone

**Fig. 3.10. The structures of ubiquinone, menaquinone (vitamin $K_2$), plastoquinone, and phylloquinone (vitamin $K_1$). Purple photosynthetic bacteria contain ubiquinone, menaquinone, or both, depending on the bacterial species; chloroplasts contain plastoquinone and phylloquinone.**

In both species, the bacteriochlorophyll and bacteriopheophytin, the iron atom and the quinones are all on two of the polypeptides, which are folded into a series of *a* helices that pass back and forth across the cell membrane. The third polypeptide resides largely on the cytoplasmic side of the membrane, but it also has one transmembrane α helix. The cytochrome subunit of the reaction center in *R. viridis* sits on the external (periplasmic) surface, of the membrane. Fig. 3.11 *b* shows the arrangement of the pigments in greater detail. Two of the four bacteriochlorophylls are packed closely together. Studies of the reaction center's optical absorption spectrum indicate that *this is the bacteriochlorophyll dimer that releases an electron when the reaction center is excited with light.* Although the structures of the plant reaction centers are not yet known in detail, photosystem II reaction centers resemble reaction centers of purple bacteria in several ways.

The amino acid sequences of their two major polypeptides are homologous to those of the two polypeptides that hold the pigments in the bacterial reaction center. Also, the reaction centers of photosystem II contain a nonheme iron atom and two molecules of plastoquinone, a quinone that is closely related to ubiquinone and they contain one or more molecules of pheophytin *a* and several molecules of chlorophyll *a* in addition to the two that probably make up P680. The reaction center of photosystem 1 is larger and more complex. It contains two large polypeptides and at least seven other smaller subunits. The reactive chlorophyll a dimer P700 resides on the two main polypeptides, along with about 60 additional molecules of chlorophyll *a,* two quinones, and an iron-sulfur center.

## Movement of Electrons from P870 to Bacteriopheophytin

Let's now consider the acceptors that extract an electron from P870 in the purple bacterial reaction center. The first acceptor the reduction of which can be well resolved kinetically is one of the two bacteriopheophytins. When reaction centers are excited with a short flash of light, an excited singlet state of the reactive bacteriochlorophylls (P870*) is formed essentially instantaneously. This state decays in about $3 \times 10^{-2}$ s, and as it does, a $P870^{\cdot+}BPh700^{\cdot+}$ radical pair is created. Here $P870^{\cdot+}$ is the radical formed by removing an electron from P870, and $BPh^{\cdot-}$ is the anionic radical formed by adding an electron to a bacteriopheophytin.

The creation of the radical-pair state can be detected spectrophotometrically by the loss of absorption bands of P870 and the bacteriopheophytin and the formation of new absorption bands attributable to the two radicals. The bacteriopheophytin that undergoes reduction is the one on the right in the figure 3.11(a) and (b). Differences in their amino acid environments make the two bacteriopheophytins spectroscopically distinguishable. The $P870^{\cdot+}BPh^{\cdot-}$ radical pair disappears in about $2 \times 10^{-10}$ s when an electron jumps from $BPh^{\cdot-}$ to one of the quinones ($Q_A$). This leaves the reaction center in the state $P870^{\cdot+}Q_A^{\cdot-}$, where $Q_A^{\cdot-}$ is the anionic semiquinone. The electron carriers that participate in these first few steps are all fixed in position in the reaction center so that the initial steps in electron transfer involve electron movements only.

One indication of this is that the electron-transfer reactions, in addition to being phenomenally fast, are almost independent of temperature. The reactions also are amazingly efficient. This can be expressed in terms of the *quantum yield* of $P870^{\cdot+}Q_A^{\cdot-}$, which is the number of moles of $P870^{\cdot+}Q_A^{\cdot-}$ formed per einstein of light absorbed. The measured quantum yield in purified reaction centers is $1.02 \pm 0.04$. *Essentially every time the reaction center is excited, an electron moves from P870 to $Q_A$.*

## A Cyclic Electron-Transport Chain Outward

From $Q_A^{\cdot-}$, an electron moves to the second quinone bound to the reaction center. In the meantime, a *cytochrome replaces the electron that was removed from P870*, preparing the reaction center to operate again. In *R. viridis,* the electron donor is the bound cytochrome with four hemes shown at the top of figure 3.11*a*. In other species, it often is a soluble *c*-type cytochrome with a single heme, resembling mitochondrial cytochrome *c*. When the reaction center is excited a second time, a second electron is pumped from P870 to the bacteriopheophytin and on to $Q_A$ and $Q_B$. This places $Q_B$ in the fully reduced form, $Q_B^{2-}$.

Uptake of two protons transforms the reduced quinone to the uncharged quinol, $QH_2$, which dissociates from the reaction center into the phospholipid bilayer of the cell membrane. The protons taken up in the formation of $QH_2$ come from the cytosol. $QH_2$ is reoxidized by a cytochrome $bc_1$ complex that is very similar to complex III of the mitochondrial respiratory chain. Electrons move through the cytochrome $bc_1$ complex to a c-type cytochrome, which then diffuses to the reaction center and provides an electron for the reduction of P870$^{+\bullet}$. As it does in mitochondria, movement of electrons through the cytochrome $bc_1$ complex results in the release of protons on the outer surface of the membrane.

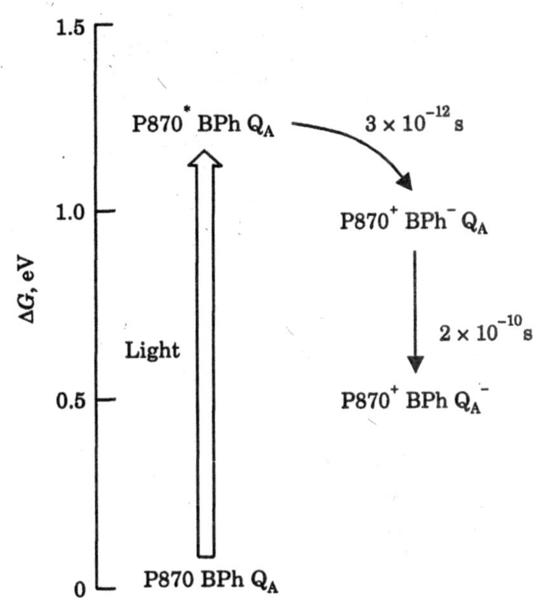


**Fig. 3.11. The initial electron-transfer steps in reaction centers of purple photosynthetic bacteria. (P870* = the first excited singlet state of P870; BPh = bacteriopheophytin; $Q_A$ = menaquinone or ubiquinone, depending on the bacterial species.) In this scheme, various states of the photosynthetic apparatus are positioned vertically according to their free energies, with the states that have the highest free energies at the top.**

The proton extrusion can be explained by the Q cycle that we discussed in connection with the mitochondrial complex. Flow of electrons from $QH_2$ back to P870 thus pumps protons outward across the cell membrane, generating a transmembrane electrochemical potential gradient. The inside of the cell becomes negatively charged relative and the cytosolic pH becomes higher than the external pH. Flow of protons back into the bacterial cell, down the electrochemical potential gradient, is mediated by an ATP-synthase resembling the ATP-synthase of the mitochondrial inner membrane. As in mitochondria, the movement of protons through the $F_0$ base-piece of the enzyme drives the formation of ATP. Note that the electron-transport system of purple photosynthetic bacteria is cyclic. Excitation of the reaction center with light creates a strong reductant (BPh$^{-\bullet}$) and a strong oxidant (P870$^{+\bullet}$), and electrons return from BPh$^{-\bullet}$ to P870$^{+\bullet}$ via quinones, the cytochrome $bc_1$ complex, and c-type cytochromes. This cyclic flow of electrons results in the formation of ATP but no net oxidation or reduction. However, biosynthesis of carbohydrates from $CO_2$ requires a reductant, such as NADPH, in addition to ATP.

Photosynthetic bacteria obtain the necessary reductant by using ATP to energize the transfer of electrons from succinate or other sub-strates to NADP+. These reactions are carried out by dehydrogenase complexes in the cell membrane.

## An Antenna System

The reactive chlorophyll or bacteriochlorophyll molecules of P700, P680, or P870 account for only a small fraction of the total pigment in photosynthetic membranes. Chloroplasts contain about 300 chlorophylls per molecule of P700 and P680, and the cell membranes of purple photosynthetic bacteria have from 25 to several hundred bacteriochlorophylls per P870 molecule, depending on the species. Most of the chlorophyll or bacteriochlorophyll is not photochemically active. Instead, it serves as an antenna. *When one of the molecules in the antenna is excited with light, it can transfer its energy to a neighboring molecule by resonance energy transfer. Energy absorbed anywhere in the antenna migrates rapidly from molecule to molecule until it is trapped by an electron-transfer reaction in a reaction center.*

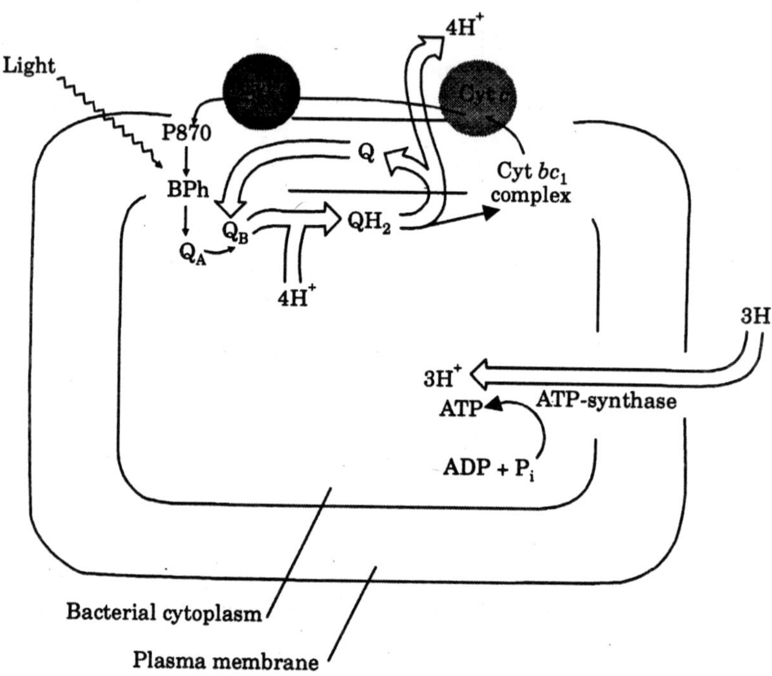


**Fig. 3.12. In purple photosynthetic bacteria, electrons return to P870$^+$ from the quinones $Q_A$ and $Q_B$ via a cyclic pathway. When $Q_B$ is reduced with two electrons, it picks up protons from the cytosol and diffuses to the cytochrome *bc*$_1$ complex. Here it transfers one electron to an iron-sulfur protein and the other to a *b*-type cytochrome and releases protons to the extracellular medium. The electron-transfer steps catalyzed by the cytochrome *bc*$_1$ complex probably include a Q cycle similar to that catalyzed by complex III of the mitochondrial respiratory chain. The c-type cytochrome that is reduced by the iron-sulfur protein in the cytochrome *bc*$_1$ complex diffuses to the reaction center, where it either reduces P870$^+$ directly or provides an electron to a bound cytochrome that reacts with P870$^+$. In the Q cycle, four protons probably are pumped out of the cell for every two electrons that return to P870. This proton translocation creates an electrochemical potential gradient across the membrane. Protons move back into the cell through an ATP-synthase, driving the formation of ATP.**

Measurements of the lifetime of fluorescence from the antenna pigments indicate that the energy is trapped in a reaction center within about $10^{-10}$ s in purple bacteria and within about $5 \times 10^{-10}$s in chloroplasts. The distinction between the antenna and the reaction centers grew out of experiments done by Robert Emerson and William Arnold in the 1930s. They measured the amount of $O_2$ that evolved when they excited suspensions of green algae with a train of short flashes of light. To obtain the maximum $O_2$ per flash, they had to allow a period of about

$2 \times 10^{-2}$ s of darkness between successive flashes. The amount of $O_2$ evolved per flash decreased if the flashes were spaced more closely together. Each flash evidently generated a product that had to be consumed before the photosynthetic apparatus was ready to work again.

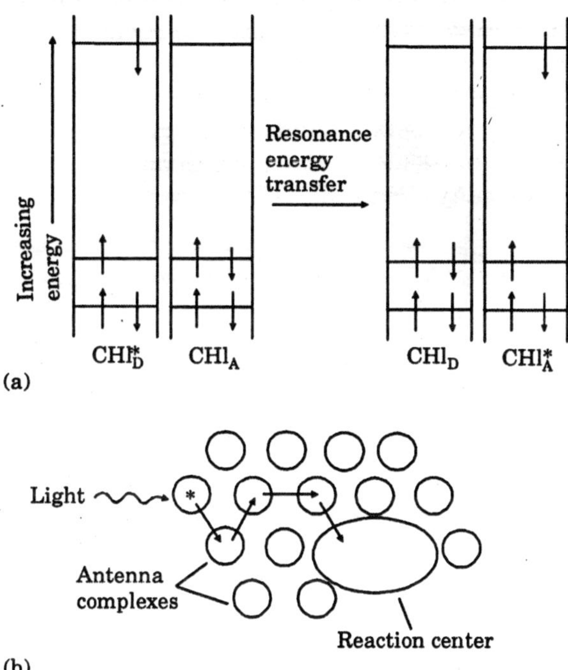


**Fig. 3.13.** ***(a)*** **A molecule in an excited state ($CHl_D^*$) can transfer its energy to another molecule ($Chl_A$). The donor molecule returns to its ground state ($Chl_D$), and the acceptor is elevated to an excited state ($CHl_A^*$). This requires that the difference in energy between $CHl_D^*$ and $Chl_D$ (the excitation energy of $Chl_D$) be the same as the difference between $Chl_A^*$ and $Chl_A$ (the excitation energy of $Chl_A$). When the energies match in this way, a resonance can occur between the two states {$CHl_D^*$ $Chl_A$} and {$Chl_D$ $Chl_A^*$}.** ***(b)*** **The energy of light absorbed by pigment-protein complexes in the antenna system hops rapidly from complex to complex by resonance energy transfer until it is trapped in an electron-transfer reaction in a reaction center.**

If the next flash arrived too soon, its energy was wasted, mainly as heat and fluorescence. We know now that *the chlorophyl complexes that undergo oxidation in the reaction centers(P700 and P680 in algae) have to return to their reduced states before they can react again. The acceptors that remove electrons from the chlorophyll complexes also have to be repxidized.* Emerson and Arnold found that the amount of $O_2$ evolved per flash increased as they excited the algae with a larger and larger number of photons on each flash but reached a plateau when they made the flashes sufficiently strong. Even with flashes of saturating intensity and optimal timing, the amount of $O_2$ was very small relative to the chlorophyll content of the algae. The cells contained about 2500 molecules of chlorophyll for each molecule of $O_2$ they evolved. This was very puzzling at the time because it was generally accepted that chlorophyll was intimately involved in the photochemistry of $CO_2$ fixation and $O_2$ evolution. Now it is clear that *most of the chlorophyll is part of the antenna system.* When the flash intensity is high, the amount of $O_2$ evolution that can occur on each flash is limited by the concentration of reaction centers, which is much smaller than the total concentration of chlorophyll. The cnlorophyll or bacteriochlorophyll molecules that comprise the antenna are bound noncovalently to integral membrane proteins. Each polypeptide typically carries only a few pigment molecules. These small complexes aggregate into larger arrays that allow excitations to hop rapidly from complex to complex. Antenna systems also contain a variety of carotenoids and other pigments. *These accessory pigments fill in the antenna's absorption spectrum in regions where chlorophylls do not absorb well.* Chlorophyll *a* absorbs red or blue light well but not green. Carotenoids, which are long, linear polyenes, have absorption bands in the green region. The energy they absorb is transferred to the chlorophyll molecules of the antenna and from there to the reaction centers. This transfer is thermodynamically downhill.

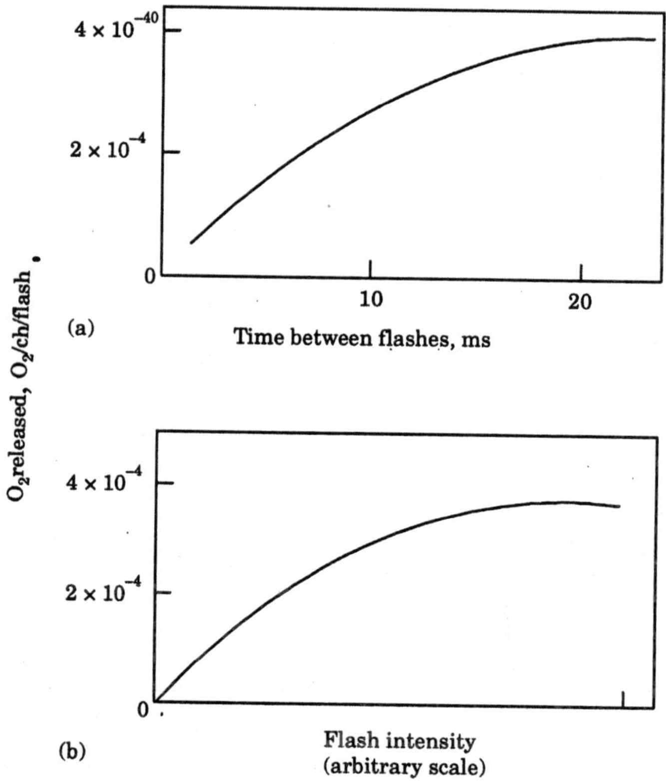


**Fig. 3.14. The amount of $O_2$ released when a suspension of algae *(Chlorella pyrenoidosa)* was excited with a train of short flashes of light, as a function of (*a*) the time between the flashes and *(b)* the intensity of the flashes. Each flash lasted about 10 μs. The total amount of $O_2$ released was measured after several thousand flashes and was divided by the number of flashes and by the amount of chlorophyll in the suspension. (What happens on the first few flashes is discussed later.) The measurements in (*a*) were made with flashes comparable to the strongest flashes used in (*b*). For the measurements shown in (*b*), the flashes were spaced 20 ms apart. Note that the maximum amount of $O_2$ released per flash was only one molecule of $O_2$ per several thousand molecules of chlorophyll. With saturating flashes, the amount of $O_2$ evolution is limited by the concentration of reaction centers, whereas most of the chlorophyll in the algae is part of the antenna system.**

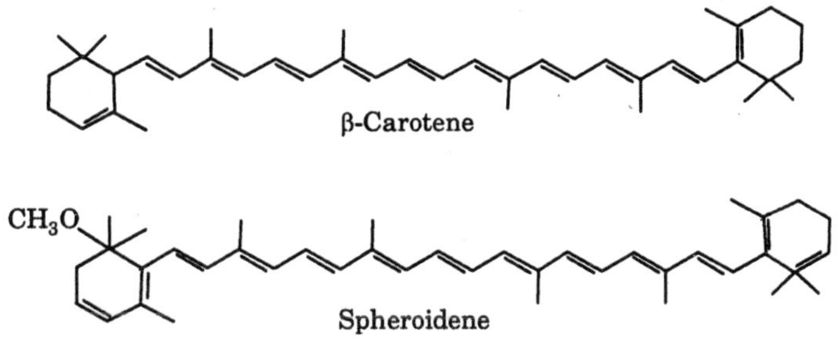

β-Carotene

$CH_3O$

Spheroidene

**Fig. 3.15. Structures of β-carotene, a major carotenoid in many types of plants, and spheroidene, a common carotenoid of photosynthetic bacteria. About 350 different natural carotenoids exist. These vary widely in colour, depending principally on the number of conjugated double bonds.**

## Two Photosystems of Chloroplasts

*We mentioned earlier that chloroplasts contain two reactive chlorophyll complexes: P700 and P680. Together with their antennae and their initial electron acceptors and donors, the reaction centers that contain P700 or P680 form two distinct assemblies called photosystem I and photosystem II. Much evidence indicates that photosystems I and II operate in series.* Excitation of P700 (photosystem I) generates a strong reductant, which transfers electrons to $NADP^+$ by way of several secondary electron carriers. Excitation of P680 (photosystem II) generates a strong oxidant, which oxidizes $H_2O$ to $O_2$. The reductant formed in photosystem II injects electrons into a chain of carriers that connect the two photosystems. This scheme, which is called the *Z scheme*, was first suggested by R. Hill and F. Bendall in 1 960. Note that the Z scheme differs from the bacterial electron-transport chain discussed previously in being linear rather than cyclic. The electron acceptors on the reducing side of photosystem II resemble those of purple bacterial reaction center.

The acceptor that removes an electron from P680 is a molecule of pheophytin *a*. The second and third acceptors are plastoquinones. As in bacterial reaction centers, electrons move one at a time from the first quinone to the second. When the second quinone becomes doubly reduced, it picks up protons from the stromal side of the thylakoid membrane and dissociates from the reaction center. The chain of carriers between the two photosystems includes the *cytochrome $b_6f$ complex* and a copper protein, *plastocyanin*. Like the mitochondria and bacterial cytochrome $bc_1$ complexes, the cytochrome $b_6f$ complex contains a cytochrome with two *b-type* hemes (cytochrome $b_6$), an iron-sulfur protein, and a *c*-type cytochrome (cytochrome *f*). As electrons move through the complex from reduced plastoquinone to cytochrome *f*, plastoquinone probably executes a Q cycle similar to the cycle we presented for UQ in mitochondria and photosynthetic bacteria.

The cytochrome $b_6f$ complex provides electrons to plastocyanin, which transfers them to $P700^{+}_{\bullet}$ in the reaction center of photosystem I. The electron carriers between P700 and $NADP^+$ and between $H_2O$ and P680 are discussed later, after we discuss some of the evidence supporting the Z scheme. The initial observations that led to the proposed Z scheme came from measurements of how the quantum yield of $O_2$ evolution depends on the excitation wavelength. The quantum yield of $O_2$ evolution is the molar ratio of $O_2$ evolved to photons absorbed.

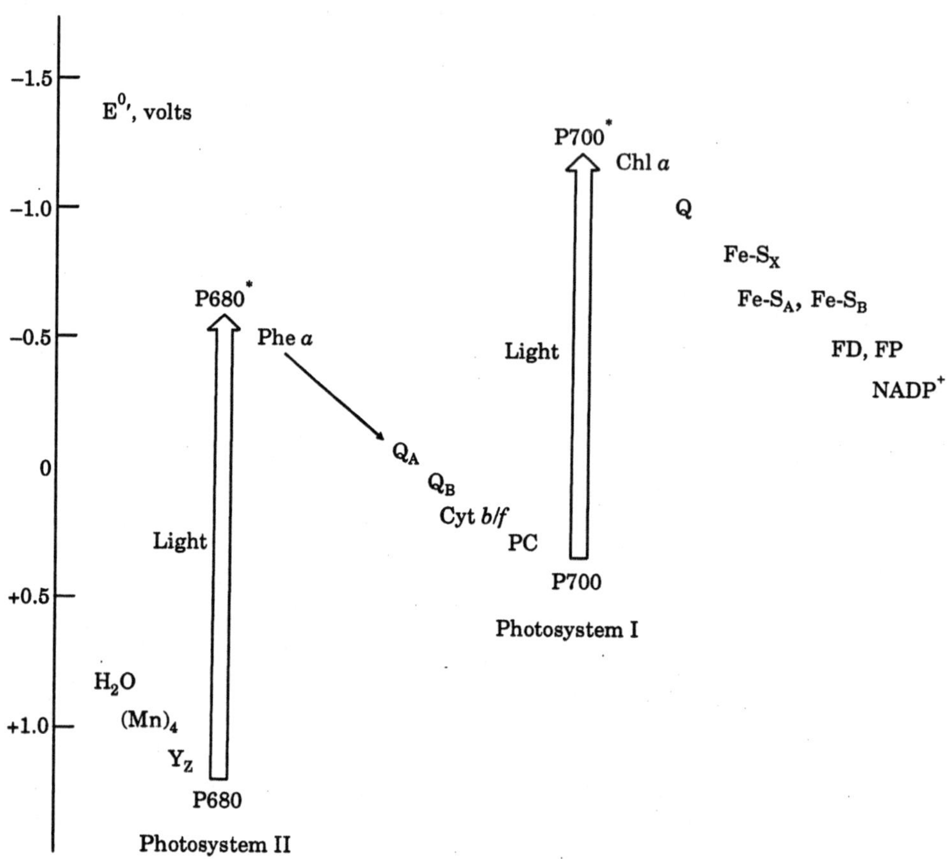


**Fig. 3.16. The *Z* scheme. [$(Mn)_4$ = a complex of four Mn atoms bound to the reaction center of photosystem II; $Y_z$ = tyrosine side chain; Phe *a* = pheophytin *a;* $Q_A$ and $Q_B$ = two molecules of plastoquinone; Cyt *b/f* = cytochrome $b_6f$ complex; PC = plastocyanin; Chl *a* = chlorophyll a; Q = phylloquinone (vitamin $K_1$); Fe-$S_x$, Fe-$S_A$, and Fe-$S_B$ = iron-sulfur centers in the reaction center of photosystem I; FD = ferredoxin; FP = flavoprotein (ferredoxin-NADP oxidoreductase)]. The sequence of electron transfer through Fe-$S_A$ and Fe-$S_B$ is not yet clear.**

In green algae, the quantum yield is nearly independent of wavelength between 400 and 675 nm but falls off drastically in the far-red region near 700 nm. This seems odd, because most of the absorbance in the far-red region is due to chlorophyll *a*. How can light absorbed by chlorophyll *a* be used less efficiently than light absorbed by carotenoids or other accessory pigments? A clue came from Emerson's finding in 1956 that 700-nm light is used more efficiently if it is superimposed on a background of weak blue-green light. It was found subsequently that the blue-green light actually does not have to be presented simultaneously with the red light.

Illumination with blue-green light improves the utilization of far-red light even if the blue-green light is turned off several seconds before the red light is turned on. These observations could be explained by the Z scheme if the absorption spectra of the antennas associated with photosystems I and II are different. Because the two photosystems must operate in series, light is used most efficiently when the flux of electrons through photosystem II is equal to that through photosystem I. If light of a particular wavelength excites one photosystem more frequently than the other, some of the light is wasted. It turns out that photosystem I can absorb and use virtually all wavelengths of light up to about 740 nm. Photo-system II, however, is not excited well by far-red light.

The pool of electron carriers between the two photosystems is, therefore, drained of electrons if algae are illuminated with far-red light in the absence of added reductants. P700 remains oxidized and is unable to respond to the light. Photosystem II does absorb blue-green light well, so illumination with green light replenishes the pool. Because electrons remain in the pool when the blue-green light is turned off, far-red light can be used effectively for a time even after a period of darkness. Subsequent experiments by Louis Duysens provided strong support for the idea that *photosystem I withdraws electrons from carriers between the two photosystems, whereas photosystem II feeds electrons to these components.*

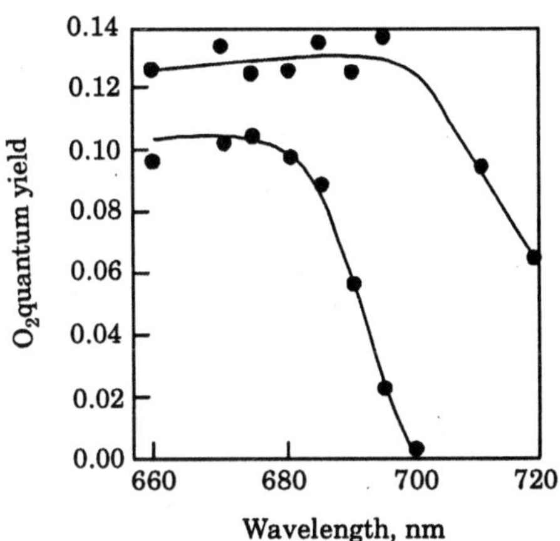


**Fig. 3.17. The quantum yield of $O_2$ evolution from a suspension of algae *(chlorella pyrenoidosa)* as a function of the excitation wavelength. Measurements were made without any supplementary light (lower curve) and with supplementary blue-green light (upper curve). The quantum yield was calculated as moles of $O_2$ evolved per einstein of light incident on the sample; $O_2$ evolution caused by the supplementary light alone was subtracted. Without the supplementary light the quantum yield falls off precipitously in the far-red region above 680 nm. The antenna of photosystem I absorbs light well in this region, but that of photosystem II does not. Supplementary blue-green light, which is absorbed well by photosystem II, increases the quantum yield with which the far-red light can be used.**

Duysens and his colleagues showed that illuminating algae with far-red light causes cytochrome *f* to become oxidized, as we expect if far-red light drives photosystem I but not photosystem II. When a supplementary green light is turned on, some of the cytochrome returns to the reduced state. This makes sense if green light can drive photosystem II. During the illumination, individual cytochrome molecules cycle repeatedly between the oxidized and reduced states, so that the cytochrome population as a whole is in a steady state of partial oxidation. Green light actually can cause either oxidation or reduction of cytochrome *f* depending on the initial conditions, so it must be able to excite either photosystem II *or* photosystem I. Far-red light, however, can only cause oxidation of the cytochrome. The electron carriers in photosystem II are located mainly in the stacked, granal regions of the thylakoid membrane, whereas those of photosystem I are mainly in the unstacked stromal lamellae.

The cytochrome $b_6f$ complex is found in both regions of the membrane. The distribution of photons between the two photosystems depends partly on the relative numbers of antenna complexes in the two regions of the membrane and can change as a result of movements of antenna complexes from one region to another. Movements of one type of antenna complex appear to be regulated by phosphorylation of the protein. If chloroplasts are illuminated with light that is absorbed better by photosystem I than by photosystem II, the mobile antenna complex moves into the stacked regions, and a larger fraction of the excitations then is passed to photosystem II. If the reaction centers of photosystem I and photosystem II are segregated into separate regions of the thylakoid membrane, how can electrons move from photosystem I to photosystem II? Evidently the plastoqumone that is reduced in photosystem II can diffuse rapidly in the membrane, just as ubiquinone does in the mitochondrial inner membrane.

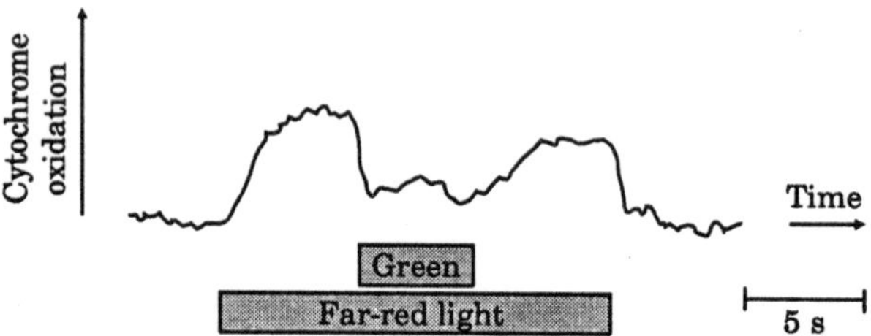


**Fig. 3.18. Illumination of a suspension of red algae (*Porphyridium cruentum*) with far-red light (680 nm) causes an oxidation of cytochrome *f*, as measured by an absorbance change at 422 nm. (The box below the trace indicates the period when the far-red light was on.) Green (562-nm) light superimposed on top of the far-red light causes the cytochrome to become more reduced. Additional measurements showed that green light also could cause cytochrome oxidation if it was turned on in the absence of the far-red light when the cytochrome was initially reduced. This result indicates that the green light can drive either photosystem I or photosystem II. Far-red light can cause only cytochrome oxidation, because it is absorbed only by photosystem I.**

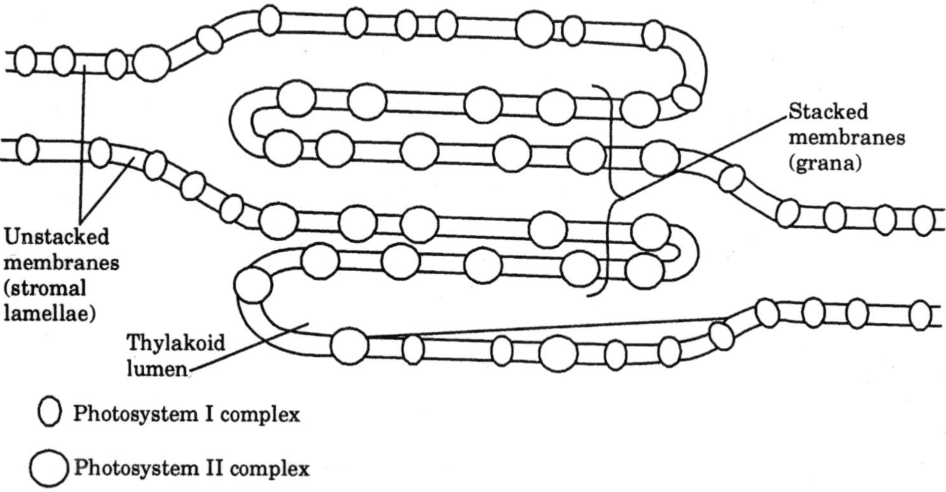


**Fig. 3.19. The photosystem II complexes are located mainly in the stacked (granal) regions of the thylakoid membrane, whereas photosystem I complexes are most abundant in the unstacked stromal lamellae. The cytochrome $b_6f$ complex is located in both regions of the membrane. Some types of antenna complexes also reside in both regions and can shift from one region to the other in response to a change in conditions. The ATP-synthase occurs mainly in the unstacked regions. These conclusions came from studies in which thylakoid membranes were fragmented gently by ultrasound and then separated by density-gradient centrifugation into fractions representing stacked and unstacked membranes.**

Plastoquinone thus carries electrons from photosystem II to the cytochrome $b_6f$ complex. Plastocyanin acts similarly as a mobile electron carrier from the cytochrome $b_6f$ complex to the reaction center of photosystem I, just as cytochrome *c* carries electrons from the mitochondrial cytochrome $bc_1$ complex to cytochrome oxidase and as a c-type cytochrome provides electrons to the reaction centers of purple bacteria.

## Photosystem

On the reducing site of photosystem I, the initial electron acceptor appears to be a molecule of chlorophyll *a*. The second acceptor probably is a quinone, phylloquinone (vitamin $K_1$). In these respects, photosystem I resembles photosystem II and purple photosynthetic bacteria, which use pheophytin *a* or bacteriopheophytin *a* followed by a quinone. From this point on, photosystem I is different: its next electron carriers consist of iron-sulfur proteins instead of additional quinones. Photosystem I contains three iron-sulfur clusters firmly associated with the reaction center.

The cysteines of Fe-Sx are provided by the two main polypeptides of the reaction center, which also bind P700 and its initial electron acceptors; Fe-$S_A$ and Fe-$S_B$ are on a separate polypeptide. The quinone that is reduced in photosystem I probably transfers an electron to Fe-$S_x$, which in turn reduces Fe-$S_A$ and Fe-$S_B$. From here, electrons move to *ferredoxin, a* soluble iron-sulfur protein found in the chloroplast stroma, then to a flavoprotein *(ferredoxin-NADP oxidoreductase)*, and finally to $NADP^+$.

## Photosystem II

Oxidation of two molecules of $H_2O$ to $O_2$ requires removing four electrons for each $O_2$ produced. According to the Z scheme each electron must traverse the photochemical reactions of both photosystem I and photosystem II, so at least eight photons (2 × 4) must be absorbed for each $O_2$ released. Experimental measurements of the number of photons that are needed, the quantum requirement, are needed on the order of 8 – 12. (The quantum requirement is the reciprocal of the quantum yield of $O_2$ evolution.) The quantum requirement exceeds eight if some of the photons that are absorbed are lost as heat or fluorescence from the antenna or if some of the electrons pumped through the photosystems are not removed to $NADP^+$ but instead cycle back into the electron-transport chain between the two photosystems. If the photosystem II reaction center transfers only one electron at a time, how does it assemble the four oxidizing equivalents needed for oxidation of $H_2O$ to $O_2$?

One possibility is that several different photosystem II reaction centers cooperate, but this seems not to happen. Instead, each reaction center progresses independently through a series of oxidation states, advancing to the next state each time it absorbs a photon. In this event $O_2$ evolution would occur only when a reaction center has accumulated four oxidizing equivalents. Support for this conclusion comes from measurements made by Pierre Joliot of the amount of $O_2$ evolved on each flash when chloroplasts are excited with a series of short flashes after a period of darkness. No $O_2$ is released on the first or second flashes but on the third flash, there is a burst of $O_2$. After this, the amount of $O_2$ released on each flash oscillates, going through a maximum every fourth flash. Bessel Kok pointed out that this pattern can be explained if the photosystem II reaction center cycles through five different oxidation states, $S_0$ through $S_4$.

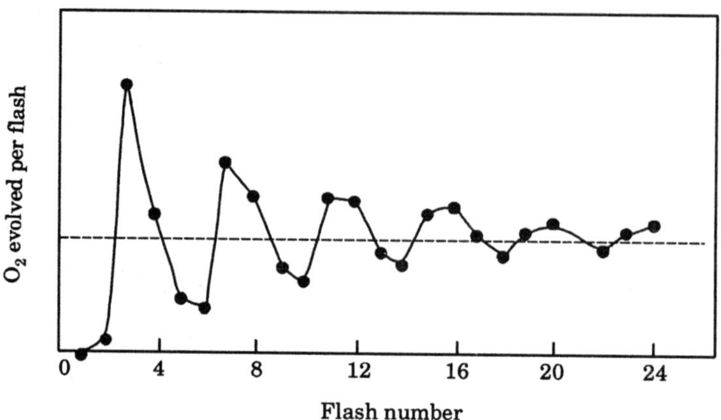


**Fig. 3.20. $O_2$ evolution from a suspension of chloroplasts that was excited with a series of 24 flashes after having been kept in the dark for several minutes. Little or no $O_2$ is evolved on the first two flashes. $O_2$ evolution peaks on the third flash and on every fourth flash thereafter. The oscillations in $O_2$ evolution are damped, and after many flashes the yields converge on the level indicated by the horizontal line.**

When the system reaches state $S_4$, $O_2$ is given off, and the reaction center returns to state $S_0$. The fact that the first burst of $O_2$ comes on the third flash instead of the fourth can be explained if the reaction centers relax mainly into $S_1$ rather than into $S_0$ during the dark period before the flashes. The Component that undergoes oxidation as photosystem II progresses from one S state to the next is a complex of four atoms of manganese bound to the two central polypeptides of the reaction center. $P680^{+}_{\bullet}$ draws electrons from the Mn complex by way of a tyrosine in one of these polypeptides.

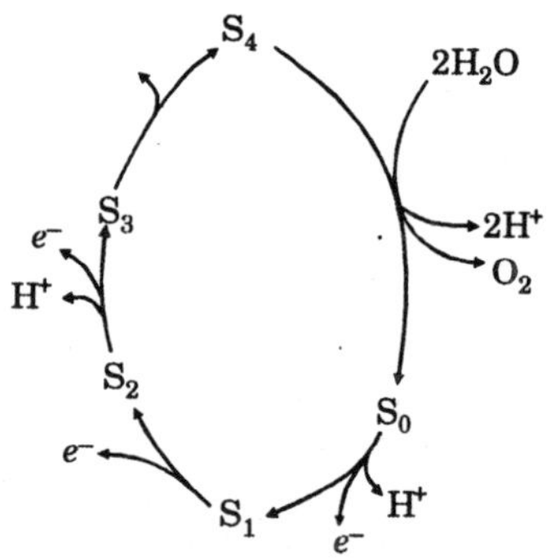

**Fig. 3.21. The five oxidation states of the $O_2$-evolving apparatus. One electron ($e^-$) is removed photochemically in each of the transitions between states $S_0$ and $S_4$. $S_4$ decays spontaneously, releasing $O_2$. Protons are released at several steps of the cycle.**

In the course of this reaction, the phenolic side chain of the tyrosine is oxidized transiently to a free radical. Although the structure of the Mn complex is not yet known, most of the transitions of the complex probably represent sequential oxidations of Mn atoms from the Mn(III) level to Mn(IV). The $E°$ of the $P680^{+}_{\bullet}/P680$ redox couple must be greater than +1 V, because the oxidized Mn complex that $P680^{+}_{\bullet}$ generates is powerful enough to oxidize water. At pH 5, which is approximately the pH of the thylakoid lumen, the $O_2/H_2O$ couple has an $E°$ of +0.94 V. $P700^{+}_{\bullet}/P700$, by contrast, has an $E°'$ of only about +0.5 V. It is surprising that $P680^{+}_{\bullet}$ is a much stronger oxidant than $P700^{+}_{\bullet}$, since they both consist of chlorophyll *a* dimers.

Apparently, the redox potential of the chlorophyll radical could depend strongly on the surroundings in the protein.

## Flow of Electrons from $H_2O$ to $NADP^+$

As in mitochondria and purple photosynthetic bacteria, the flow of electrons through the chloroplast's cytochrome $b_6f$ complex pumps protons across the thylakoid membrane. When plastoquinone undergoes reduction in photosystem II and in the cytochrome $b_6f$ complex, protons are taken up from the stromal side of the membrane. When the plastoquinone is reoxidized, protons are released into the thylakoid lumen. This lowers the pH in the lumen, and makes the inside positively charged with respect to the stroma. Four protons probably move across the membrane for each pair of electrons that proceed to photosystem I. Two more protons are released in the lumen for each molecule of $H_2O$ that is oxidized to $\frac{1}{2}O_2$, and one proton is taken up from the stroma for each molecule of $NADP^+$ that is reduced to NADPH. Proton movement back out through the thylakoid membrane is conducted by an ATP-synthase, and this movement drives the formation of ATP.

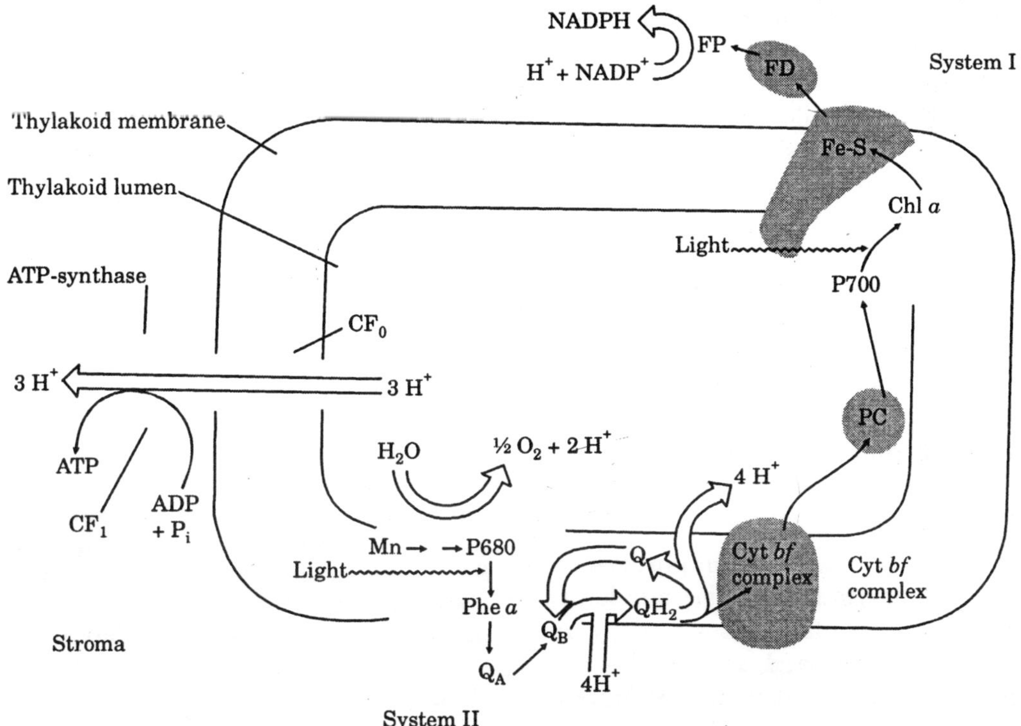


**Fig. 3.22. Transport of two electrons from photosystem II through the cytochrome $b_6f$ complex to photosystem I results in the movement of four protons from the chloroplast stroma to the thylakoid lumen. The proton translocation probably occurs in a Q cycle resembling. Two more protons are released in the lumen for each molecule of $H_2O$ that is oxidized to $O_2$, and one additional proton is removed from the stroma for each molecule of $NADP^+$ reduced to NADPH. (Two protons are taken up when the flavoprotein is reduced by photosystem I; one ends up on NADPH, and the other is returned to the solution.) The flow of protons from the thylakoid lumen back to the stroma through an ATP-synthase ($CF_0$-$CF_1$) drives the formation of ATP. The abbreviations used in this figure are the same.**

The chloroplast ATP-synthase is structurally very similar to the mitochondria ATP-synthase Its head-piece($CF_1$) contains the catalytic sites; its hydrophobic base-piece ($CF_0$) includes the proton-conducting channel. Observations in chloroplasts played a key role in the development of the chemiosmotic theory of oxidative phosphorylation.

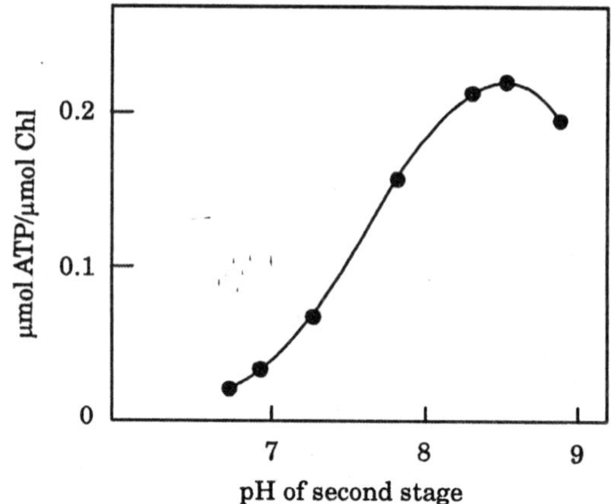


**Fig. 3.23. Chloroplasts can use an electrochemical potential gradient for protons to form ATP in the dark. In a two-step experiment, spinach chloroplasts first were equilibrated at an acidic pH (pH 4.0). The pH of the solution then was raised quickly to a value between 6.7 and 8.8, as indicated on the abscissa of the graph, and ADP and $P_i$ were added. Because the pH in the thylakoid lumen was buffered at pH 4, the sudden increase in the external pH created a pH gradient of between 2.7 and 4.8 pH units across the thylakoid membrane. Proton efflux through the ATP-synthase can drive the formation of ATP. Under optimal conditions, one molecule of ATP was formed for approximately every five chlorophylls in the sample (see the ordinate scale on the graph). For comparison, chloroplasts contain only about one molecule of P700 and one of P680 per 400 chlorophylls. ATP formation caused by acid-base transition of spinach chloroplasts.**

André Jagendorf and his colleagues discovered that if chloroplasts are illuminated in the absence of ADP, they developed the capacity to form ATP when ADP was added later, after the light was turned off. The amount of ATP synthesized was much greater than the number of electron-transport assemblies in the thylakoid membranes, so the energy to drive the phosphorylation could not have been stored in an energized form of one of the electron carriers. Protons were taken up from the solution outside the thylakoid membranes during the illumination, and the ability to form ATP correlated with the magnitude of the pH gradient across the membrane. If a pH gradient in the right direction was set up across the thylakoid membrane by raising the pH on the outside, chloroplasts could make ATP without any illumination at all. Formation of ATP by photosynthetic systems is often called photophosphorylation.

Although photophosphorylation and oxidative phosphorylation are very similar, they do differ in a few details. The respiratory chain pumps protons outward across the inner membrane, from the mitochondrial matrix to the intermembrane space the chloroplast electron-transport chain pumps protons into the thylakoid lumen. The electrochemical potential gradient for protons across the mitochondrial membrane consists mainly of an electric potential ($\Delta\psi$); in chloroplasts it consists mainly of a pH gradient ($\Delta$pH), which can be greater than 3 pH units during strong illumination. In both cases, however, the catalytic head-piece of the ATP-synthase is on the side of the membrane that becomes more alkaline.

The relative importance of $\Delta$pH and $\Delta\psi$ probably depends on the permeability of the membrane to other ions such as $Cl^-$. Movement of $Cl^-$ into the thylakoid lumen reduces

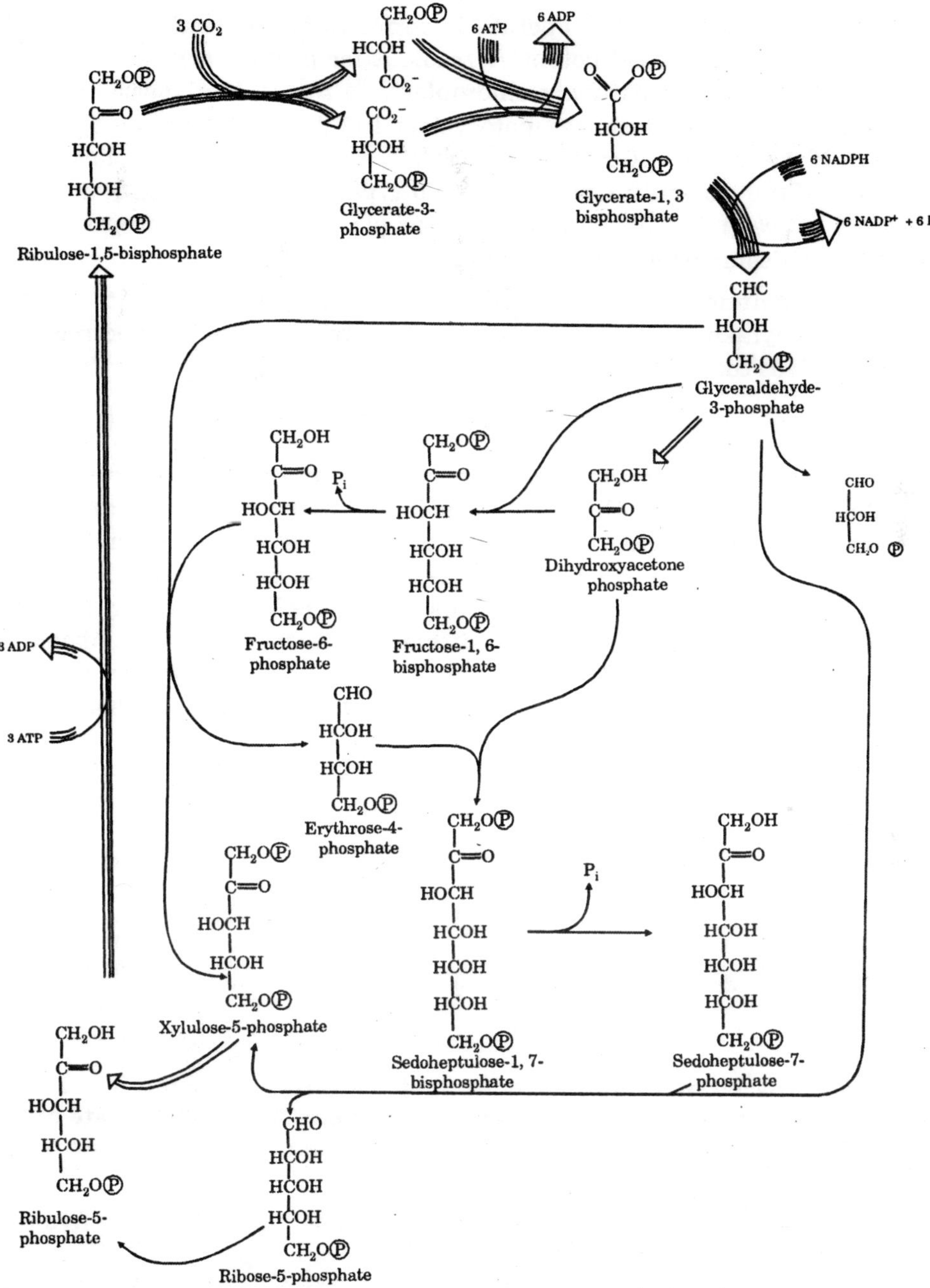


**Fig. 3.24. The reductive pentose cycle, or Calvin cycle. The number of arrows drawn at each step in the diagram indicates the number of molecules proceeding through that step for every three molecules of $CO_2$ that enter the cycle. The entry of three molecules of $CO_2$ results in the formation of one molecule of glyceraldehyde-3-phosphate (box on right), and requires the oxidation of six molecules of NADPH to $NADP^+$ and the breakdown of nine molecules of ATP to ADP.**

$\Delta\psi$ allowing for the buildup of a larger pH gradient. In the Z scheme, photosystem II, the cytochrome $b_6f$ complex and photosystem I operate in series to move electrons from $H_2O$ to $NADP^+$ and to create an electrochemical potential gradient for protons across the thylakoid membrane. In addition to this linear pathway, chloroplasts in some plant species may use a cyclic electron-transfer scheme that includes photosystem I and the cytochrome $b_6f$ complex but not photosystem II. If the concentration of $NADP^+$ is low, electrons ejected by photosystem I can be passed from one of the iron-sulfur proteins to a *b*-type cytochrome, and from there to the plastoquinone pool between two photosystems. From plastoquinone, electrons return to P700 via the cytochrome $b_6f$ complex and plastocyanin.

The Q cycle that is included within this cyclic electron pathway results in proton translocation across the thylakoid membrane and thus contributes to the electrochemical potential gradient that drives the ATP-synthase.

## Carbon Fixation

*The ATP and NADPH generated by the photosynthetic electron-transfer reactions are used for fixation of $CO_2$.* Reactions in which $CO_2$ is incorporated into carbohydrates were discovered in the 1950s by Melvin Calvin and his coworkers in some of the first biochemical studies employing radioactive tracers. A key experiment was to expose algae to a brief period of illumination in the presence of $^{14}CO_2$ and then to disrupt the cells and search for organic molecules that had become labeled with $^{14}C$. The material that turned out to be labeled most rapidly was glycerate-3-phosphate, and almost all of its $^{14}C$ was in the carboxyl group. Further experiments showed that the glycerate-3-phosphate is formed from ribulose- 1, 5-bisphosphate.

Ribulose-1, 5 bisphosphate ($CH_2O$Ⓟ–C=O–HCOH–HCOH–$CH_2O$Ⓟ) —Ribulose bisphosphate carboxylase ($H^+$)→ [Enolate ($CH_2O$Ⓟ–C–$O^-$ = C–OH–HCOH–$CH_2O$Ⓟ)] —$CO_2$→ [2-Carboxy-3-ketoarabinitol-bisphosphate ($CH_2O$Ⓟ–HO–C–$CO_2^-$–C=OH–HCOH–$CH_2O$Ⓟ)] —Ribulose bisphosphate carboxylase ($OH^-$)→ 2 Glycerate-3-phosphate ($CO_2^-$–HCOH–$CH_2O$Ⓟ)

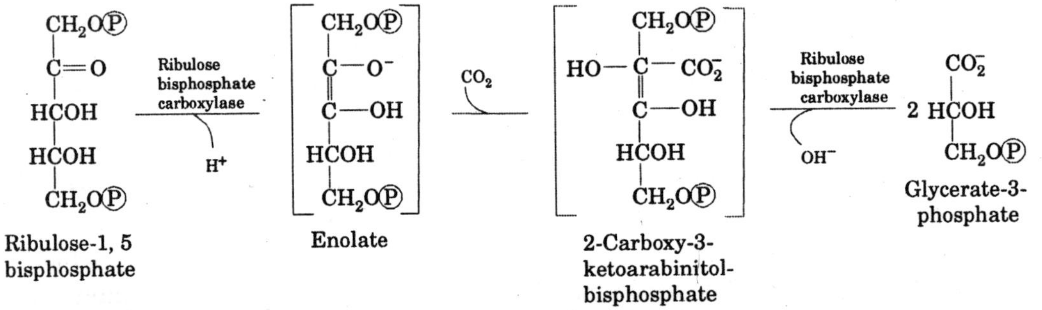

**Fig. 3.25. The reaction catalyzed by ribulose bisphosphate carboxylase involves 2-carboxy-3-ketoarabinitol-1, 5-bisphosphate as an enzyme-bound intermediate. The intermediate probably forms by the addition of $CO_2$ to the enolate of ribulose-1, 5-bisphosphate. The substrate is known to be $CO_2$ rather than bicarbonate.**

A molecule of $CO_2$ is incorporated in the process, so that one molecule of the five-carbon sugar ribulose-1, 5-bisphosphate yields two molecules of glycerate-3-phosphate. The reaction is catalyzed by ribulose bisphosphate carboxylase in the chloroplast stroma. Other enzymes in the stroma can convert glycerate-3-phosphate back to ribulose-1, 5-bisphosphatc. The reactions of this reductive pentose cycle, or Calvin cycle. 3-Phosphoglycerate is first phosphorylated to glycerate-1, 3-bisphosphate at the expense of ATP and then reduced to glyceraldehyde-3-phosphate by NADPH. (Note that the nucleotide specificity in the reductive step differs from that of the cytosolic glyceraldehyde-3-phosphate dehydrogenase, which uses $NAD^+$ and NADH.) Glyceraldehyde-3-phosphate and its isomerization product dihydroxyacetone phosphate can combine to form fructose-1, 6-bisphosphate under the influence of aldolase. Fructose-6-phosphate, formed by hydrolysis of the fructose-1, 6-bisphosphate, combines with another molecule of glyceraldehyde-3-phosphate, generating xylulose-5-phosphate and erythrose-4-phosphate.

The enzyme that catalyzes this reaction is similar to the transketolase of the pentose phosphate pathway. The erythrose-4-phosphate that is formed goes on to combine with dihydroxyacetone phosphate to give sedoheptulose-1, 7-bisphosphate in a second reaction catalyzed by aldolase. After hydrolysis to sedoheptulose-7-phosphate, the seven-carbon sugar reacts with glyceraldehyde-3-phosphate in another transketolase reaction, forming ribose-5-phosphate and a second molecule of xylulose-5-phosphate. Xylulose-5-phosphate and ribose-5-phosphate both can be isomerized to ribulose-5-phosphate. Finally, ribulose-5-phosphate is phosphorylated to ribulose-1, 5-bisphosphate at the expense of an additional ATP, completing the cycle. In figure 3.25 note that three molecules of ribulose-1, 5-bisphosphate are regenerated for every three that are carboxylated.

In the process, there is a net gain of one molecule of glyceraldehyde-3-phosphate. The metabolic cost is nine molecules of ATP converted to ADP and six molecules of NADPH oxidized to $NADP^+$, or three molecules of ATP and two of NADPH. The glyceraldehyde-3-phosphate that is produced is exported from the chloroplast to the cytosol or converted to hexoses for storage in the chloroplast as starch.

## Photorespiration and the C-4 Cycle

Ribulose-bisphosphate carboxylase typically accounts for more than half the soluble protein in a leaf. It is the world's most abundant enzyme and probably the most abundant protein. The enzyme found in plants has eight copies each of two types of subunits. The larger subunit contains the catalytic site; the role of the smaller subunit is unclear. In addition to serving as a substrate, $CO_2$ activates ribulose-bisphosphate carboxylase by binding to the $\epsilon$-amino group of a lysyl residue in the large subunit to form a carbamate (lysine—NH—$CO_2^-$). This process requires $Mg^{2+}$, which binds to the carboxyl group of the carbamate.

The $Mg^{2+}$ in turn forms part of the binding site for a second molecule of $CO_2$, which acts as the substrate in the carboxylase reaction. The reactive molecule of $CO_2$ probably adds to the enolate of ribulose-1, 5-bisphosphate to form 2-carboxy-3-ketoarabinitol-1, 5-bisphosphate as an intermediate. The same active site of ribulose bisphosphate carboxylase also catalyzes a competing reaction in which $O_2$ replaces $CO_2$ as a substrate. The products of this *oxygenase* reaction are 3-phosphoglycerate and a two-carbon acid, 2-phosphoglycolate. Phosphoglycolate is oxidized to $CO_2$ by $O_2$ in additional reactions involving enzymes in the cytosol, mitochondria, and another organelle, the peroxisome.

This *photorespiration* is not coupled to oxidative phosphorylation, and it appears to constitute a severe drain on chloroplast metabolism. In some plants, as much as one-third of the $CO_2$ that is fixed is released again by photo-respiration. If photorespiration has any benefit to the plant, it is not obvious. The outcome of the competition between the carboxylase and oxygenase reactions depends on the concentrations of $CO_2$ and $O_2$. The Km of the activated enzyme for $CO_2$ is about 20 μM, and that for $O_2$ is about 200 μM, so $CO_2$ is the referred substrate.

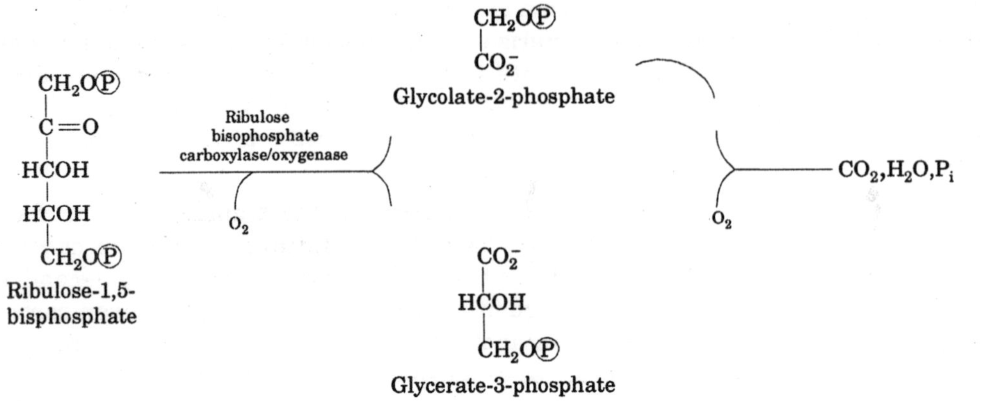

$CH_2O(P)$
|
$C{=}O$
|
HCOH
|
HCOH
|
$CH_2O(P)$
Ribulose-1,5-bisphosphate

Ribulose bisphosphate carboxylase/oxygenase
$O_2$

$CH_2O(P)$
|
$CO_2^-$
Glycolate-2-phosphate

$O_2$
$CO_2, H_2O, P_i$

$CO_2^-$
|
HCOH
|
$CH_2O(P)$
Glycerate-3-phosphate

**Fig. 3.26. Photorespiration results from the oxygenation reaction catalyzed by ribulose bisphosphate carboxylase/oxygenase. 2-Phosphoglycolate generated by the reaction moves from the chloroplast to the cytosol, where other enzymes break it down to $CO_2$, $H_2O$, and $P_i$. The oxygenation reaction, like carboxylation, does not require light directly. It occurs mainly during illumination, however, because the formation of the substrate, ribulose-1, 5-bisphosphate, requires ATP and NADPH.**

In illuminated chloroplasts, however, the concentration of $O_2$ is elevated as a result of the photosynthetic oxidation of $H_2O$. The concentrations of the two gases are frequently on the order of the $K_m$ values, making $O_2$ a serious competitor. Plants that live in dry climates face the additional dilemma that opening the stomata in the leaves to improve the exchange of $CO_2$ and $O_2$ with the atmosphere can result in dehydration. If photorespiration is only a liability, natural selection presumably has favoured evolution of a ribulose bisphosphate carboxylase with minimal activity as an oxygenase. Nonetheless, the oxygenase activity of the enzyme in higher plants is only slightly below that of the enzymes in photosynthetic bacteria.

*Some species of plants have evolved an alternative strategy for favouring the carboxylase over the oxygenase: They increase the concentration of $CO_2$ in the region of the enzyme.* These plants, which include corn, sugar cane, and numerous tropical species, have a layer of specialized mesophyll cells at the outer surface of the leaf. Mesophyll cells use $CO_2$ from the air to convert phosphoenolpyruvate to oxaloacetate. Oxaloacetate is reduced to malate, which then moves to the *bundle sheath cells* that surround the vascular structures in the interior of the leaf. Here malate is 'decarboxylated to pyruvate in an oxidative reaction that reduces $NADP^+$ to NADPH.

The pyruvate returns to the mesophyll cells, where it is phosphorylated to phosphoenolpyruvate. This phosphorylation is driven by splitting of ATP to AMP and pyrophosphate and subsequent hydrolysis of the pyrophosphate to phosphate. This cyclic series of reactions delivers $CO_2$ and reducing power (NADPH) to the bundle sheath cells, at a cost of hydrolyzing one ATP to one AMP. The bundle sheath cells fix the $CO_2$ by the ribulose bisphosphate carboxylase reaction and the reductive pentose cycle. Plants that have this mechanism for concentrating $CO_2$ can grow considerably more rapidly than species that do not, particularly in hot, dry climates. This auxiliary cycle is called the C-4 cycle because it involves the four-carbon acids malate and oxaloacetate.

Currently interest has arisen in the possibility of increasing agricultural productivity by using genetic engineering techniques to introduce the C-4 pathway into plant species that lack it.

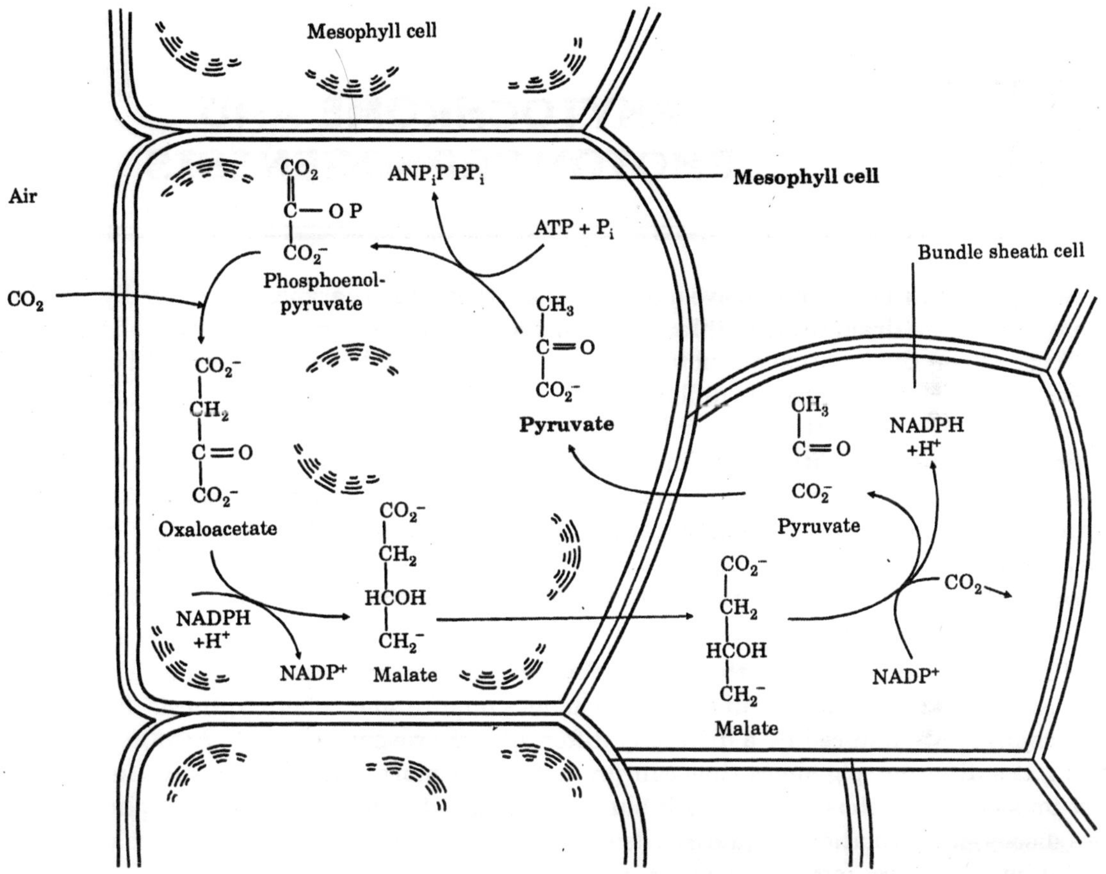


**Fig. 3.27. The C-4 Pathway requires the cooperation of two types of cells. Mesophyll cell (left)take up $CO_2$ from the air and export malate to the bundle sheath cells (right). The bundle sheath cells return pyruvate to the mesophyII cells and fix the $CO_2$ using ribulose bisphosphate canboxylase and the reductive pentose cycle.**

# CHAPTER 4 PHYTOCHROME AND PHOTOMORPHOGENESIS

Seedlings grown in darkness ave a pale, almost ethereal appearance. This spindly, "etiolated" form of growth is dramatically different from the stockier, green appearance of light-grown seedlings. Given the key role of photosynthesis in plant metabolism, one would be tempted to attribute much of this contrast to differences in the availability of light-derived metabolic energy. However, it takes very little light or time to initiate the transformation from the etiolated to the green appearance. Within 10 min after applying a single flash of relatively dim light to a dark-grown pea seedling, one can measure a decrease in the rate of stem extension, the beginning of apical hook straightening, and the initiation of the synthesis of pigments characteristic of green plants. Light has acted as a signal to induce a change in the form of the seedling from one that facilitates growth beneath the soil to one more adaptive to growth above ground in the light.

Photosynthesis cannot be the driving force of this transformation because chlorophyll is not present during this time. Full de-etiolation does require some photosynthesis, but the initial rapid changes are induced by a distinctly different light response called *phototnorphogenesis.* Among the different pigments that can promote photomorphogenic responses in plants, the most important are those that absorb blue light and red light. The blue-light photoreceptors were discussed in relation to guard cells and phototropism. The focus in this chapter will be the red-light photoreceptor, *phytochrome.*

## Biochemical Properties of Phytochrome

Although phytochrome was not identified as a unique chemical species and named until 1959, various red light-induced morphogenic responses in plants had been well documented since the 1930s. The list of such responses is now enormous and includes one or more responses at almost every stage in the life history of a wide range of different green plants. A key breakthrough in the history of phytochrome was the discovery that the effects of red light (650-680 nm) on morphogenesis could be reversed by a subsequent irradiation with light of longer wavelengths (710-740 nm), called far-red light.

The first demonstrated case of red/far-red light photoreversibility was lettuce seed germination: red light promoted germination, and far-red light given immediately after the red light reversed its effects and inhibited germination. It was predicted that the photo receptor for this photoreversible system must itself be photoreversible; that is, it would change its own absorption properties (change colour) reversibly after red and far-red light treatments. An instrument to detect photoreversible colour changes in plants was designed, and in 1959 this prediction was verified by observing such reversible changes in both plants and plant extracts.

By using photoreversibility as a direct qualitative and quantitative assay of this pigment, scientists were soon able to purify and chemically characterize phytochrome as a protein pigment containing a specific light-absorbing molecule, or chromophore. Because knowledge of the photochemical and biochemical properties of purified phytochrome has shed light on how phytochrome functions in vivo, we will review these properties in this section.

## Forms of Phytochrome

In unirradiated plants, phytochrome is present in a red light-absorbing form referred to as Pr. This form, blue in colour, is converted by red light to a far- red light -absorbing form called Pfr, which is blue-green in colour. The Pfr form, in turn, can be converted back to Pr by far-red light. This quality of photoreversibility is the most distinctive property of phytochrome, and it may be expressed in an abbreviated form as follows:

$$\text{Pr} \underset{\text{Far-red light}}{\overset{\text{Red light}}{\rightleftharpoons}} \text{PFr}$$

**Table 4.1. Typical photoreversible responses induced by phytochrome in a variety of higher and lower plants.**

| *Group* | *Genus* | *Stage of development* | *Pfr effects* |
|---|---|---|---|
| Angiosperm | *Lactuca* (Lettuce) | Seed | Promotes germination |
| | *Avena* (oat) | Seeding (etiolated) | Promotes de-etiolation (*e.g.*, leaf unrolling) |
| | *Sinapis* (mustard) | Seedling | Promotes formation of leaf primordia, development of primary leaves |
| | *Pisum* (pea) | Adult | Inhibits internode elongation |
| | Xanthium (cocklebur) | Adult | Inhibits flowering (photoperiodic response) |
| Gymnosperm | *Pinus* (pine) | Seedling | Enhances rate of chlorophyll accumulation |
| Pteridophyta | *Onoclea* (sensitive fern) | Young gametophyte | Promotes growth |
| Bryophyta | *Polytrichum* | Germling | Promotes plastid replication |
| Chlorophyta | *Mougeotia* | Mature gameotophyte | Promotes chloroplast orientation to directional dim light |

The interconversion of the Pr and Pfr forms can be measured in vivo or in vitro. In fact, most of the spectral properties of carefully purified phytochrome are the same as those observed in vivo. The Pr form of phytochrome absorbs very little in the far-red region of the spectrum, but the spectra of Pr and Pfr have significant overlap in the red region. Because red light is absorbed by both forms, when Pr, molecules are given red light most of them will absorb it and be converted to Pfr, but then some of the Pfr will also absorb the red light and be converted back to Pr.

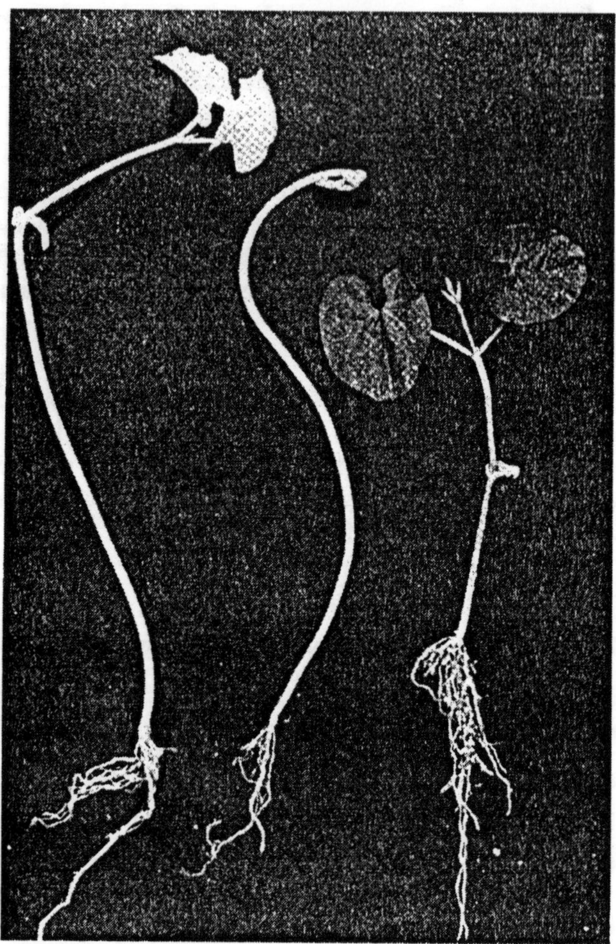

**Fig. 4.1. Bean *(Phaseolus vulgaris)* seedlings grown for six days either in continuous white light (right), in total darkness (center), or in darkness but with 5 min of dim red light given once a day (left). Note that 5 min of dim red light is sufficient to prevent some of the symptoms of etoliation, such as reduced leaf size and maintenance of the apical hook.**

The proportion of phytochrome in the Pfr form after saturating red-light irradiation is only 85%. Similarly, the very small amount of far-red light absorbed by Pr makes it impossible to convert Pfr entirely to Pr by broad-spectrum far-red light, though an equilibrium of 97% Pr and 3% Pfr can be achieved. This equilibrium is termed the photostationary state. Note that both forms of phytochrome also absorb in the blue region of the spectrum. However, most blue-light effects on plants are thought to be due to the action of a separate blue-light photoreceptor rather than phytochrome.

## Intermediates Between Pr and Pfr

The photoconversions of Pr to Pfr and of Pfr to Pr are not one-step processes. There are forms with intermediate absorbances between Pr and Pfr. By irradiating phytochrome with very brief flashes of light, absorption changes that occur in less than a millisecond can be observed. This technique shows that after Pr absorbs a flash of red light, several short-lived spectral forms are generated in a reproducible sequence before Pfr is formed. A different set of intermediate spectral forms occur in the photoconversion of Pfr to Pr.

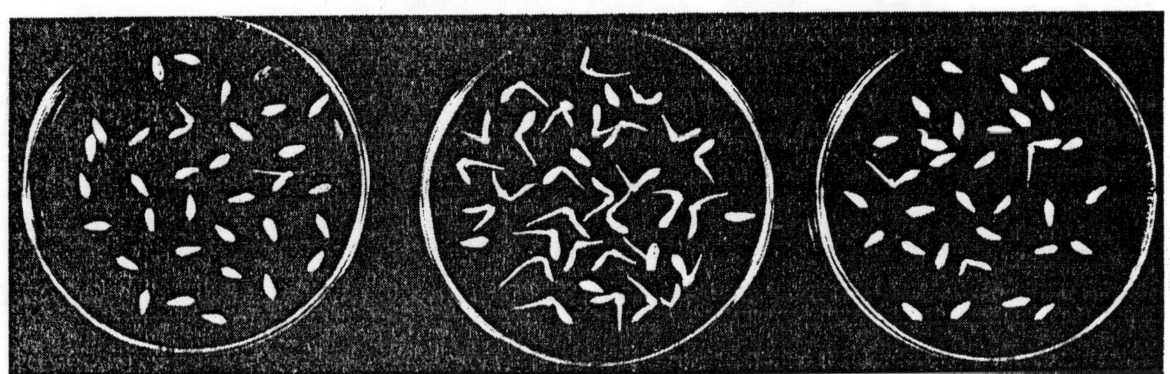

**Fig. 4.2. The promotive effect of red light (R) on lettuce seed germination is reversed by far-red light (FR). Seeds irradiated by red light followed by far-red light (R-FR) have about the same germination rate as those that have been kept in darkness (D). Lettuce seed germination is a typical photoreversible response controlled by phytochrome.**

Sunlight includes a mixture of all visible wavelengths. Under such white-light conditions, both Pr and Pfr are excited, and phytochrome cycles continuously between Pr and Pfr. In this situation, the intermediate forms of phytochrome accumulate and make up a significant fraction of the total? phytochrome. It is possible that such intermediates could even play a role in initiating or amplifying phytochrome responses under natural sunlight conditions, but the question has as yet not been resolved.

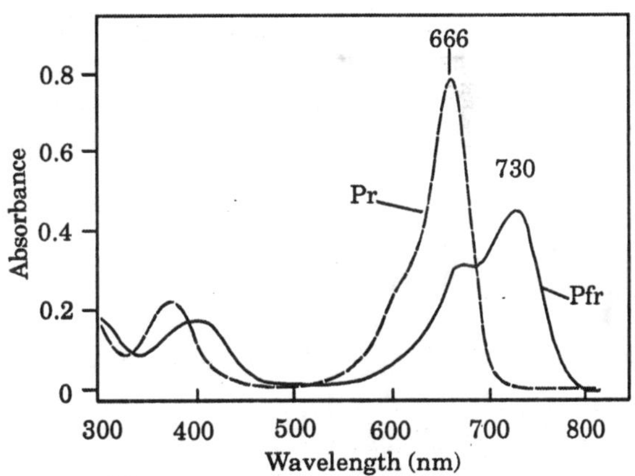


**Fig. 4.3. The absorption spectra of the Pr (dashed line) and Pfr (solid line) forms of phytochrome overlap. Because red light (640-690 nm) is absorbed by both Pr and Pfr, it does not generate pure Pfr but rather sets up a photoequilibrium of about 85% Pfr and 15% Pr. Similarly, far-red light (710-750 nm) does not convert all of phytochrome to Pr, but generates a photo-equilibrium of about 97% Pr and 3% Pfr.**

## Physiologically Active form of Pfr

Because phytochrome responses are induced by red light, they could, in theory, result either from the appearance of Pfr or from the disappearance of Pr. In most cases studied, a quantitative relationship holds between the magnitude of the response and the amount of Pfr generated by light, but no such relationship holds for the amount of Pr generated by response-inducing light. Evidence such as this has led to the conclusion that, in most cases, Pfr is the physiologically active form of phytochrome. In the cases in which it has been shown that a phytochrome response is not quantitatively related to the absolute amount of Pfr, it has been proposed that the ratio between Pfr and Pr determines the magnitude of the response.

## The Amount of Pfr

Light is not the only factor regulating the appearance or disappearance of Pfr. Both in vivo and in vitro, the Pfr of certain plants spontaneously changes to Pr in darkness by a reaction called dark reversion. The rate of this reaction is dependent on temperature and pH and can be greatly accelerated, even to a time course of seconds, by certain chemicals, especially reducing agents. Phytochrome extracted from seedlings of cereals, such as oat and rye, exhibits dark reversion only if it is partially fragmented by enzymes called proteases. It is possible that the dark reversion of Pfr observed in vivo in some plants is also dependent on partial proteolysis.

Whatever the mechanism of dark reversion, the phenomenon of a timed disappearance of so potent an effector as Pfr has led to the hypothesis that reversion may have a regulatory function, particularly in relation to flowering. Unfortunately, the physiological significance of dark reversion remains obscure. Both forms of phytochrome are targets of proteases present in cells and, like all other cellular proteins, undergo breakdown or destruction during the normal cellular turnover process. In most plants examined, the photoconversion of Pr to Pfr accelerates the destruction of phytochrome, suggesting that Pfr is more labile to proteolysis than Pr. There is evidence that Pfr may become marked or tagged for proteolytic destruction by the ubiquitin system. As discussed already ubiquitin is a small polypeptide that binds covalently to proteins and serves as a recognition site for a large proteolytic complex.

The disappearance of Pfr by this process would be quite distinct from dark reversion. In addition to dark reversion and degradation, phytochrome levels are also regulated by the rate of synthesis. Phytochrome is synthesized in the Pr form in the dark. The same light stimulus that converts it to Pfr and thus accelerates its degradation also decreases its synthesis. As we shall see at the end of the chapter, the synthesis of phytochrome is itself phytochrome-regulated. The contributions of all these factors to Pr and Pfr levels are summarized in the diagram 4.4.

## Types of Phytochrome

The diversity and amount of biochemical information available on phytochrome are impressive. Obtaining this information required a method for purifying phytochrome from plant tissue, which was developed through research pioneered by H. W. Siegelman. After surveying many different plants and plant tissues, Siegelman found that the young etiolated seedlings of cereals, such as oats and rye, were particularly rich sources of phytochrome. Some initial success in purifying phytochrome from etiolated oats was reported in 1964 and many workers subsequently used dark-grown oat seedlings for phytochrome extractions.

As a result. most of the biochemical information on phytochrome is derived from studies on oat seedlings, although there is a considerable amount of data on phytochrome isolated from dark-grown rye and pea seedlings as well. Although on a fresh-weight basis there is 50 times more phytochrome in etiolated tissue than in equivalent green tissue of the same age, the overall content of phytochrome in etiolated plants is low, about 0.2% of the total extractable protein. Thus, to purify even milligram quantities usually requires over a kilogram fresh weight of starting material. Etiolated seedlings are also a rich source of proteases, and phytochrome is especially susceptible to proteolysis. To minimize this problem extractions must proceed rapidly, at a low temperature, and in buffers containing protease inhibitors. The stability of phytochrome during its purification is also favoured by using green-light illumination, which is absorbed only poorly by phytochrome, instead of white light. Very little phytochrome is extractable from green plants, and some of it is different from the majority of the phytochrome present in etiolated plants.

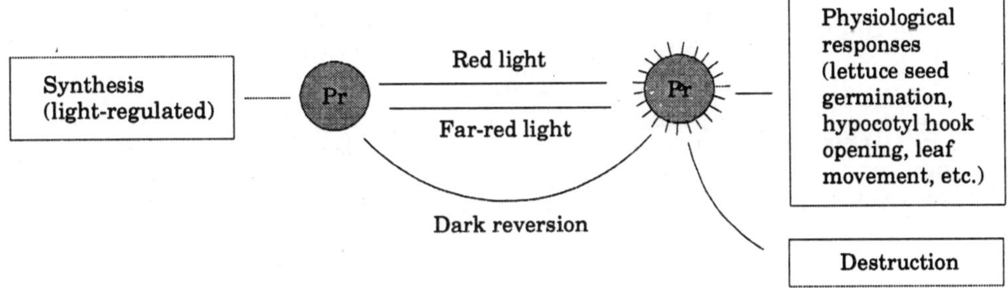


**Fig. 4.4. A summary of processes regulating the turnover of phytochrome in cells.**

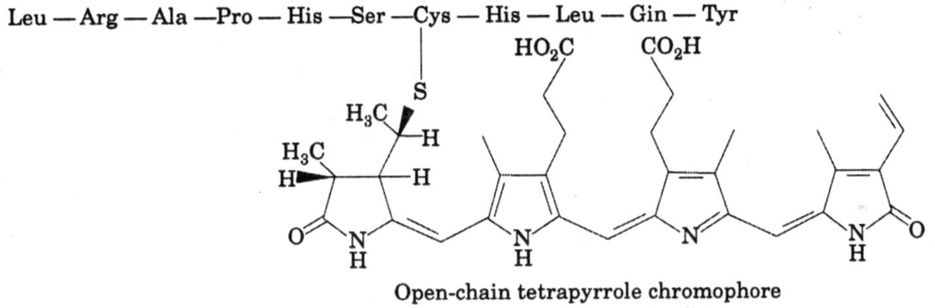

Leu — Arg — Ala — Pro — His — Ser — Cys — His — Leu — Gin — Tyr

S

$HO_2C$ $CO_2H$

$H_3C$ H H $H_3C$ H

O N H N H N N H O

Open-chain tetrapyrrole chromophore

**Fig. 4.5. Structure of the Pr chromophore and the peptide region bound to the chromophore through a thioether linkage. Amino acid abbreviations: Leu, leucine; Arg, arginine; Ala, alanine; Pro, proline; His, histidine; Ser, serine; Cys, cysteine; Gln, glutamine; Tyr, tyrosine.**

Recent research has shown that there are, in fact, two different phytochromes, with different (though similar) amino acid sequences. Current terminology refers to them as type I and type II phytochrome. Type I is about nine times more prevalent in dark-grown pea seedlings than Type II, whereas in light-grown pea seedlings the amounts of the two types are about equal. There is evidence that the two types of phytochrome regulate distinctly different functions in plants. So far, almost all of the research on phytochrome has been done on type I, and very little is known about the structure and function of Type II. Thus, unless otherwise noted, most of the phytochrome referred to in the rest of this chapter can be assumed to be type I.

## Phytochrome Dimer

The chromophore of phytochrome is an open-chain tetrapyrrole and is covalently bound to the protein portion of phytochrome (phytochrome apoprotein) through a thioether linkage. There appears to be only one chromophore per monomer of protein, and the primary sequence of amino acids immediately surrounding the thioether linkage to the chromatophore has been determined . The Pr and Pfr chromophores differ mainly in the number of their resonating double bonds. In addition, the Pfr configuration has the methene bridge of its A ring bound to one or two amino acid residues in the protein structure. Because phytochrome is unusually susceptible to proteolysis during purification, its published chemical properties have undergone several revisions as progressively larger versions of the molecule have been isolated.

The currently accepted monomer molecular mass of intact.type I oat phytochrome is 124,000 Da, and the value for phytochrome from other species is reported to be close to this. The chemical characterizations of purified phytochrome reviewed here are all of the 124,000 Da version, which is apparently undegraded and most closely resembles the phytochrome prevalent in the living cells of dark -grown plants. The gene for phytochrome (type I) has been isolated, cloned, and sequenced, and the primary sequence of amino acids has been deduced from the DNA sequence. Native phytochrome has a molecular mass of about 240,000 Da and thus consists of a dimer of two equivalent subunits.

The Pr form has been visualized by spraying a solution of the protein onto mica flakes, drying it under vacuum, and then "shadowing" the dried molecules by using platinum atoms to enhance its three-dimensional structure. After further treatment, the preparation can be viewed in an electron microscope. As phytochrome molecules appear as a collection of spots, the number depending on the orientation of the protein. Based on this information and more accurate measurements using x-ray scattering, the model has been proposed, (Tokutomi et al., 1989). Each of the subunits (I and II) is composed of two separate domains, an N-terminal chromophore binding domain (A) and a C-terminal nonchromophoric domain (B). Each domain is represented by a disk of the appropriate size and thickness. The shape of phytochrome may be important for its functions in the cell.

## Conformational Changes

The amino acid composition of phytochrome shows a higher percentage of hydrophobic amino acids than is typical of water-soluble proteins. Recent data indicate that the photoconversion of Pr to Pfr exposes more of the hydrophobic region of the molecule, consistent with the idea that Pfr, particularly the nonchromophoric domains, may function in association with membranes. Amino acid side chains of the Pr and Pfr forms of phytochrome are differentially reactive to various reagents, suggesting that the protein undergoes some conformational change during the phototransformation.

This shape change must be subtle because it cannot be detected by spectral assays, such as circular dichroism or nuclear magnetic resonance, that reveal general aspects of a protein's shape. Some of the differential reactivity of Pr.and Pfr may be due to a rearrangement of the chromophore relative to the protein rather than to a conformational change of the protein itself. Another indication that Pr and Pfr differ conformationally comes from immunological studies. In this regard, monoclonal antibodies (mAbs) have been especially useful.

These are antibodies produced by cells that have all been cloned from a single antibody-producing cell. Because each antibody-producing cell produces only one type of antibody molecule, each mAb raised against a protein is specific for a particular site on that protein.

Thus, mAbs to phytochrome react only with specific regions of the molecule. Several workers who have raised mAbs to phytochrome have found that certain of these antibodies can discriminate between the Pr and Pfr forms. Such results support the hypothesis that conformational changes occur in defined regions of the phytochrome protein during its photoconversion. Monoclonal antibodies have also been useful for comparisons of phytochromes from different species.

Such studies have revealed that, overall, phytochrome is not a highly conserved protein; that is, it does not have many sequences that have remained unchanged during its evolutionary history. However, the relatively rarer mAbs that do cross-react with phytochrome molecules derived from distantly related species are likely to be directed against the (presumably) few highly conserved domains of the molecule (Cordonnier et al., 1986).

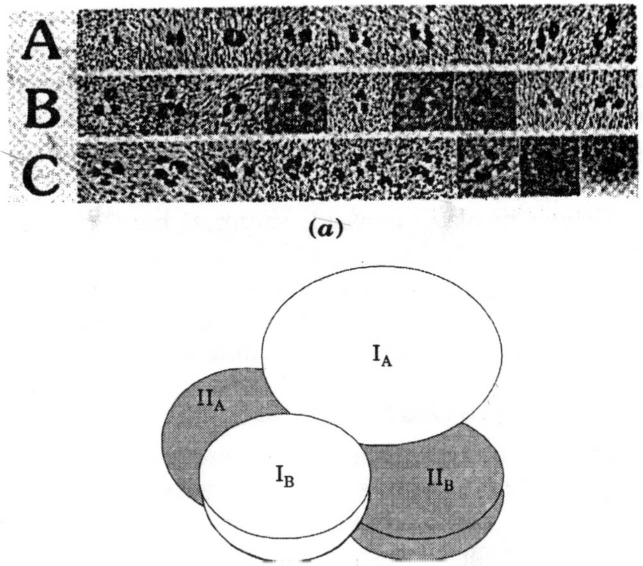

A
B
C

*(a)*

$I_A$ $II_A$ $I_B$ $II_B$

*(b)*

**Fig. 4.6. Visualization of phytochrome. (a) Phytochrome from peas, purified and then sprayed onto a mica flake. After drying, the specimen was shadowed with platinum at an angle of 6° to enhance the three-dimensional structure. A carbon replica was prepared and viewed in.the electron microscope. Various images were produced, including pairs of spots (A), three spots (B), and clusters of spots (C). The variation in the pattern in (a) is believed to be caused by different orientations of the protein on the mica, (b) The structure of the phytochrome dimer based on a type of x-ray scattering that does not require crystallization. Each monomer consists of a chromophore binding domain (A) and a smaller nonchromophoric domain (B). The monomers are labeled I and II. The molecule as a whole has an ellipsoidal rather than a globular shape. Phytochrome may interact with membranes via conformational changes in the nonchromophoric B domains.**

## Phytochrome Localization

Valuable insights into the function of a protein can be gained from learning where it is localized in the organism — for example, whether it is present only in roots or, on the subcellular level, only in nuclei or mitochondria. It is not surprising, then, that significant effort has been expended to determine the location of phytochrome both throughout the plant and within individual cells. In this section we will review the results of these studies.

## Spectrophotometrica Detection

The photoreversible spectral changes that phytochrome undergoes after irradiations with red and far-red light uniquely identify the presence of phytochrome and permit its quantitation.

These spectral changes can be detected in vivo; in fact, the initial discovery of phytochrome was made possible by the availability of a spectrophotometer that could quantitate changes in the red and far-red regions of the spectrum even against the background interference of living tissue. The presence of chlorophyll makes it very difficult to detect phytochrome spectrally in green tissue. But in etiolated plants, phytochrome can easily be detected and measured by in vivo spectroscopy.

Using this assay method, scientists have confirmed the presence of phytochrome in many different tissues of monocots and dicots and in gymnosperms, ferns, mosses, and algae. Dark-grown seedlings have been studied in greatest detail. Here the highest phytochrome levels are usually found in meristematic regions or in regions that were recently meristematic, such as the bud and first node of peas and the tip and node regions of the coleoptile in oats.

## Immunocytochemical Methods

Physical and chemical fixation methods are available that can immobilize phytochrome in cells while preserving its cross-reactivity with antibodies. There are also many techniques for visualizing where antibodies have bound to phytochrome in fixed tissue. Such immunocytochemical methods have been used to localize phytochrome in plant cells and tissues. The early studies revealed a tissue-level distribution similar to that shown by spectrophotometric methods but added information on the location of phytochrome in roots (it is concentrated in the cap region).

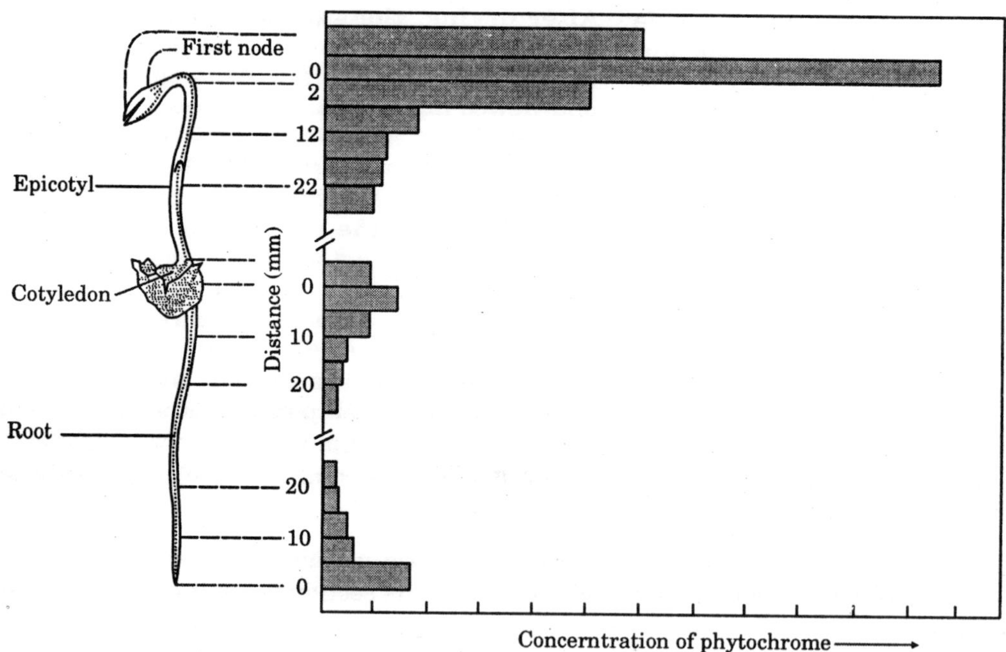


**Fig. 4.7. Distribution of phytochrome in an etiolated pea seedling, as measured spectrophotometrically. Notice that phytochrome is most heavily concentrated in the apical meristems of the root and epicotyl, where the most dramatic developmental changes are occurring.**

Moreover, they showed for the first time what spectrophotometric methods could not resolve: the sub-cellular distribution. Phytochrome was found both in the cytosol and in association with organelles, including plastids and nuclei. One of the more interesting results from the early immunocytochemical studies was the finding that the photoactivation of phytochrome induced a dramatic change in its subcellular distribution. When phytochrome is converted to Pfr, its distribution in coleoptile parenchymal cells of oat and rice seedlings changes within minutes from mainly diffuse through the cytosol to mainly sequestered in discrete regions. When Pfr is photoconverted back to Pr by far-red light, phytochrome slowly returns to the original diffuse distribution.

In pea seedlings, phytochrome seems to become inaccessible to antibodies immediately after the red-light treatment, and the sequestered distribution appears only after a subsequent 10 min incubation in the dark. These results could indicate that Pfr associates tightly with some intracellular receptor that is important for its function soon after its production in the cell. Recent ultrastructural studies have revealed that most of the sequestered Pfr in oat cells is associated with discrete structures that are globular to oval in shape, are about 300 nm in size, and have no morphologically identifiable membranes surrounding them.

The function or significance of these structures is as yet unknown, though one hypothesis is that they represent intracellular sites where the Pfr-dependent dark destruction of phytochrome occurs. In summary, the immunocytochemical results indicate that phytochrome is most concentrated in young, undifferentiated tissue and, on the subcellular level, undergoes a distribution change from diffuse cytosolic to sequestered upon its conversion to Pfr.

The strong; correlation between phytochrome and cells that have the potential for dynamic developmental changes is certainly consistent with the important role of phytochrome in controlling such changes. In addition, the dramatic shift in the subcellular distribution of phytochrome upon conversion to Pfr is in accord with the idea that Pr and Pfr have radically different roles in the cell. Learning more about the distribution of Pfr may help to clarify how it alters plant cell development so significantly.

## Photoreversible Responses

Because phytochrome regulates plastid development and modulates gene expression, there is no doubt that it controls the function of at least some organelles in vivo. But does it *directly* associate with these organelles in order to alter their activity? The most reliable data available on the intracellular sites of phytochrome are the in situ results reviewed above. These data, however, are equivocal on the question of whether phytochrome is associated with specific organelles.

With some antibody preparations and some tissues, it appears that phytochrome is associated with plastids, mitochondria, and nuclei. With other antibody preparations and tissues, no evidence of phytochrome association with any subcellular membrane is evident. A less direct way to address this problem is to investigate whether phytochrome copurifies with any organelles or regulates any organellar function in vitro.

The answer in both cases is consistently yes. Photoreversible effects on organellar enzyme or other functions have been reported in isolated plastids, peroxisomes, nuclei, and mitochondria. Furthermore, phytochrome has been detected in association with these purified organelles by spectral and/or immunochemical methods. However, it has not been determined whether the association of phytochrome with these organelles occurred before they were isolated (i.e., in vivo) or only during (and only as a result of) the extraction procedure. Thus, the hypothesis that phytochrome directly associates with organelles as part of its cellular functioning must be tested further for confirmation or disproof.

## Phytochrome-Induced Responses

The variety of different phytochrome responses in intact plants is great, in terms of both the *kinds* of responses and the quantity of light needed to induce the responses. A review of this variety will show how diversely the effects of a single photoevent—the absorption of light by Pr—are amplified throughout the plant. The kinds of phytochrome-induced responses may be logically grouped, for ease of discussion, into rapid biochemical events and slower morphological events, including movement and growth. Some of the early biochemical responses affect later morphological responses. These rapid responses will be treated in detail in a later section.

## Lag Time and Escape Time

Morphological responses to the photoactivation of phytochrome may be observed visually after a lag time as brief as a few minutes (e.g., the initiation of chloroplast rotation in *Mougeotia)* to as long as a few weeks (e.g., flowering in many photoperiodic species. The more rapid of these responses are usually reversible movements of organelles or reversible dimensional changes (swelling, shrinking) in cells, but even some growth responses occur remarkably fast. Red-light inhibition of the stem elongation rate of light-grown *Chenopodium album* is observed within 8 min after increasing its relative level of Pfr Information about the lag time for a phytochrome response helps scientists to evaluate the kind of biochemical events that could precede and play a causal role in the induction of that response.

The shorter the lag time, the more limited the range of biochemical events that could have been involved. Variety in phytochrome responses can also be seen in the phenomenon called escape from photoreversibility. Red light-induced events are reversible by far-red light for only a limited period of time, after which the response is said to have "escaped" from reversal control by light. A currently favoured model to explain this phenomenon assumes that phytochrome-controlled morphological responses are the result of a step-by-step sequence of linked biochemical reactions in the responding cells. Each of these sequences has a point of no return beyond which it proceeds irrevocably to the response. The escape time for different responses ranges from less than a minute to, remarkably, hours.

## Quantity of Light

Phytochrome responses may also be distinguished by the amount of red light required to induce them. This amount is referred to as fluence and is measured in moles of quanta per square meter. The total fluence of light given to a sample is a function of two factors: the photon flux density, or number of photons striking a given area per unit time (N), and the irradiation time ($t$). For every phytochrome response there is a characteristic range of light fluences over which the magnitude of the response is proportional to the fluence. This range varies greatly for different phytochrome responses. Some responses can be initiated by fluences as low as 0.1 nmol $m^{-2}$ (one-tenth of the amount of light emitted from a firefly in a single flash). These are classified as *very low fluence (VLF) responses.*

Red light can affect mesocotyl and coleoptile growth in etiolated oat seedlings at fluences this low. Other responses cannot be initiated until the fluences reach 1.0 µmol $m^{-2}$. These are referred to as *low fluence (LF) responses,* and they include most of the classical photoreversible responses, such as the promotion of lettuce seed germination, that are mentioned in Table 4.1. The amount of light needed to induce VLF responses converts less than 0.02% of the total phytochrome to Pfr. Since the far-red light that would normally reverse a red-light effect converts 97% of the Pfr to Pr about 3% of the phytochrome remains as Pfr— significantly more than is needed to induce VLF responses. Thus, far-red light cannot reverse VLF responses.

It is the red-light action spectrum, rather than photoreversibility, that suggests that phytochrome is the photoreceptor for VLF responses. Still another group of phytochrome responses can be distinguished by the quantity of light needed to promote them. These are the high fluence (HF) responses, several of which are listed in Table 4.2. These responses are elicited by prolonged or continuous irradiation, requiring exposures for hours, not minutes, and fluences in excess of 10 µmol $m^{-2}$. They have action spectra with peaks in the far-red and blue regions and they are not photoreversible.

**Table 4.2. Some plant photomorphogenic responses induced by high irradiances.**

| |
|---|
| Anthocyanin synthesis in various dicot seedlings and in apple skin segments |
| Inhibition of hypocotyl elongation in mustard, lettuce, and petunia seedlings |
| Induction of flowering in *Hyoscyamus* (henbane) |
| Plumular hook opening in lettuce |
| Enlargement of cotyledons in mustard |
| Ethylene production in sorghum |

Experimental results suggest that phytochrome may be one of the photoreceptors that promote this response, although a blue-light photoreceptor and, in some cases, one or more of the chlorophylls in photosystem I or II may also be involved (Mancinelli, 1980). Perhaps it should not be surprising that a reaction which requires so much light over so long an exposure time would involve more than one photoreceptor.

## Adaptation to Changes in Light Conditions

The presence of a red/far-red reversible pigment in all green plants, from algae to dicots, suggests that these wavelengths of light provide information that helps plants adjust to their environment. What environmental conditions result in changes in the relative levels of these two wavelengths of light in natural radiation ? The ratio of red light (R) to far-red light (FR) varies remarkably in different environments. This ratio can be defined as

$$R/FR = \frac{\text{photon fluence rate in 10-nm band centered on 660 nm}}{\text{photon fluence rate in 10-nm band centered on 730 nm}}$$

Table 4.3 summarizes both the total photons (400-800 nm) and the R/FR values in six natural environments. Both parameters vary greatly in different environments. Compared with direct daylight, there is relatively more FR in sunset light, under 5 mm of soil, and especially under the canopy of other plants (as on the floor of a forest).

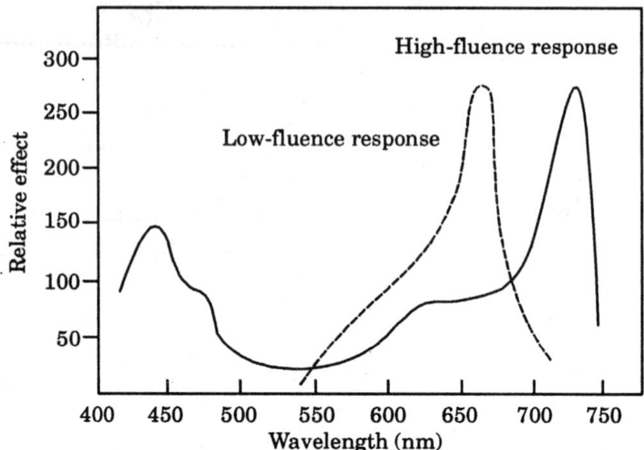


**Fig. 4.8. A comparison of the action spectra of typical high fluence (solid line) and low fluence (dashed line) responses. The high-fluence responses have peaks in the blue and far-red regions of the spectrum. Low fluence responses have a single peak at about 660 nm. High fluence responses are not photoreversible, although there is evidence that phytochrome is one of the pigments involved.**

The canopy phenomenon results from the fact that green leaves absorb red light because of their high chlorophyll content but are relatively transparent to far-red light. Phytochrome, then, can serve as an indicator of the degree of shading of a plant by other plants. As shading increases, R/FR decreases and the ratio of Pfr to the total phytochrome (Pfr/$P_{toal}$ decreases. When simulated natural radiation was used to vary the FR content, it was found that the higher the FR content (i.e., the lower the Pfr/$P_{total}$) the higher the stem extension rate for "sun plants," that is, plants that normally grow in an open-field habitat (Fig. 4.9 solid line).

**Table 4.3. Ecologically important light parameters. The light intensity factor (400-800 nm) is given as the photon flux density, and phytochrome-active light is given as the R/FR ratio.***

| | *Photon flux density ($\mu mol\ m^{-2}\ s^{-1}$)* | *R/FR* |
|---|---|---|
| Daylight | 1900 | 1.19 |
| Sunset | 26.5 | 0.96 |
| Moonlight | 0.005 | 0.94 |
| Ivy canopy | 17.7 | 0.13 |
| Lakes, at a depth of 1 m | | |
| Black Loch | 680 | 17.2 |
| Loch Leven | 300 | 3.1 |
| Loch Borralie | 1200 | 1.2 |
| At a depth of 5-mm soil | 8.6 | 0.88 |

*Absolute values taken from spectroradiometer scans; the values should be taken to indicate the relationships between the various natural conditions and not as actual environmental means.

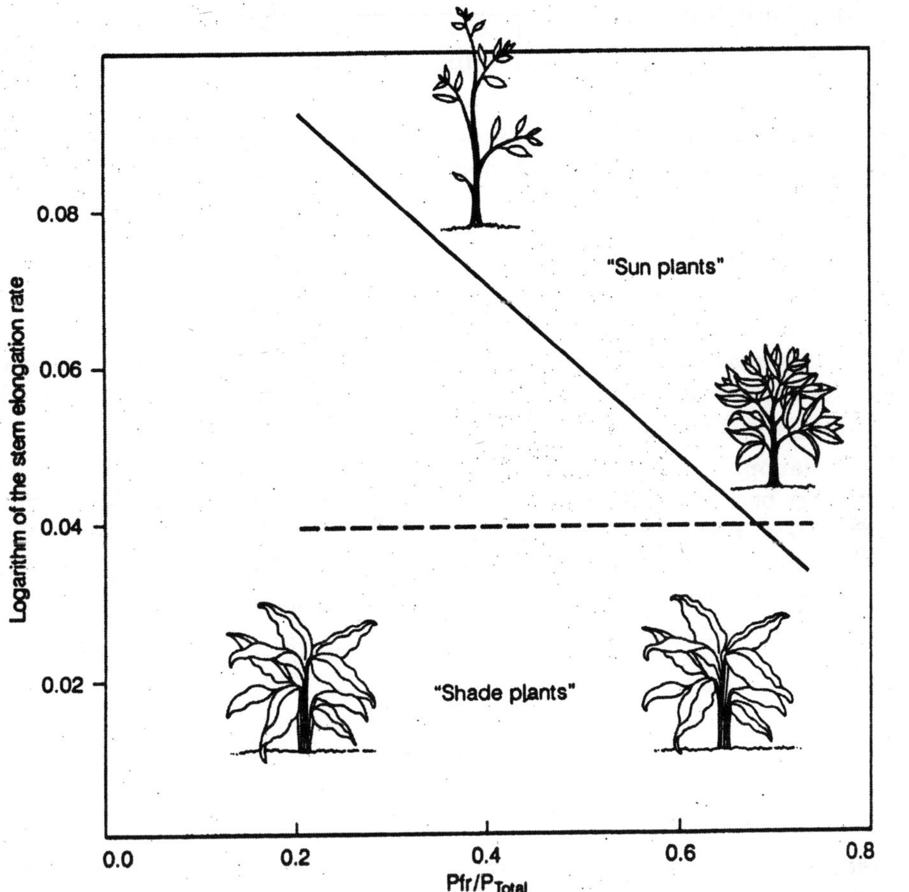


**Fig. 4.9. The relative ratio of Pfr to total phytochrome (Pfr/$P_{total}$) has an effect on the stem elongation rate of species that normally grow in a more sunlight environment (solid line) but not on species that normally grow in a more shaded environment (dashed line). This result supports the idea that phytochrome plays a role in shade perception for plants not adapted to a shaded environment.**

Simulated canopy shading induced these plants to allocate more of their resources growing taller. This correlation did not hold for the woodland understory plants that were tested. These "shade plants," which normally grow in a shaded environment, showed little or no alteration in their stem extension rate as they were exposed to higher R/FR values). Thus, there appeared to be a systematic relationship between phytochrome-controlled growth and species habitat. Such results are taken as an indication of the involvement of phytochrome in shade perception. For a "shade-avoiding" plant there is a clear adaptive value in allocating its resources toward more rapid extension growth when it is shaded by another plant.

In this way it can enhance its chances of growing above the canopy and acquiring a greater share of unfiltered, photosynthetically active light. The price for favouring internode expansion is usually reduced leaf area and reduced branching, but, at least in the short run, this evolutionary adaptation seems to work. Light quality also plays a role in regulating the germination of some seeds. Large-seeded species, with ample food reserves to sustain prolonged seedling growth in darkness (e.g., underground), generally do not have a light requirement for germination.

However, such a requirement is often observed in the small seeds of herbaceous and grassland species, many of which remain dormant, even while hydrated, if they are buried below the depth to which light penetrates. Even when such seeds are on or near the soil surface, their level of shading from the vegetation canopy (*i.e.*, the R/FR they receive) is likely to affect their germination. There are well-documented instances in which the FR enrichment imparted by a leaf canopy inhibits germination in a range of small-seeded species. For seeds of the tropical species *Cecropia obtusifolia* and *Piper auritum* planted on the floor of a deeply shaded forest, this inhibition can be reversed by placing a narrow-band red filter immediately above the seeds. This filter permits the red component of the canopy-shaded light to pass through while almost completely blocking the farred component.

Although the canopy transmits very little red light, it is enough to stimulate the seeds to germinate, probably because most of the inhibitory far-red light is excluded by the filter and the R/FR ratio is very high. These seeds would also be more likely to germinate in spaces receiving sunlight through gaps in the canopy than in densely shaded spaces. The sunlight would help to ensure that the seedlings became photosynthetically self-sustaining before their seed food reserves were exhausted.

## Daily Rhythms

Various metabolic processes in plants, such as $O_2$ evolution and respiration, cycle alternately through high-activity and low-activity phases with a regular periodicity of about 24 h. These rhythmic changes are referred to as circadian rhythms, from the Latin *circa diem,* meaning "approximately a day." The period of a rhythm is the time elapsed between successive peaks or troughs in the cycle, and because the rhythm persists in the absence of external controlling factors it is considered to be endogenous. Not surprisingly, metabolic circadian rhythms are often expressed in visible forms that affect the whole plant or large parts of it. A well-described example is the leaf "sleep movements" of certain plants in the legume family. These movements are the rhythmic expression of circadian ion fluxes into and out of motor cells at the base of the leaves.

The swelling or shrinking of these cells in a structure called the pulvinus controls the up-and-down movement of the leaves. There is abundant evidence that the molecular basis of rhythms involves, at some level, the regular cycle of ions or metabolites into and out of organelles and/or the cytosol of cells. The pulsed application of agents, such as lipid solvents, that alter membrane permeability shifts the phase of circadian rhythms. Such agents either advance or delay the daily cycle, depending on the point in the cycle at which they are applied.

Phase-response curves, in which the magnitude of the advance or delay of a cycle is plotted against the time at which an agent was applied, are often used to illustrate this phenomenon. Light is a strong modulator of rhythms in both plants and animals. Although circadian rhythms that persist under controlled laboratory conditions usually have periods one or more hours longer or shorter than 24 h, in nature their periods tend to be uniformly closer to 24 h because of the synchronizing effects of light at day-break.

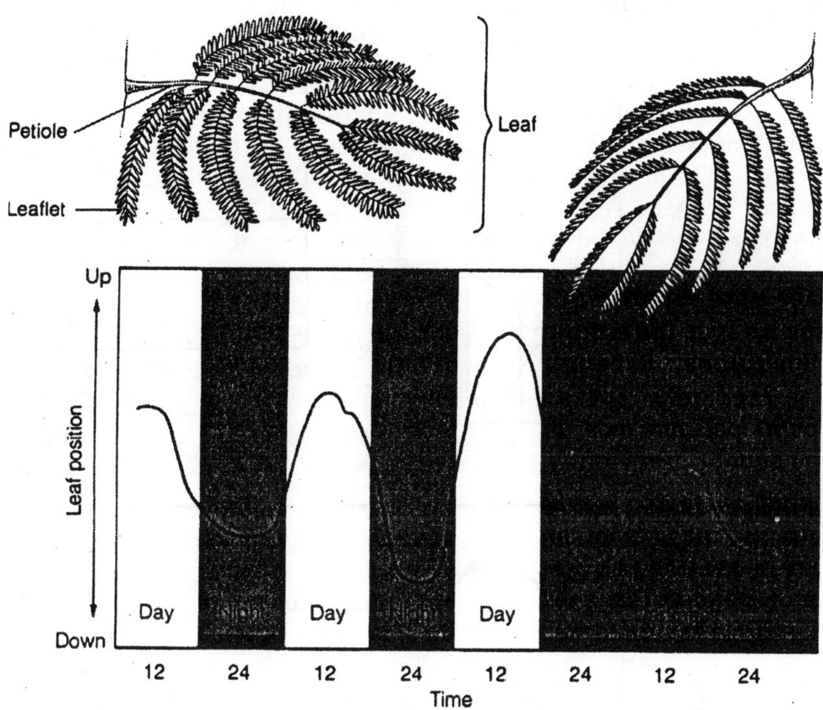


**Fig. 4.10. Circadian rhythm in the diurnal movement of *Albizzia* leaves. The leaves are elevated in the morning and lowered in the evening. The rhythm persists at a lower amplitude for a limited time in total darkness. In addition to the raising and lowering of leaves, the leaflets undergo opening and closing responses.**

Pulses of light, like pulses of membrane-active agents, generate characteristic phase-response curves. A key photoreceptor for these light responses is phytochrome. It is no coincidence that phase-response curves generated by pulses of light resemble those generated by membrane reagents. As indicated below, phytochrome itself appears to modulate the permeability of membranes.

## Cellular and Molecular Mode of Action

Phytochrome-regulated changes in plants all begin with absorption of light by the pigment. The absorbed ligh somehow alters the molecular properties of phytochrome so that it can induce a sequence of cellular reactions that ultimately cause a change in growth, development, or position of an organ. Such a progression of causally linked events is termed a transduction sequence. Every environmental stimulus that affects plant or animal metabolism does so through one or more such sequences. Identifying even the first few steps of such a sequence is a daunting task. Although it may be easy enough to document an early biochemical change induced by a stimulus, it is far more difficult to show that this change is causally related to a specific visible alteration in the plant. Furthermore, though we should not complicate our hypotheses more than necessary, there is no reason to propose that all phytochrome phenomena result from the same early sequence of biochemical reactions. From the numerous explanations for phytochrome effects, certain hypotheses have emerged as particularly plausible.

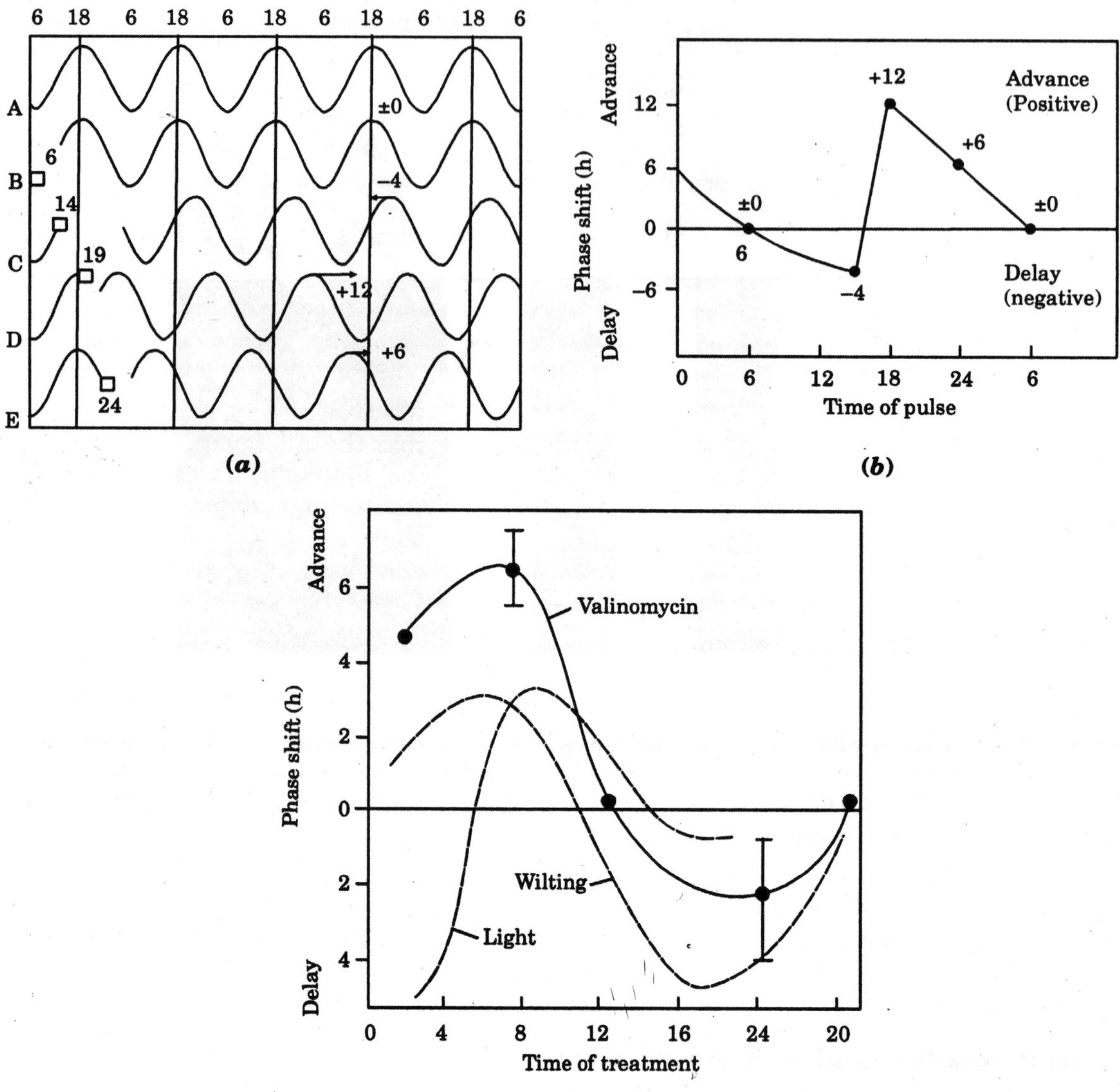


**Fig. 4.11. The phase-response curve of circadian rhythms. (a) Method for obtaining data for phase-response curves. The control response is curve A. The numbers at the top of the figure indicate the clock time, which is arbitrary. At various times during the cycle a pulse is given (shown by the boxes and adjacent numbers indicating the time), resulting in a shift of the curve by the amount indicated at the right of each curve. The rhythm may either be delayed (curve C) or advanced (curves D and E) in response to the pulse. (b) A plot of the data in (c). Effects of two environmental stimuli (light and water stress) and one agent that affects membrane permeability (the potassium ionophore valinomycin) on phase shifts of the circadian leaf movements of the bean *Phaseolus coccineus*. All the agents advance or delay the rhythm, depending on when during the rhythm they were applied. Zero indicates the last night peak (when the leaves are lowered) of the circadian cycle before treatment. These results are consistent with the idea that ion movements are an important factor in circadian timekeeping.**

According to one of these hypotheses, many responses result from changes that phytochrome induces in membranes. Since other well-known light transduction pathways, such as photosynthesis and, in animals, vision, involve the mediation of membranes, it would be surprising if the phytochrome system were an exception. Phytochrome has repeatedly been shown to modulate membrane functions, though how it alters membrane properties is still controversial. The evidence that phytochrome-mediated changes in membrane properties may regulate photomorphogenesis is summarized in the next two subsections.

## Phytochrome Can Associate with Membranes

As discussed earlier, immunocytochemical results give equivocal answers to the question of whether phytochrome is associated with organelles or other cellular membranes in situ. The evidence in support of the hypothesis that phytochrome can bind to membranes is indirect, but it comes from both in vivo and in vitro studies. In vitro, the Pfr form of phytochrome binds tightly to pure lipid bilayers.

Unlike Pr, it also facilitates ion movements across these bilayers. Although high salt concentrations can inhibit the binding of Pfr to spherical lipid bilayers, called liposomes, they cannot dissociate Pfr from liposomes once it has bound to them; detergents are required to produce this dissociation. These results suggest that the initial attachments of Pfr to membranes may be through ionic bonds but the final association is maintained by hydrophobic forces. There is also in vitro evidence that Pfr binds to biological membranes.

Mitochondrial preparations isolated from red light–irradiated oat seedlings have significant levels of phytochrome (Pfr) associated with them, whereas those from unirradiated plants contain no detectable phytochrome. The association of Pfr with isolated mitochondria does not appear to be a random surface-sticking phenomenon, because the Pfr is not accessible to digestion by enzymes (proteases) that would be expected to remove proteins bound only on organelle surfaces. Furthermore, the phytochrome associated with purified mitochondria can photoreversibly regulate some of their functions. Distinct from the question of whether phytochrome binds directly to membranes is the hypothesis that phytochrome regulates membrane functions. Abundant in vivo data support this contention.

The most convincing results are those that demonstrate rapid photoreversible modulation of the electrical potential across the plasma membrane. Such modulation has been directly measured individual cells and has also been inferred from the rapid effects of red and far-red light on the surface potential of roots and oat (*Avena*) coleoptiles. The lag between the production of Pfr and the onset of measurable potential changes varies from organism to organism; for example, it is about 1.7 s for depolarization (inside becomes less negative relative to outside the plasma membrane) in the giant alga *Nitella* and 4.5 s for hyperpolarization in *Avena*.

The longer lag in *Avena* suggests that some secondary reaction (rather than Pfr itself) may induce the potential changes. As yet, this question has not been resolved. Changes in the bioelectric potential of cells imply changes in the flux of ions across the plasma membrane. Phytochrome rapidly modulates fluxes of $Ca^{2+}$ into and out of oat cells and, with a longer lag time, induces the flux of other ions in other cells. There is evidence linking the rapid red light-induced membrane depolarization in *Nitella* with a $Ca^{2+}$ influx at the plasma membrane. Other in vivo data indicating that phytochrome regulates membrane functions are those demonstrating the photoregulation of leaf and leaflet movements in *Mimosa, Albizzia, Samanea,* and other legumes.

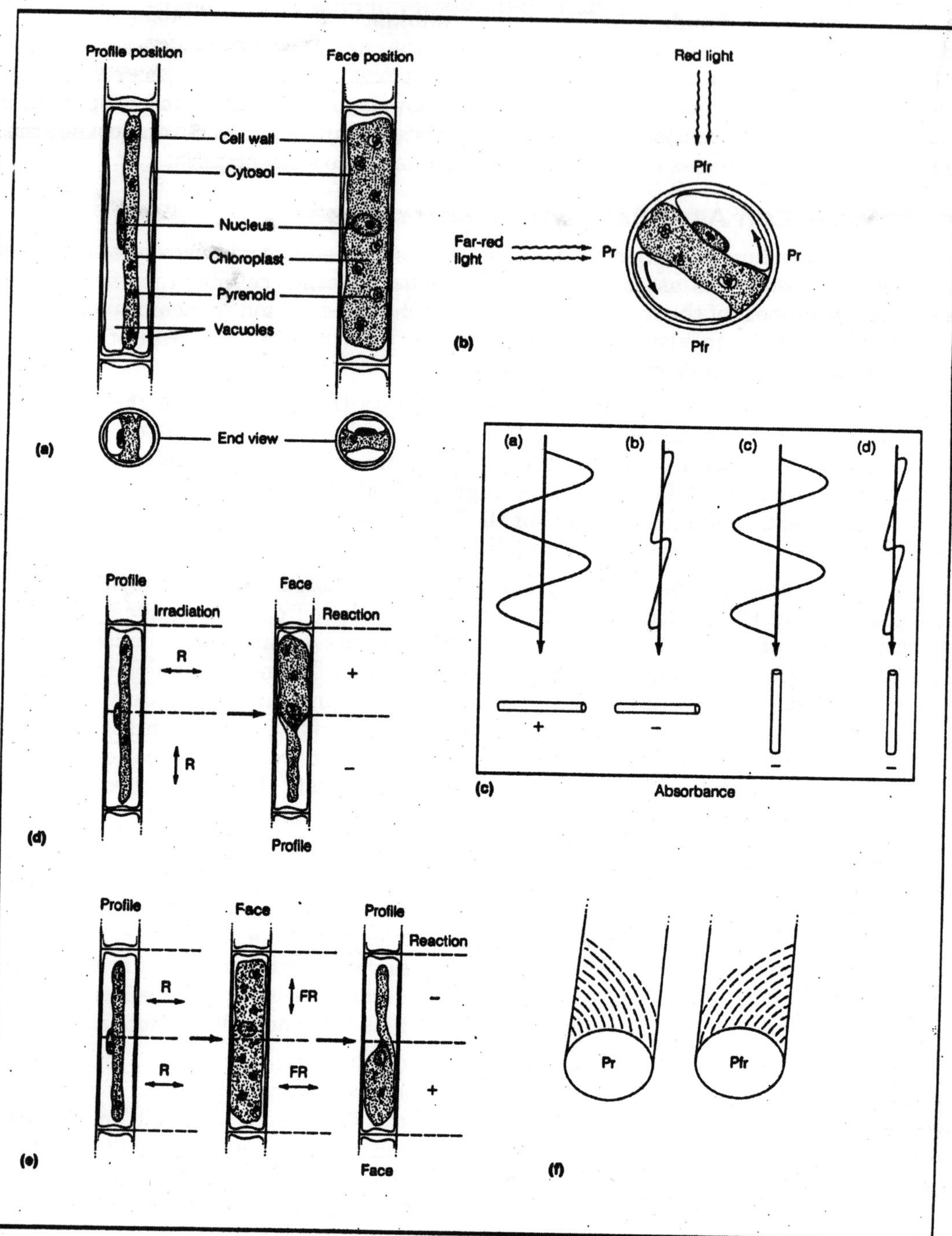


Fig. 4.12. Promotion of Choloroplast.

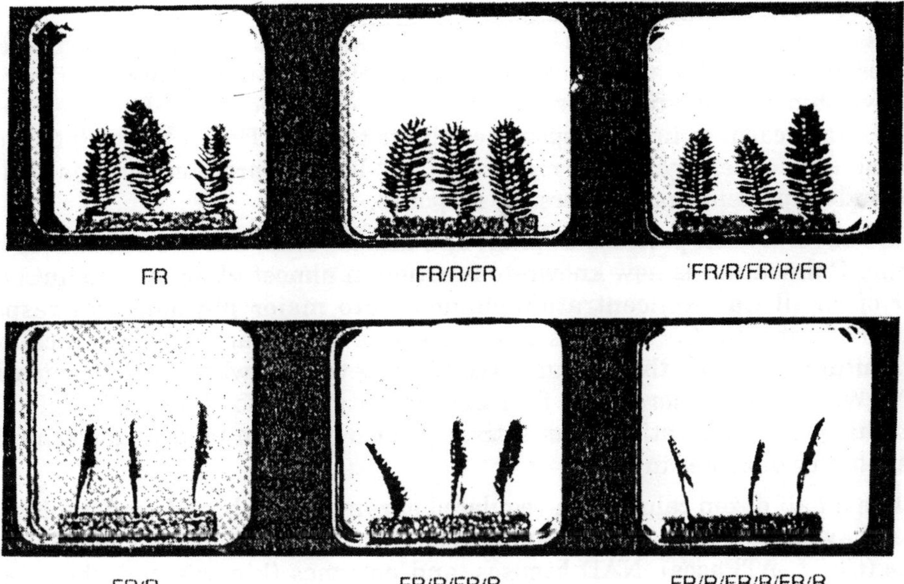


**Fig. 4.13. Phytochrome regulation of leaflet movement in *Mimosa pudica L.* The leaflets were first opened in white light and then transferred to darkness. Just before transfer to darkness they were exposed to a succession of red (R) or far-red (FR) pulses of 2-min duration. The leaflets on the top row all received a pulse of far-red light as the final treatment and remained open for 30 min after transfer to darkness. In contrast, the leaflets on the bottom row received red light as their final treatment and were fully closed after 30 min in the dark. The results show the requirement for Pfr for leaf closure in the dark and the photoreversibility of the effect.**

As we have seen in the section on circadian rhythms, these movements result from the selective swelling and shrinking of pulvinar motor cells located at the bases of the leaves and leaflets. The swelling and shrinking, in turn, are regulated by the rapid entry and exit of ions, especially $K^+$. Phytochrome modulates these salt movements, presumably by affecting the permeability of the membranes across which these ions are transported. Whether phytochrome directly or indirectly modulates the permeability of membranes remains an open question.

A more important question is, what sort of permeability change could amplify a dim light signal into major changes in growth and development? If we interpret the photoregulated bioelectric potential changes as indicating that light induces membrane permeability changes, we should ask what ion fluxes are modulated by phytochrome and what developmental impact such fluxes could have. Considering that the different ion transport systems in cells are closely coupled in their functions, it seems unlikely that phytochrome could affect the transport of any one ion without affecting others simultaneously.

In fact, there are many reports of phytochrome-regulated proton fluxes, $K^+$ fluxes, and $Ca^{2+}$ fluxes. Among these ions, $Ca^{2+}$ has been considered a good candidate for transducing the photoactivation of phytochrome into significant physiological changes in plants. Since the $Ca^{2+}$ studies are at an early stage, their discussion in the next section should be taken as only one example of how phytochrome-regulated membrane changes could affect plant growth and development.

## Calcium and Calmodulin

As discussed already calcium ions have long been known to be important for the regulation of sensory responses in animal cells. More recently, their role in regulating the responses of plants to environmental changes has also come to be appreciated. Hence, it was logical to examine the role of calcium in the response of plants to light. The $Ca^{2+}$-regulated responses in both plant and animal cells generally require a $Ca^{2+}$-binding protein, and it is this complex of $Ca^{2+}$ and a binding protein that triggers the response.

There are several such proteins in plant and animal cells, but the one that is best known is calmodulin. Calmodulin is now known to function in almost all plant and animal cells as a transducer of small $Ca^{2+}$ concentration changes into major physiological responses. The concentration of unbound $Ca^{2+}$ in cells is usually very low (near $10^{-7}$ M), but calmodulin has such a high affinity for $Ca^{2+}$ that it can begin binding the ion when its concentration rises to only $10^{-6}$ M. When one or more of its four high-affinity sites for $Ca^{2+}$ is occupied by the ion, calmodulin changes conformation to an activated form that can bind to and stimulate a wide range of regulatory enzymes in cells.

More than a half-dozen calmodulin-regulated enzymes have been found in plants and more than a dozen have been documented in animal cells. These include plasma membrane–localized $Ca^{2+}$ pumps ($Ca^{2+}$-ATPases), NAD kinases, and enzymes (kinases and phosphatases) that regulate the phosphorylation (and thus the activity) of other enzymes. All of these enzymes play central roles in the regulation of various metabolic activities in plants.

Calmodulin is present both in the cytosol and in organelles, including plastids, mitochondria, and nuclei. Because of its high affinity for $Ca^{2+}$, Calmodulin and the target enzymes it controls can be activated by very small changes in the $Ca^{2+}$ concentration of the subcellular environment. This property of the calmodulin system makes it attractive for amplifying the photoconversion of a few phytochrome molecules into major cellular changes and for modulating other regulatory systems, such as hormonal responses, as well. A current model of how $Ca^{2+}$ could mediate phytochrome responses predicts that (1) Pfr should promote an increase in the $Ca^{2+}$ concentration of one or more subcellular compartments, (2) some Pfr-stimulated enzymes will also be stimulated by calmodulin, and (3) calmodulin inhibitors should interfere with some red light-induced responses.

Indirectly, the model also predicts that some phytochrome responses should be inducible in darkness by chemical agents, known as *ionophores,* that promote $Ca^{2+}$ uptake into cells. All of these predictions have been tested at some level, and the preliminary results, which appear to favour the model, are reviewed briefly here. An obvious first question is, does phytochrome promote $Ca^{2+}$ entry into plant cells? As assayed by autoradiography of $^{45}Ca^{2+}$ in *Mougeotia* (4.14) and by quantitation of total $Ca^{2+}$ in germinating fern spores, the answer is yes. To investigate whether the red light-stimulated entry of $Ca^{2+}$ into cells played any role in modulating phytochrome responses, workers have tested whether chemically induced uptake of $Ca^{2+}$ into cells could mimic Pfr effects. Pfr is a strong promoter of fern spore germination. Treatment of unirradiated fern spores with the $Ca^{2+}$ ionophore A23187 can substitute for Pfr in inducing the germination of the spores.

Although this ionophore can promote the entry of other divalent cations (such as $Mg^{2+}$), it stimulates germination only if $Ca^{2+}$ is in the medium bathing the spores. In a similar experiment, ionophore A23187 can substitute for photoactivated phytochrome in promoting chloroplast rotation in *Mougeotia* (Fig. 4.15), but again only if $Ca^{2+}$ is in the medium. In this experiment, the $K^+$ ionophore valinomycin is totally ineffective as an inducer, so the response is not merely due to nonspecific membrane disturbances resulting from ionophore incorporation. Finally, calcium and A23187 can substitute for red light in regulating leaflet closure in *Albizzia*. Is calmodulin the agent that transduces $Ca^{2+}$ entry into physiological responses? This question has been addressed by testing whether certain chemicals that bind to $Ca^{2+}$-activated calmodulin and prevent it from activating target enzymes are able to block $Ca^{2+}$-inducible phytochrome responses.

The phenothiazines and calmidozolium are examples of such chemicals. These inhibitors can bind to $Ca^{2+}$-binding proteins other than calmodulin and they can have nonspecific side effects unrelated to calmodulin-binding proteins, so the results of studies using these inhibitors must be interpreted with caution.

Nonetheless, both the fern spore germination and chloroplast rotation responses referred to in the preceding paragraphs are blocked by these calmodulin inhibitors. There are preliminary reports that other phytochrome responses are blocked by calmodulin inhibitors as well. One of the more rapid visible responses to the photoactivation of phytochrome can be logically linked to preceding $Ca^{2+}$-and calmodulin-dependent cellular events. This response is the red-light inhibition of internode extension rates referred to earlier.

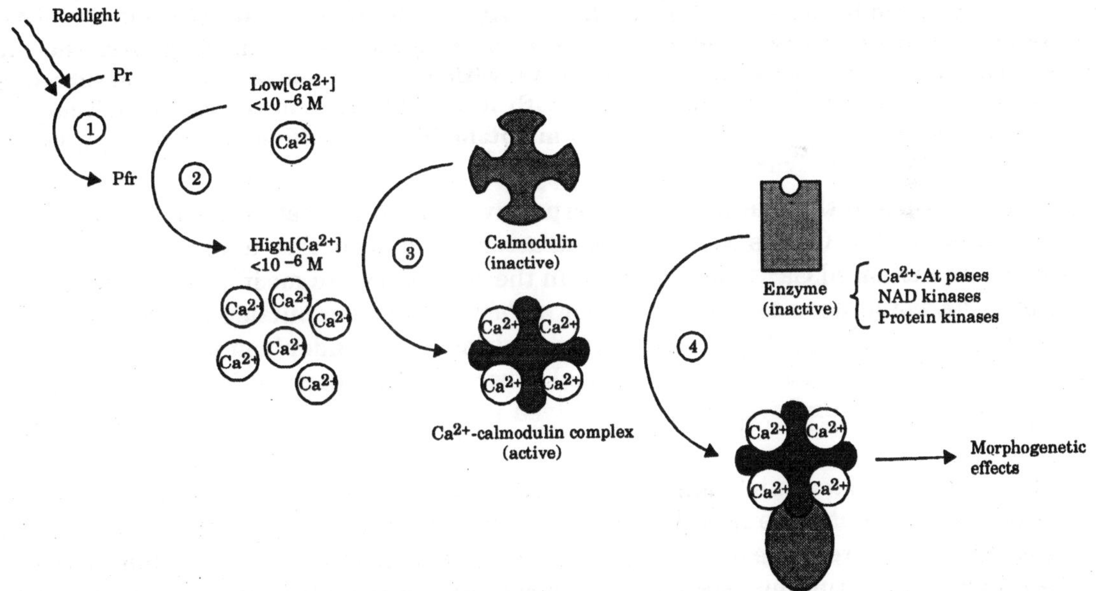


**Fig. 4.14. Proposed model for the cascade of events initiated by the activation of phytochrome. Step 1: Phytochrome activated by red light (650-680 nm) irradiation. Step 2: Pfr-mediated increase in the cytosolic $Ca^{2+}$ concentration. Step 3: Activation of calmodulin by formation of a $Ca^{2+}$-calmodulin complex. Step 4: Binding of calcium-activated calmodulin to specific enzymes, thus activating them. These events could take place in the cytosol or within organelles, such as plastids, mitochondria, or nuclei.**

A major enzyme under calmodulin control is a plasma membrane $Ca^{2+}$-ATPase, which pumps $Ca^{2+}$ out into the wall of plant cells. Pfr stimulates $Ca^{2+}$ efflux from oat cells. Increasing the $Ca^{2+}$ concentration in plant cell walls inhibits some (as yet unidentified) biochemical process needed for cell wall loosening and cell extension. The induction of a transitory increase in cytosolic $Ca^{2+}$ concentration by Pfr, then, could be expected to promote (1) the activation of calmodulin, (2) stimulation of the calmodulin-modulated $Ca^{2+}$-ATPase, (3) increased $Ca^{2+}$ concentration in the walls of irradiated cells, and (4) decreased cell extension rates. This model is obviously speculative, but it helps to illustrate how Pfr-induced $Ca^{2+}$ fluxes could help lead to Pfr-induced growth changes.

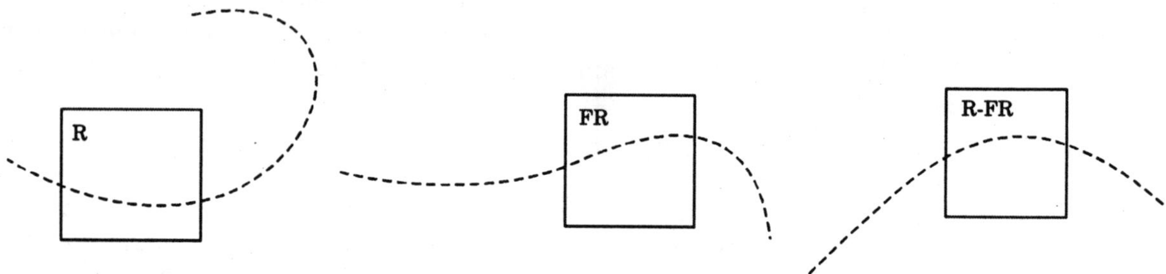


**Fig. 4.15. Photoreversible uptake of $^{45}Ca^{2+}$ by cells of the filamentous alga *Mougeotia*. Filamentous of *Mougeodia* were transferred to a medium containing $^{45}Ca^{2+}$ after a portion of their cells (enclosed by the box) was irradiated with R, FR, or R followed by FR. The relative amount of $^{45}Ca^{2+}$ taken up by the irradiated cells was estimated by autoradiography; the darker the cell, the more $^{45}Ca^{2+}$ it took up. Cells irradiated with R took up more $^{45}Ca^{2+}$ than surrounding cells, whereas those irradiated with FR or with R-FR did not. These results indicated that phytochrome activation could result in calcium uptake by *Mougeotia* cells, consistent with step 2 of the model shown in Figure 4.13.**

It should be noted that calmodulin is responsive only to increases in the concentration of *free* $Ca^{2+}$. Most cellular $Ca^{2+}$ is bound to cell structures and is not free to react with calmodulin. To promote an increased $Ca^{2+}$ concentration in the cytosol, where most of the cell's calmodulin is present, Pfr would have to promote either a $Ca^{2+}$ influx from outside the cell or release of $Ca^{2+}$ from organelles (e.g., the vacuole, mitochondria, or the endoplasmic reticulum) into the cytosol. Ruth Satter and her colleagues at the University of Connecticut have provided evidence that phosphoinositide turnover may be involved in the control of leaf movement by light in *Samanea saman*.

White light caused a rise in the levels of inositol triphosphate ($IP_3$) and diacylglycerol (DAG) after less than a minute of exposure. $IP_3$ could then release free $Ca^{2+}$ from organelles. It is very difficult to measure the level of free $Ca^{2+}$ in plant cells and even more difficult to determine whether a particular stimulus, such as light, has altered this level. Because the $Ca^{2+}$ entering plant cells is so rapidly bound to or sequestered in organelles, the evidence that Pfr promotes $Ca^{2+}$ uptake into cells or causes an increase in $IP_3$ does not indicate whether Pfr also promotes an increase of free $Ca^{2+}$ in the cytosol. Answering this question experimentally will be crucial for assessing the validity of the hypothesis that $Ca^{2+}$ and calcium-binding proteins such as calmodulin mediate some phytochrome responses. Until then, this idea must be considered only a plausible and testable hypothesis.

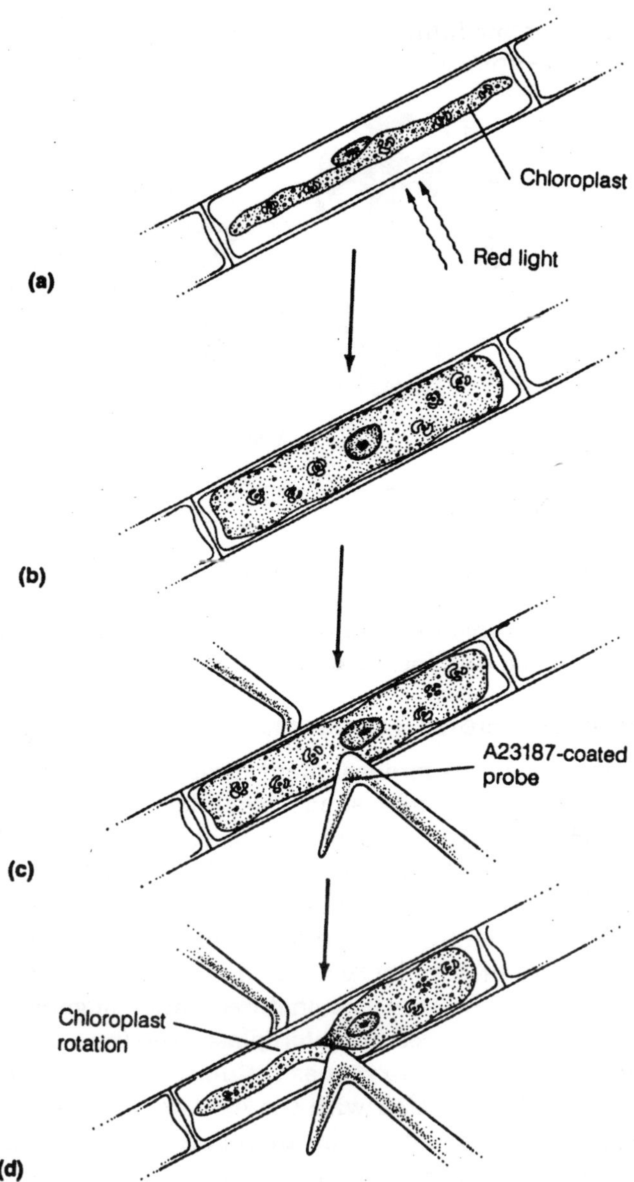


**Fig. 4.16. (left) The promotion of Chloroplast rotation in *Mougeotia* by light or by positional application of a calcium ionophore, A23187. (*a*) A *Mougeotia* cell receives red-light irradiation through the edge surface of its Chloroplast, which induces its Chloroplast to turn to face view (*b*) as seen by an observer looking down from above. Chloroplast rotation can also be induced in the dark by placing A23187-coated probes against the two sides of the cell adjacent to the edge surface of the Chloroplast (c, d′). This result supports the hypothesis that calcium functions as a chemical messenger to couple the stimulus of Pfr production with physiological responses in plants.**

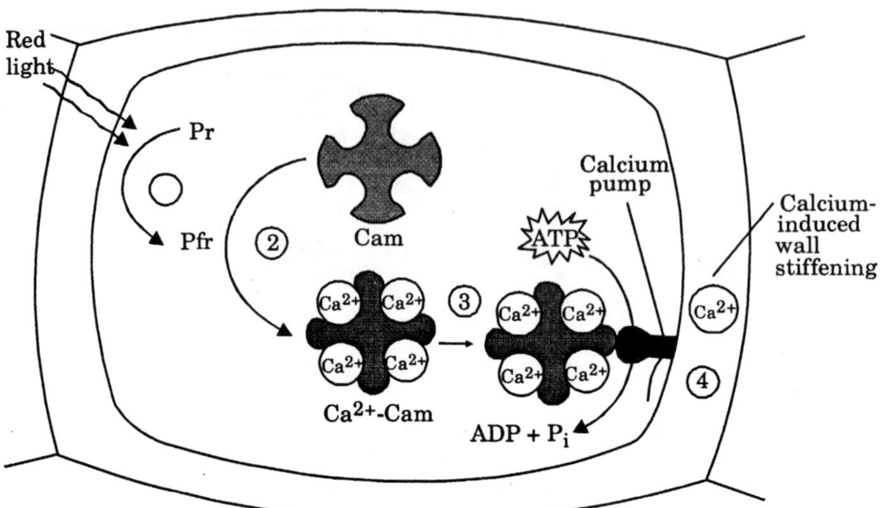


**Fig. 4.17. (right) A model for phytochrome regulation of wall extensibility based on calmodulin. Step 1: Red light converts Pr to Pfr. Step 2: $Ca^{2+}$-calmodulin complexes ($Ca^{2+}$-CaM) are formed. Step 3: The $Ca^{2+}$-CaM complex activates a calcium pump on the plasma membrane. Step 4: The calcium concentration of the cell wall increases, inhibiting cell wall-loosening processes.**

## Phytochrome Regulates Gene Transcription

As the term photomorphogenesis implies, plant development is profoundly influenced by light. Etiolation symptoms include spindly stems, small leaves (in dicots), and the absence of chlorophyll. Complete reversal of these symptoms by light involves major long-term alterations in metabolism that can be brought about only by changes in gene expression. Historically, the hypothesis that phytochrome regulates morphogenesis primarily through its modulation of gene expression preceded the hypothesis that phytochrome affects primarily membrane properties. Whether the control of one is mediated by regulation of the other, or whether the two modes of action are independent, is not yet clear. The photoregulation of gene expression has focused on the nuclear genes encoding messages for chloroplast proteins, the small subunit (SSU) of ribulose-1, 5-bisphosphate carboxylase/oxygenase (Rubisco) and the major light harvesting chlorophyll *a/b* protein (LHCP) associated with the light harvesting complex of photosystem II. These two proteins play an important role in chloroplast development and greening; hence, their regulation by phytochrome has been studied in detail.

The genes for both of these proteins — *rbcS* and *cab*, respectively—are present in multiple copies in the genome; that is, they, belong to multigene families. Working with the duckweed *Lemna,* Tobin and her colleagues at the University of California, Los Angeles have shown by Northern blotting that the abundance of the mRNAs for these two multigene families increases linearly as a function of time soon after exposure of dark-grown seedlings to light. Only a 1-min treatment is needed to increase the amount of mRNA, and the effect is reversed by far-red light. That phytochrome works at the transcriptional level was demonstrated through in vitro transcription in isolated nuclei (nuclear run-off experiments).

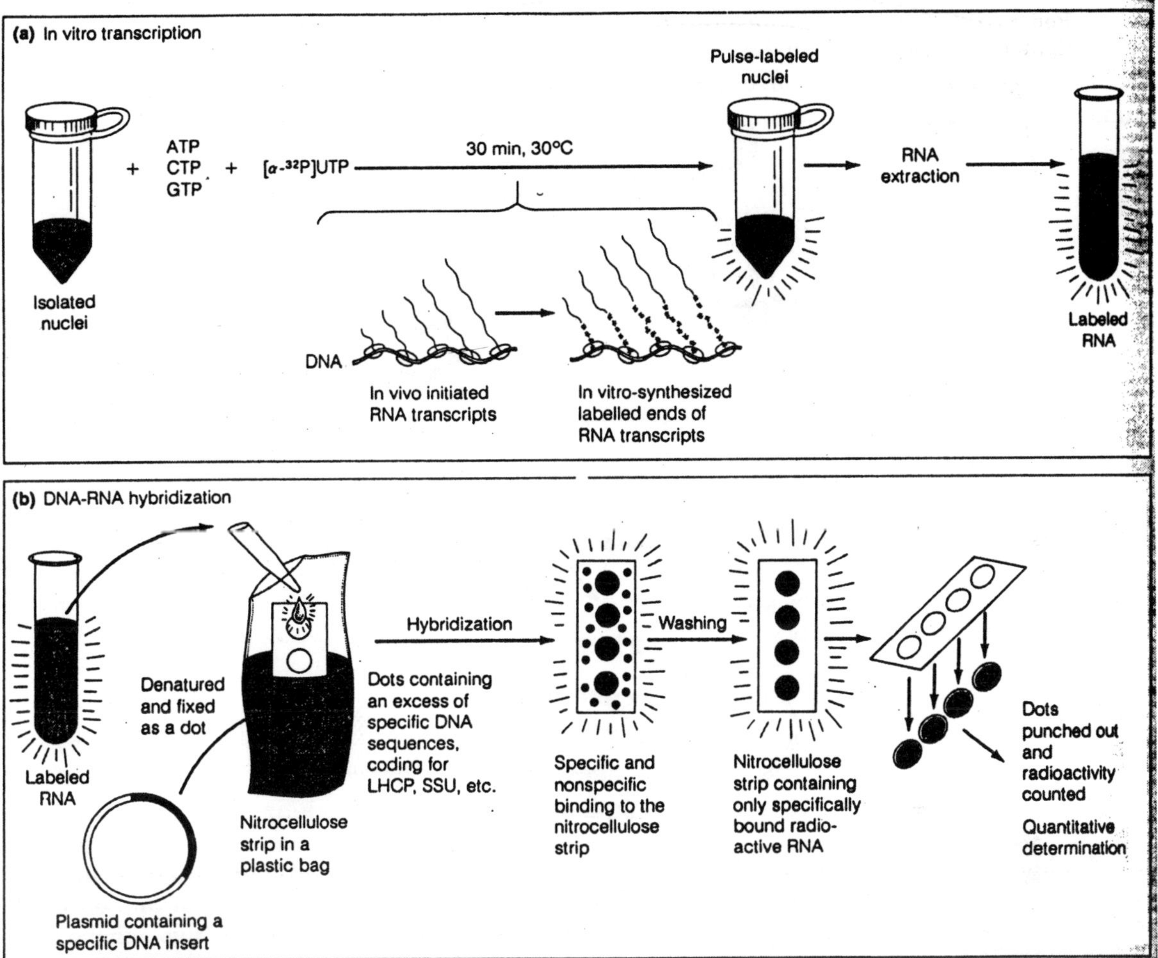


**Fig. 4.18. Nuclear run-off assay, (a) In vitro transcription. Isolated nuclei are incubated in a small tube with unlabeled ATP, CTP, and GTP, plus labeled UTP. During a 30-min incubation transcripts already initiated prior to nuclear isolation incorporate labeled UTP. The end-labeled RNA transcripts are purified, (b) DNA-RNA hybridization. Plasmids containing cDNA inserts for the gene of interest are spotted onto a strip of nitrocellulose paper. The strip is incubated in a plastic bag containing buffer and a drop of the labeled RNA. The mRNA complementary to the cDNA hybridizes to the spot, and the remainder of unbound RNA is removed by washing. The spots can then be punched out and the radioactivity quantified by counting. Alternatively, the radioactivity in each spot can be estimated by exposure to x-ray film. The size of the developed spot is roughly proportional to the radioactivity.**

Nuclei were isolated from tissue exposed to various light regimes and then incubated in the presence of radiolabeled ribonucleotide triphosphates. The labeled transcripts were hybridized to nitrocellulose filters containing cDNA copies of the *rbcS* or *cab* genes, and the bound radioactivity was counted. Such assays have shown that phytochrome responding to red light enhances the rate of transcription of the rbcS and cab genes. Similar results have now been obtained for a number of species, so the effect appears to be widespread. Light-induced rbcS and cab mRNAs are translated on free polysomes (i.e., polysomes not attached to endoplasmic reticulum) as high-molecular-weight precursors containing a transit peptide at the amino-terminal end of the protein.

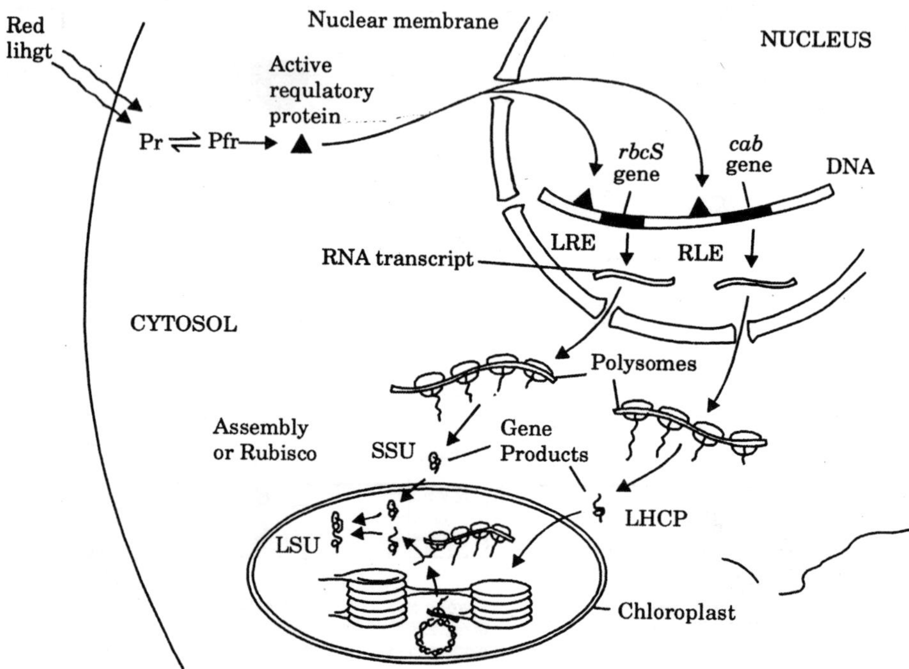


**Fig. 4.19. Model for phytochrome regulation of *rbcS* and *cab* genes. Red light converts to Pr to Pfr, initiating a sequence of biochemical events that leads to the activation of one or more regulatory proteins in the cytosol. The regulatory proteins migrate to the nucleus, where they bind to specific light-regulated elements (LREs) in the promoter region of the *rbcS* and *cab* genes. Transcription is stimulated, leading to enhanced synthesis of the gene products, the small subunit (SSU) of Rubisco and the light harvesting chlorophyll *a/b* protein (LHCP). These proteins contain transit peptides that facilitate their entry .into the chloroplast. Once inside the chloroplast, SSU combines with LSU (the large subunit of Rubisco) to form the holoenzyme. LHCP is incorporated into photosystem II on the thylakoid membrane.**

The transit peptide contains hydrophobic sequences and serves as a targeting signal for transport into chloroplasts. That is, it facilitates binding to and transport across the double membrane of the chloroplast. Upon entry into the chloroplast, the transit peptides are cleaved by a specific processing protease, allowing the proteins to fold and assemble into their active conformations.

## The Promoter Region of the *rbcS* Gene-Is Light-Regulated

Promoters are DNA sequences located upstream from a gene—that is, on the 5′ flanking side—which function in the regulation of transcription. Eukaryotic promoters typically are composed of two functionally different regions: a short sequence that determines the transcription start site (the so-called TATA box, named for its most abundant nucleotides) and a more distant region that controls the developmental expression of the gene. Such regulatory sequences in the promoter are called cis-acting DNA elements. Recently, the promoter regions of the *rbcS* gene from peas have been analyzed to identify cis-acting elements that might be involved in light regulation. To study light-regulated elements (LREs), Chua and colleagues at Rockefeller University transferred the *rbcS* gene (including the promoter region) of pea into petunia and tobacco plants, producing transgenic plants. This was accomplished with the tumor-inducing (Ti) plasmid of the soil bacterium *Agrobacterium tumefaciens,* which can be used as a vector for introducing foreign genes into plant cells.

Fortunately, the cDNA probe for the *rbcS* gene of peas did not hybridize very efficiently with the Rubisco small subunit mRNA of either petunia or tobacco. Thus, it was possible to use the cDNA from peas to monitor the expression of the pea gene in the transgenic plants. The expression of the foreign *rbcS* gene in petunia and tobacco showed remarkable fidelity during development. That is, expression was highest in the leaves of the transgenic plant and was stimulated by red light and inhibited by far-red light. By selectively removing ("deleting") specific regions of the promoter region before transferring the gene to tobacco, it was possible to identify two short sequences of DNA in the 5′ upstream region that regulated both light induction and organ specificity of *rbcs* gene transcription.

When a hybrid gene was constructed from the LRE region of the *rbcS* gene and the TATA box and coding sequence of a non-light-regulated foreign gene, the foreign gene became light-regulated when introduced into tobacco. It is generally thought that cis-acting regulatory elements such as the LREs are regulated by proteins encoded elsewhere in the genome, called trans-acting factors. Trans-acting factors are DNA-binding proteins that bind to the cis-acting DNA sequences and regulate transcription of the gene?

As discussed in relation to auxins and gibberellins DNA-binding proteins thought to play a role in hormone-regulated transcription have been identified using the gel retardation assay. A protein factor called GT-1 has been isolated from nuclear extracts that bind specifically to the LREs of the *rbcS* genes of several species. The binding site for the GT-1 protein has been identified. It is located about 150 bp upstream of the transcription start site of many *rbcS* genes. Recently, Lam and Chua (1990) demonstrated that the GT-1 sequence alone has sufficient information to turn a promoter that is not regulated by light into a light-regulated one. Thus, the GT-1 sequence "is likely to be a part of the molecular light switch for *rbcS* activation".

Other possible trans-acting factors have been identified, and it is likely that the regulation of transcription by light is exceedingly complex, involving a host of such factors. Based on the foregoing results, the following model has been proposed. Red light causes the conversion of Pr to Pfr. Pfr either directly or indirectly (e.g., through $Ca^{2+}$ and calmodulin) activates one or more regulatory proteins (phytochrome itself has no DNA-binding activity). The activated regulatory protein (s) then binds to the LRE sequences or other potential regulatory sites and stimulates transcription of the gene as illustrated. It is assumed that most phytochrome-stimulated genes show this basic pattern of regulation.

## Phytochrome Regulates the Expression of its Own Gene

Phytochrome is encoded by a multigene family (e.g., the oat nuclear genome contains at least four phytochrome genes). When Pr is converted to Pfr by a pulse of red light, the steady-state phytochrome mRNA level decreases exponentially with a half-life of only 15 min, an extremely fast rèsponse by higher plant standards. This decrease has been shown by the nuclear run-off assay to be due to a sharp decrease in the rate of transcription. Thus phytochrome genes are autoregulated.

As noted earlier in the chapter, the repression of phytochrome gene expression is a contributing factor to the rapid decrease in total phytochrome when dark-grown plants are transferred to the light. How does phytochrome repress the transcription of its own genes? Presumably, a trans-acting factor binds to a specific LRE in the promoter region, inhibiting transcription. A highly conserved sequence has now been identified in the promoter region of the phytochrome gene in rice (*Oryza sativa*). The sequence is a potential LRE, although this function has not yet been, demonstrated by deletion analysis, as described above for the *rbcS* gene. Surprisingly, GT-1, the factor that binds to the LRE of the *rbcS* genes, also appears to bind to the conserved sequence in the phytochrome gene.

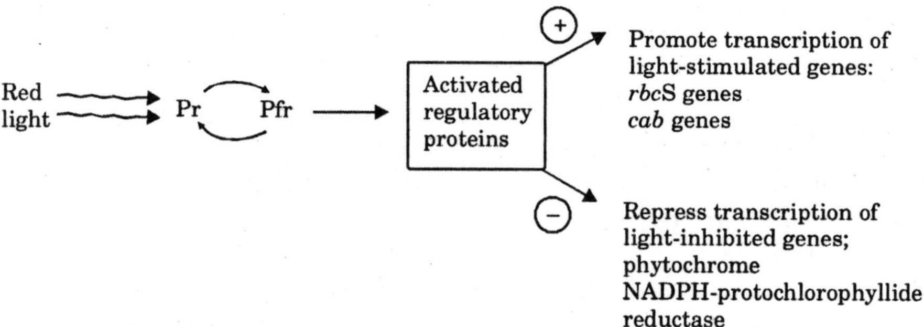


**Fig. 4.20. Scheme for phytochrome regulated transcription of genes. Phytochrome in the Pfr form causes, directly or indirectly, the activation of one or more regulatory proteins. These regulatory proteins either promote the transcription of light-stimulated genes or repress the transcription of light-inhibited genes.**

This raises the interesting possibility that the same DNA-binding protein *that promotes* the transcription of light-inducible genes may *repress* the transcription of light-inhibited genes, such as the gene for phytochrome. However, as noted earlier, the situation is undoubtedly much more complex, and much research remains to be done.

## Concluding Remarks

Photomorphogenesis refers to the dramatic effects of light on plant development and cellular metabolism. Red light exerts the strongest influence, and the effects of red light are often reversible by far-red light. Phytochrome is the pigment involved in most photomorphogenetic phenomena. Phytochrome is a large dimeric protein made up of two equivalent subunits. The monomer has a molecular mass of 124,000 Da and is covalently bound to a chromophore molecule, an open-chain tetrapyrrole. Phytochrome exists in two forms, a red light-absorbing form (Pr) and a far-red light-absorbing form (Pfr). Absorption of red light by Pr converts it to Pfr, and absorption of far-red light by Pfr converts it to Pr.

However, the absorption spectra of the two forms overlap in the red region of the spectrum, leading to an equilibrium between the two forms called a photostationary state. Pfr is considered to be the active form in most cases; that is, it gives rise to the physiological response. Other factors in addition to light regulate the steady-state level of Pfr; these include dark reversion, degradation, and the rate of synthesis. Spectroscopic and immunological studies have shown that phytochrome is most abundant in etiolated tissues and is concentrated in meristematic regions.

During photoconversion to Pfr in higher plants, the phytochrome distribution within the cell changes from a diffuse to a sequestered pattern. Thus far, no organelle has been specifically associated with Pfr in higher plant cells. In the alga *Mougeotia,* phytochrome appears to be located on the plasma membrane, where it mediates chloroplast rotation by undergoing a 90° shift in orientation. Phytochrome responses have been classified into very low fluence (VLF), low fluence (LF), and high fluence (HF) responses. These three types of responses differ not only in their fluence requirements but also in other parameters such as their escape times, action spectra, and photoreversibility.

Phytochrome plays an important role in the detection of shade in plants adapted to high levels of sunlight. Phytochrome also regulates circadian rhythms, such as the sleep movements of leaves. Many of these phenomena can be traced to alterations of membrane properties. One way in which phytochrome might exert its effects is by increasing free calcium levels. The formation of a calcium-calmodulin complex might activate enzymes important for cell regulation. In addition, phytochrome is known to regulate the transcription of a number of genes. Many of the genes involved in greening, such as the nuclear-encoded genes for the small subunit of Rubisco and the chlorophyll *a*/*b* protein of the light harvesting complex, are activated by red light.

Gene activation is thought to be mediated by regulatory proteins that bind to the promoter region of the gene! Red light also represses the transcription of a number of genes, including the gene for phytochrome. The repression of the gene for phytochrome is also presumed to be mediated by DNA-binding regulatory proteins.

## Mougeotida: A Chloroplast with a Twist

Mougeotia is a filamentous green alga whose cells contain a single ribbon (Fig. 4.20 Part a) like chloroplast nestled between two large vacuoles and surrounded by a layer of cytoplasm. The chloroplast can rotate about its long axis and can respond to incident radiation by orienting *perpendicular* to the direction of light (part *b*). Phytochrome is implicated in the response because red light elicits the greatest rotation and the effect of red light can be reversed by far-red light. Far-red light can also induce rotation, except that the chloroplast oreints *parallel* to a source of far-red light (part *b*). Using microbeams of red and far-red light, Wolfgang Haupt and his co-workers in Germany showed that the photoreceptor for chlroplast rotation is localized near the periphery of the cytoplasm in the vicinity of the plasma membrane.

When a microbeam of red light was directed at the cell surface under a microscope, the edge of the chlroplast adjacent tot eh microbeam rotated 90°, even though the chlroplast itself was not illuminated. The plasma membrane location of phytochrome in *Mougeotia* was deduced from studies using microbeams of plane-polarized red and far-red light. Phytochrome is a dichroic pigment; that is, it has a preferred electrical vector orientation for the absorption of light.

The greatest absorption occurs when the electric vector is parallel to the absorbing surface of the pigment. if we think of the pigment as a rod and the electric vector as a narrow slit of light, more light will be absorbed when the light slit is oriented parallel to the long axis of the pigment than in any other orientation (part *c*). Haupt and his co-workers discovered that the ability of plane-polarized red light to cause rotation depended on the plane of polarization relative to the long axis of the cell. This phenomenon is depicted in the experiment shown in part *d*. Half of a cell was illuminated with plane-polarized red light vibrating transverse to the cell axis, while the other half was illuminated with light vibrating parallel to the cell axis.

The chloropast rotated from the profile to the face position only in response to red light vibrating in the transverse direction, producing a twist in the chloroplast. It was concluded that phytochrome has a defined orientation in the cell. Later studies have specifically shown that Pr molecules are arranged in left-handed spirals around the cell. When happens to the orientation of phytochrome after it is converted from Pr to Pfr? Part *c* shows the results of a photoreversibility experiment.

The chloroplast is first treated with red light vibrating in a plane that induces rotation from a profile to a face position. Half of the cell is then immediately treated with far-red light vibrating in the same plane, while the other half is treated with far-red light vibrating perpendicular to the plane of the red light. The red-light effect is reversed only by the far-red light vibrating in a plane perpendicular to that of the red light, suggesting that phyotochrome molecules change their orientation by about 90° during photoconversion. The effects described above are best accounted for by a model in which phytochrome is located on the plasma membrane of *Mougheotia* (part *f*). It is difficult to imagine how Pr and Pfr could maintain their respective orientations without being associated with the membranes.

Rotation occurs because the chloroplast moves away from a localised region of Pfr and toward a region of Pr (part *b*). Thus, chloroplast movement occurs in response to a gradient of Pfr to Pr around the periphery of the cell. Does the reorientation of phytochrome molecules during photoconversion play a role in higher plants as well? The diffuse distribution of phytochrome in higher plant cells is inconsistent with such a model, but only further research will tell.

# CHAPTER 5 AMINO ACID BIOSYNTHESIS

*The carbon skeletons for amino acid biosynthesis originate from the central metabolic pathways; the first steip in amino acid synthesis usually involves addition of the α amino group.*

Amino acids are best known as the building blocks of protein, and indeed that is a main function of the 20 L-amino acids that are commonly found in proteins. In addition to this role, amino acids serve as precursors to many important small molecule compounds, including nucleotides, poiphyrins, parts of lipid molecules, and precursors for several coenzymes. Amino acids also provide the vehicles for converting nitrogen and sulfur from inorganic to organic forms.

As an alternative energy source amino acid catabolism can be coupled to the regeneration of ATP from ADP or AMP. In this Chapter we focus on the biosynthesis of amino acids and their role in bringing inorganic nitrogen and sulfur into the biological world. In addition nonprotein amino acids are discussed briefly.

## Central Metabolic Pathways

Inspection of the amino acid biosynthetic pathways shows that all amino acids arise from a few intermediates in the central metabolic pathways. Amino acids derived from a common intermediate are said to be in the same family. For example, the serine family of amino acids, which includes serine, glycine, and cysteine, all arise from glycerate-3-phosphate. The carbon flow from the central metabolic pathways to amino acids is a regulated process that provides amino acids in the amounts needed for maintenance and growth. The flow is regulated mostly by end-product inhibition, so that the amino acid (end product) inhibits the first enzyme in the pathway for its synthesis.

## Amino Acid Biosynthesis

*Escherichia coli* has served as the organism of choice for examining the pathways for amino acid biosynthesis for two reasons: (1) *E. coli* synthesizes all 20 amino acids commonly found in proteins, and (2) *E. coli* is an organism ideally suited for genetic and biochemical studies.

### *Genetic Complementation Analysis*

The investigation of an amino acid biosynthetic pathway in *E. coli* begins with the accumulation of mutants that are deficient in the capacity to synthesize that amino acid.

Mutants of this sort are known as auxotrophs. Each auxotroph bears a mutation in one of the genes that encodes an enzyme required for a step in the biosynthetic pathway of the amino acid. It is possible to determine how many steps are in a particular pathway by the process of complementation analysis.

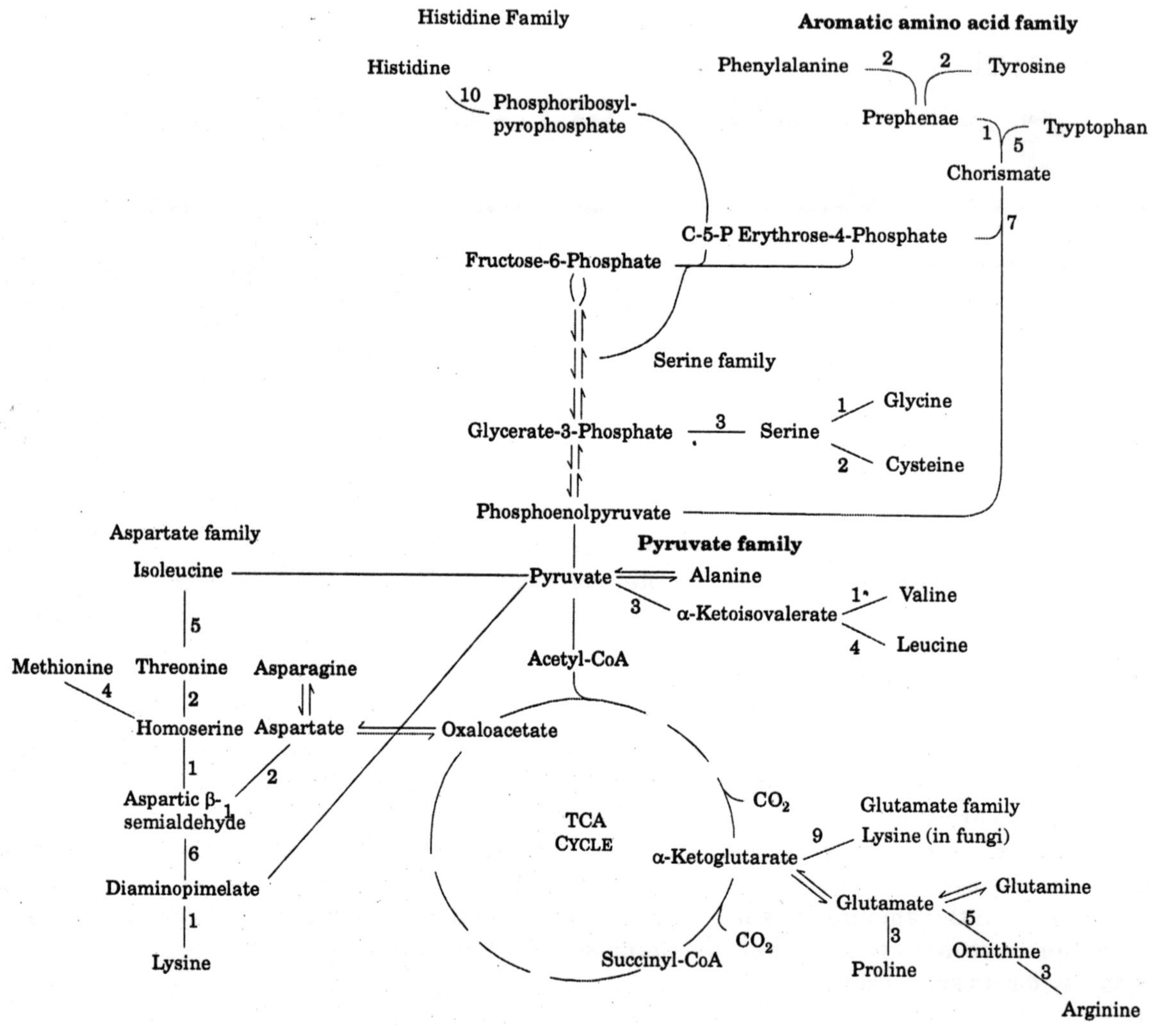


**Fig. 5.1. Outline of the biosynthesis of the 20 amino acids found in proteins. The *de novo* biosynthesis of amino acids starts with carbon compounds found in the central metabolic pathways. The central metabolic pathways are drawn in black, and the additional pathways are drawn in red. Some key intermediates are illustrated, and the number of steps in each pathway is indicated alongside the conversion arrow. All common amino acids are emphasized by boxes. Dashed arrows from pyruvate to both diaminopimelate and isoleucine reflect the fact that pyruvate contributes some of the side-chain carbon atoms for each of these amino acids. Note that lysine is unique in that two completely different pathways exist for its biosynthesis. The six amino acid familiec are screened.**

## *Biochemistry of Auxotrophs*

Complementation group analysis sets the stage for the biochemist. First it tells him or her how many proteins must be isolated and characterized. Additional aid to the biochemist is provided by the fact that mutations belonging to complementation groups block specific steps in a pathway. As a result no amino acid is synthesized, and specific intermediates accumulate. Owing to their abundance, intermediates accumulating in such mutants can frequently be isolated and their structure determined. By piecing together information of this type from studies on different mutants, an overall picture of the pathway emerges. Finding intermediates in a pathway is only half the problem.

The biochemist must also isolate and characterize the enzymes of the pathway. To this end, extracts are made from either normal or mutant cells. These extracts are tested in conjunction with different intermediates to see whether the extracts can carry out the conversion of one intermediate to the next in the pathway. This conversion is used as an assay for isolating the enzyme by progressive subfractionation of the extract to the point at which the enzyme of interest is reasonably pure.

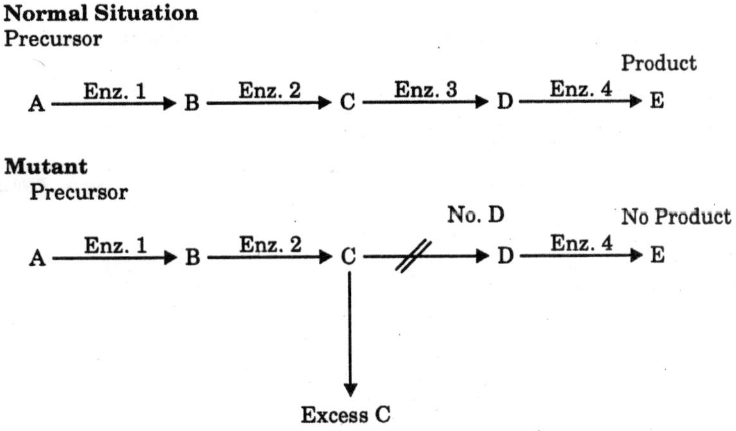

**Normal Situation**
Precursor
Product
A —Enz. 1→ B —Enz. 2→ C —Enz. 3→ D —Enz. 4→ E

**Mutant**
Precursor
No. D
No Product
A —Enz. 1→ B —Enz. 2→ C —//→ D —Enz. 4→ E
Excess C

**Fig. 5.2. Immediate consequences of a point mutation in a biosynthetic pathway. When the mutation leads to an inactive enzyme, the chain of reactions leading to the end product in the pathway is broken, and frequently large amounts of intermediate are produced, accumulate in the cell, and may leak to the environment.**

In the final stages of biochemical analysis the structure and catalytic activities of the enzyme are determined. Sometimes the investigation of a pathway is complicated by the fact that common intermediates are used in the synthesis of more than one amino acid. In such cases intermediates may be converted into more than one product by different enzymes, each product serving as an intermediate for a particular amino acid. Mutants containing defects in enzymes required for the synthesis of more than one amino acid usually require the addition of two or more amino acids to reestablish cell growth.

## Nitrogen Fixation

A common element in all protein-bound amino acids is the α-amino group. *Directly or indirectly, the α-amino groups of most amino acids are derived from ammonia by way of the amino groups of L-glutamine.* Ammonia itself can be incorporated directly into α-ketoglutarate to make glutamate, or into glutamate to make glutamine. The uptake of $NH_3$ by glutamate represents a key step in the process, whereby reduced nitrogen becomes incorporated into organic molecules.

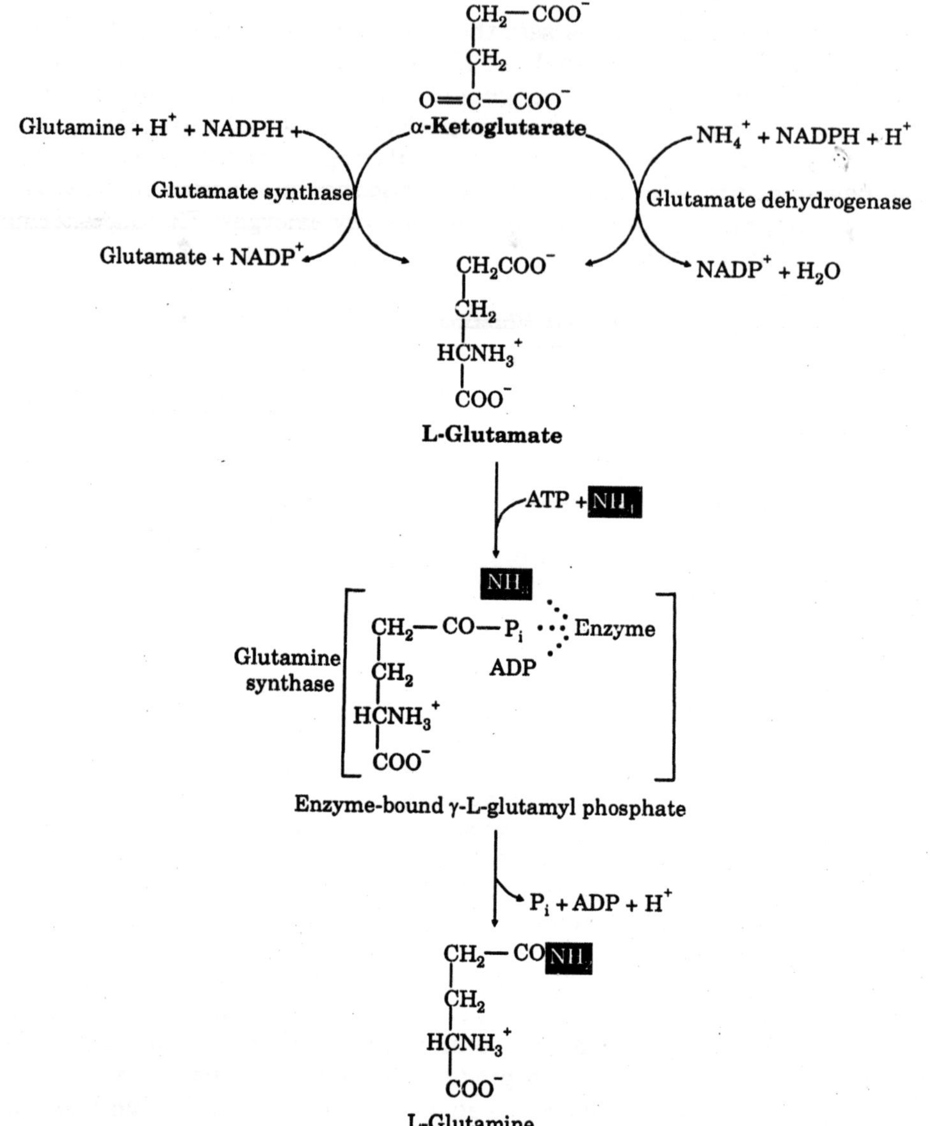


**Fig. 5.3. The conversion of ammonia into the α-amino group of glutamate and into the amide group of glutamine. The direct amination of α-ketoglutarate by $NH_4^+$ occurs only under conditions of high $NH_4^+$ concentrations, which are rarely found in nature.**

Glutamate serves as the starting material for the synthesis of glutamine, proline, and arginine. Appropriately, these four amino acids are described as belonging to the glutamate family. In fungi, lysine is also included in this family. In this section we discuss the pathways for glutamate and glutamine synthesis. We also discuss some other aspects of the nitrogen cycle. *Consideration of the arginine pathway is deferred until the next chapter.*

## *The Direct Amination of α-Ketoglutarate Leads to Glutamate*

The most direct route to glutamate (and therefore, amino group formation) is that exhibited by many bacteria when grown in a medium containing an ammonium salt as the sole nitrogen source. The reaction entails a reductive amination catalyzed by glutamate dehydrogenase. In *E. coli* this enzyme is specific for NADPH as the hydrogen donor, as might be expected for a biosynthetic reaction involving a reductive step. Studies of the glutamate dehydrogenases from green plants indicate that they require a very high concentration of $NH_3$ to be effective.

In fact an alternative pathway is responsible for glutamate biosynthesis in most plants. Thus, in green plants, as well as many bacteria and fungi, the formation of the amino group of glutamate usually originates from the amide group of glutamine. The reaction is catalyzed by glutamate synthase.

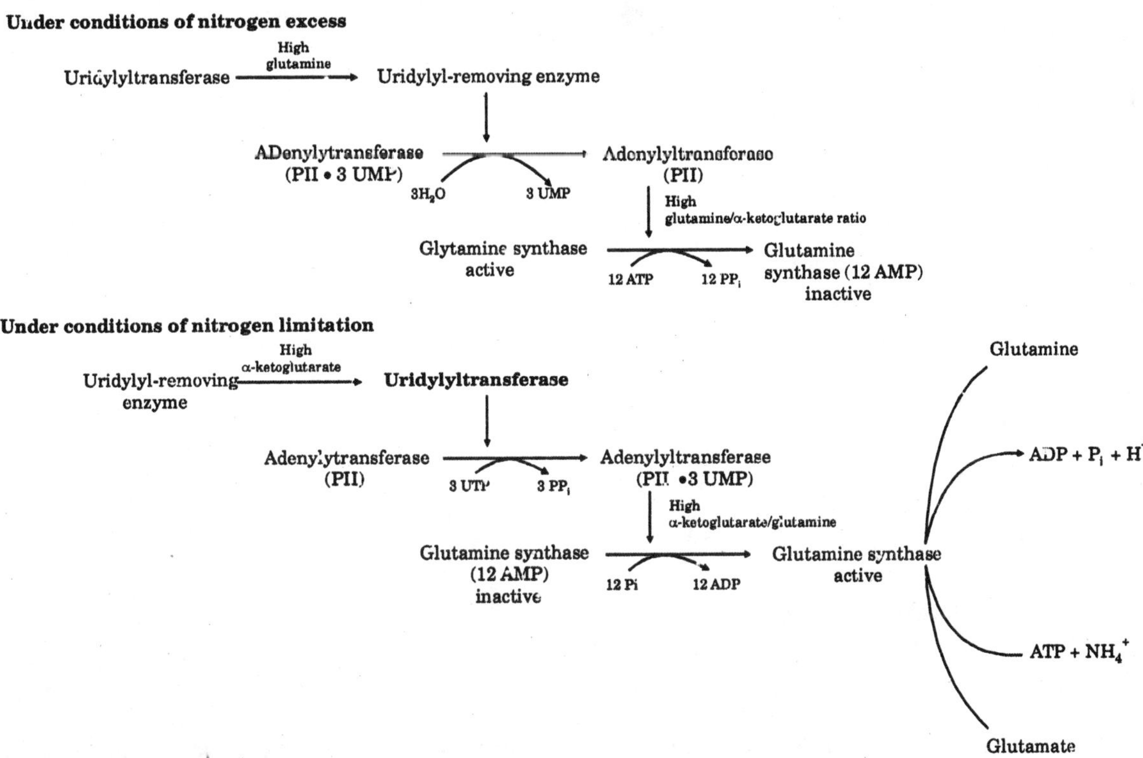


**Fig. 5.4. Regulation of glutamine synthase in *E. coli*. The activity of glutamine synthase is inhibited by adenylylation. Both adenylylation and deadenylylation are regulated by a cascade of controls that are responsive to the concentrations of glutamine and α-ketoglutarate. The concentrations of these two compounds are an indication of the nitrogen supply. Thus, under conditions of nitrogen excess (*top part of figure*) the concentration of glutamine is high, and the ratio of glutamine to α-ketoglutarate is high. This leads to inactivation of the glutamine synthase. Under conditions of nitrogen limitation (*bottom part of figure*) the α-ketoglutarate concentration is high, and the ratio of α-ketoglutarate to glutamine is high, so the glutamine synthase is activated.**

## *Amidation of Glutamate to Glutamine is an Elaborately Regulated Process*

Glutamine, which is so commonly used in amination reactions, is regenerated by a reaction in which ammonia is directly added to glutamate; the reaction is catalyzed by *glutamine synthase*. This two-step reaction starts with activation of the γ-carboxyl group of glutamate and yields a γ-glutamyl-enzyme complex this is followed by γ-glutamyl transfer of $NH_3$. Since glutamine synthase is the main enzyme initiating the flow of ammonia nitrogen into organic components, it is not surprising to find that it is a highly regulated enzyme. The *E. coli* enzyme has been studied extensively by Stadtman and Ginsberg. They have shown that the enzyme is composed of 12 identical subunits ($M_r$ 50,000) arranged in two hexameric rings.

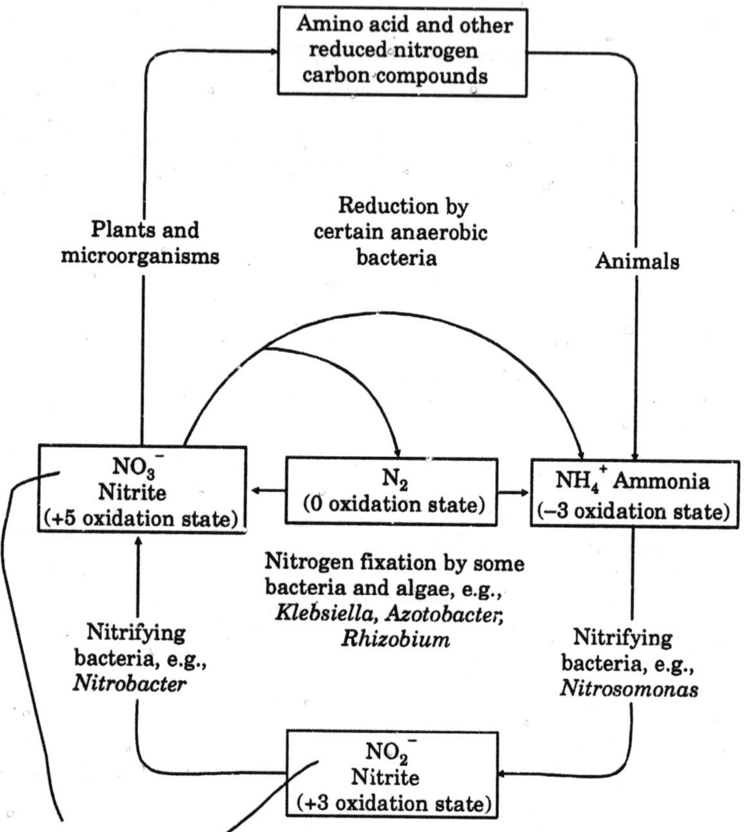


**Fig. 5.5. The nitrogen cycle depicts the flow of nitrogen in the biological world. The proteins of animals and plants are cleaved by many microorganisms to free amino acids from which ammonia (or ammonium ion) is released by deamination. Urea, the main nitrogen excretion product of animals, is hydrolyzed to $NH_3$ and $CO_2$. *Nitrosomonas* soil bacteria obtain their energy by oxidizing $NH_3$ to nitrite, $NO_2^-$. *Nitrobacter* obtain their energy by oxidizing nitrite, $NO_2^-$, to nitrate, $NO_3^-$. Plants and many microorganisms reduce nitrate for incorporation into amino acids, completing the cycle. Other microorganisms reduce nitrate partly to $NH_3$ and partly to $N_2$, which is lost to the atmosphere. Atmospheric nitrogen, $N_2$, can be recaptured, reduced, and converted into organic substances by a limited number of nitrogen-fixing bacteria and algae.**

Activity of the enzyme is negatively regulated by eight nitrogenous compounds: Carbamyl phosphate, glucosamine-6-phosphate, tryptophan, alanine, glycine, histidine, cytidine triphosphate, and AMP. With the exception of glycine and alanine these compounds receive the amide nitrogen directly from glutamine during their biosynthesis. Therefore most of these compounds can be thought of as end products originating from glutamine.

Although glycine and alanine are not direct end products, they are "indicators" of the sufficiency of the nitrogen supply of the cell. Thus, the regulation of glutamine synthase can be regarded as an example of end-product inhibition.

Even more important in the regulation of *E. coli* glutamine synthase activity is the reversible ATP-dependent adenylylation of a specific tyrosyl residue on each subunit. As the enzyme becomes progressively more adenylylated (up to the fully adenylylated form which contains 12 AMP groups per enzyme molecule), the enzyme becomes progressively less active. The adenylylation reaction and the deadenylylation reaction are regulated by the nitrogen supply in the cell.

The immediate small-molecule effectors of this regulatory system are glutamine and α-ketoglutarate. A high-glurtamine α-ketoglutarate ratio signals nitrogen excess, whereas a high-α-ketoglutarate-glutamine ratio signals nitrogen limitation. Nitrogen excess leads to inactivation of glutamine synthase, just as nitrogen limitation leads to its activation.

The activation of glutamine synthase favours fixation of ammonia in a condensation with glutamate to form glutamine. Both activation (deadenylylation) and the inactivation (adenylylation) are controlled by a cascade of regulatory interactions. The cascades involved here are unusual in that the same two regulatory proteins, *uridylyltransferase and adenylyltransferase* function in both cascades. Both of these regulatory proteins have two enzymatically distinct sites arranged so that only one site is active at any given time. Let us first consider the situation under conditions of nitrogen excess. The first regulatory protein in the cascade is converted into a uridylyl-removing enzyme. This enzyme hydrolyzes UMP from a PII · 3 UMP protein that acts in concert with adenylyltransferase to form adenylate glutamine synthase. The resulting adenylylated enzyme is inactive. Under conditions of nitrogen limitation, the first regulatory enzyme in the cascade is converted into a uridylyltransferase. This enzyme uridylates the PII protein.`

The PII · 3 UMP protein and adenylyltransferase deadenylylates the glutamine synthase, 12 AMP. The control of glutamine synthase by adenylylation and deadenylylation may be limited to certain Gram negative bacteria. However, physiologically analogous modifications of other glutamine synthases probably exist. For example, glutamine synthases in eukaryotic cells are octameric structures that respond to regulation by dissociation to a tetrameric form under conditions in which the $NH_4^+$ supply is restricted.

## Nitrogen Cycle

Glutamine synthase is just one of the links in the nitrogen cycle whereby nitrogen passes through several chemical forms. The passage of nitrogen from one form to another involves a chain of widely distributed organisms. $NH_4^+$ is the form in which nitrogen is incorporated into organic materials, but it is often less available to plants or bacteria for biosynthesis than other forms of nitrogen. When present for any length of time in the free state $NH_4^+$ is likely to be oxidized by nitrifying bacteria (such as *Nitrosomonas* and *Nitrobacter*) to nitrite ($NO_2^-$) and nitrate ($NO_3^-$).

Reduction of nitrate by plants and bacteria to $NH_4^+$ is vital to maintaining the nitrogen cycle. Nitrogen fixation whereby atmospheric nitrogen is reduced to $NH_3$ occurs in a limited number of microorganisms. Some of these microorganisms exist in the free state, and some exist as symbionts in the root nodules of certain plants.

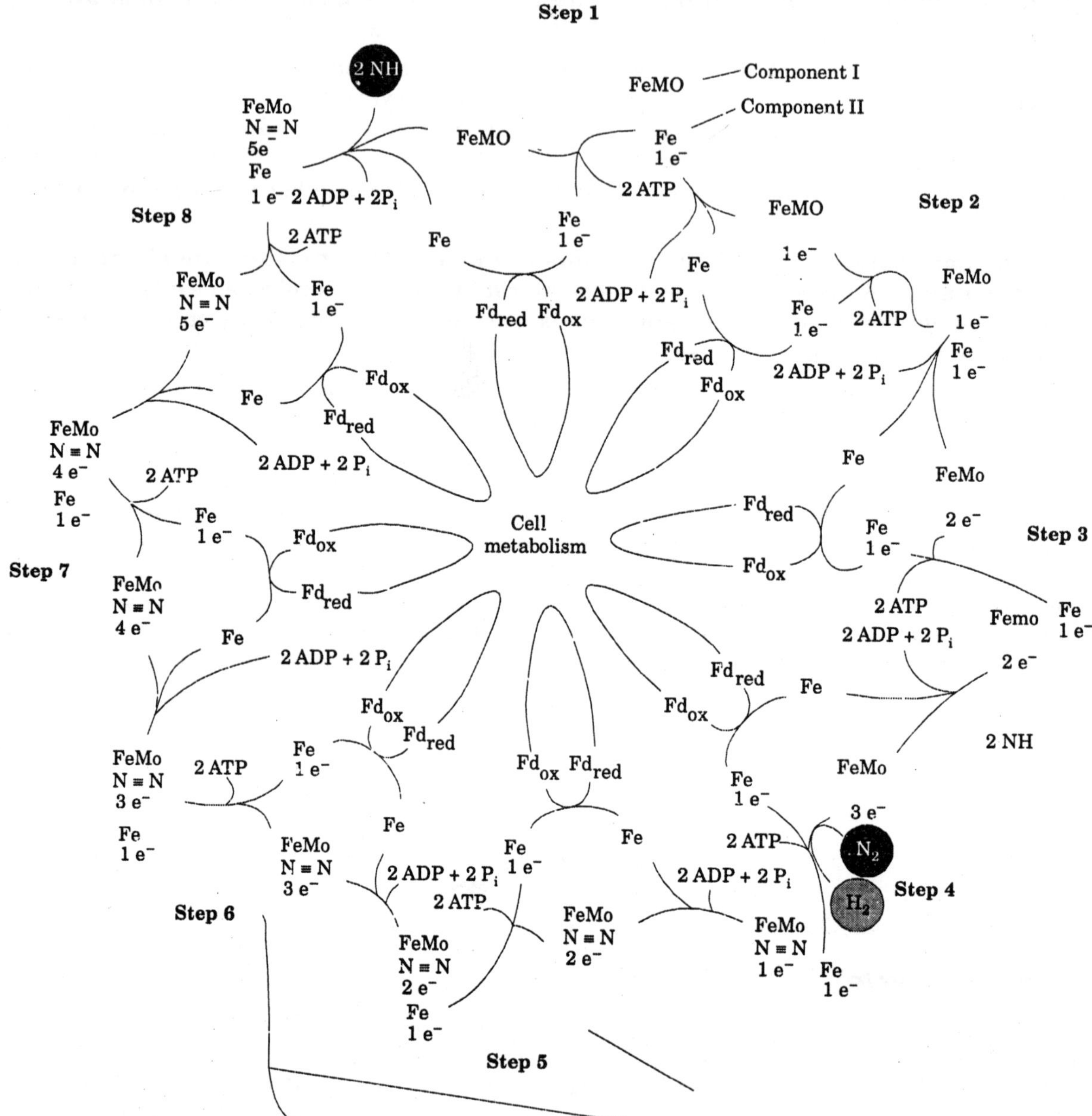


**Fig. 5.6. The eight-step nitrogenase cycle. Nitrogenase contains two complex protein components I and compoenent II. The nitrogenase cycle starts when component II (an Fe—protein) binds to component I (usually an Fe—Mo—protein). The binding requires two ATP molecules. During the bound state component II transfers an electron to component I. This is followed by the hydrolysis of the two ATPs and the separation of the two components that begin a new cycle. In each step of the cycle the Fe—protein component II picks up asingle electron from reduced ferrodoxin or flavodoxin and binds to two ATP molecules. After three single electron ($e^-$) transfers, an $H_2$ molecule is released from component II, and a $N_2$ molecule binds in its place. Following five more electron transfers, the $N_2$ molecule is fully reduced to 2 $NH_3$'s, which are released.**

## *Nitrogen Fixation Involves an Enzyme Complex Called Nitrogenase*

The conversion of gaseous nitrogen ($N_2$) into ammonia ($NH_3$) is a most intricate reaction catalyzed by nitrogenase, a two-component enzyme. In *Clostridium pasteurianum* component I is a tetrameric protein with a molecular weight of 220,000. The four subunits in component I are of two kinds: Two with molecular weights of 50,000 and two with molecular weights of 60,000. In addition component I has four iron-sulfur centers, each containing four iron atoms and four sulfur atoms; and two extractable iron-molybdenum-sulfur "cofactors" that contain eight iron atoms, six labile sulfur atoms, and one molybdenum atom.

Component II from *C. pasteurianum* contains two identical subunits, each with a molecular weight of 29,000. These proteins surround an iron-sulfur center that contains four iron atoms and four sulfur atoms. Although it might be expected that only six reducing equivalents would be required for each mole of $N_2$ (valence state 0) reduced to 2 $NH_3$ (valence state –3), in fact, the reaction catalyzed by nitrogenase converts two protons ($H^+$) to hydrogen ($H_2$) for each molecule of $N_2$ reduced. As a result the overall equation for the conversion of nitrogen to ammonia requires eight reducing equivalents.

$$N_2 + 8H^+ + 8e^- + 16\,ATP \longrightarrow 2NH_3 + H_2 + 16\,ADP + 16\,P_i \quad ...(5.1)$$

Why is hydrogen production a necessary step in the reduction of dinitrogen? It appears that it is necessary for the binding of nitrogen. The reduction of nitrogen by nitrogenase can be viewed as a cycle of eight electron transfers, each of which occurs in the cyclic transfer of a single electron from reduced ferrodoxin or flavodoxin to the Fe protein (component II) *and* the regeneration of the reduced electron donor by cell metabolism (e.g., pyruvate oxidation). On reduction, the Fe — protein binds two molecules of ATP, after which it associates with the Fe — Mo — protein (component I). The transfer of an electron to the Fe — Mo — protein is accompanied by the hydrolysis of the two bound ATP molecules and the dissociation of components I and II.

Overall, the sequential functioning of eight Fe—protein reduction–oxidation cycles is required for one Fe—Mo protein cycle. The first three transfers, which result in the production of one $H_2$ molecule, somehow pave the way for the binding of dinitrogen at the active site. The remaining transfers are necessary to reduce a single dinitrogen to two ammonias.

## *Nitrate and Nitrite Reduction Play Two Roles*

As pointed out above, $NH_3$ tends to be oxidized to nitrites or nitrates, if it is not immediately incorporated into organic matter. Nitrate reduction play two physiological roles. One, exhibited and nitrate primarily by bacteria, is a dissimilatory role in which *nitrate and nitrite serve as terminal electron acceptors*. The reduction serves to oxidize reducing equivalents such as NADH that are generated during oxidation of substrates. *The other role played by these reductions is an assimilatory one in which nitrate is first converted to nitrite and thence to $NH_3$*. These reactions are observed in bacteria, fungi, and plants. Generally, the assimilatory nitrate and nitrite reductases are soluble enzymes that utilize reduced pyridine nucleotides or reduced ferrodoxin. In contrast, the dissimilatory nitrate reductases are membrane-bound terminal electron acceptors that are tightly linked to cytochrome $b_1$ pigments.

Such complexes allow one or more sites of energy conservation (ATP generation) coupled with electron transport. The ammonium ion ($NH_4^+$), produced by fermentative bacteria that use nitrite as an oxidant, or produced during the decomposition of organic materials, is an important source of nitrogen for many plants and bacteria. Nevertheless, under vigorous aerobic conditions much of the $NH_4^+$ so produced is converted back to nitrite and nitrate by nitrifying bacteria.

## Sulfur Fixation

As already noted the serine family includes three amino acids: Serine, glycine, and cysteine. We focus on cysteine synthesis, which funnels sulfur into the biochemical world and supplies the cysteine needed for biosynthesis.

### *Cysteine Biosynthesis*

The biosynthesis of L-cysteine entails the sulfhydryl transfer to an activated form of serine. This pathway to L-cysteine has been most throughly studied in *E. coli*. In the first step an acetyl group is transferred from acetyl-CoA to serine to yield *O*-acetylserine. The reaction is catalyzed by *serine transacetylase.* The formation of cysteine itself is catalyzed by *O*-acetylserine sulfhydrylase. In some organisms sulfur incorporation involves homocysteine as an intermediate. In such cases cysteine formation occurs by a transsulfuration reaction, with the intermediate formation of L, L-cystathionine.

Cystathionine is formed in a simple condensation reaction from serine and homocysteine by cystathionine-β synthase.

The subsequent cleavage of cystathionine to yield cysteine, α-ketobutyrate and $NH_4^+$ is catalyzed by γ-cystathionase, a pyridoxal-phosphate-containing enzyme. This transsulfuration pathway is one of the routes used for methionine catabolism.

### *Sulfate Conversion of into Sulfide*

Most sulfur in nature exists in the inorganic form of sulfate ion, $SO_4^{2-}$. Sulfate must be reduced to $H_2S$ before it can be incorporated into amino acids. The eight-electron reduction of sulfur from $SO_4^{2-}$ to $H_2S$ starts with activation of sulfate. In a reaction catalyzed by *adenylylsulfate pyrophosphorylase,* inorganic sulfate is converted to adenosine-5′-phosphosulfate (APS). In *E. coli* and certain other bacteria APS is phosphorylated at the 3′—OH by APS kinase to yield 3′-phosphoadenosine-5′-phosphosulfate (PAPS). The reduction of the sulfonyl moiety of APS (in yeast and plants) or of PAPS (in *E. coli)* to sulfite occurs by transfer of the sulfonyl group to a thiol acceptor, such as thioredoxin to yield an —S—$SO_3$ derivative.

Reaction with a second thiol group (on thioredoxin or a molecule of glutathione) yields sulfite and an oxidized acceptor. The six-electron reduction of sulfite to sulfide is catalyzed by *sulfite reductase,* a multisubunit complex composed of a flavoprotein and a heme iron-sulfur protein.

## Synthesis of Isoleucine

The aspartate (oxaloacetate) family of amino acids includes aspartate, asparagine, methionine, lysine, threonine, and isoleucine. The pyruvate family includes alanine, valine, leucine, and also lysine and isoleucine. Threonine is a precursor of isoleucine. It is converted into isoleucine by a group of enzymes that are also used in the synthesis of valine.

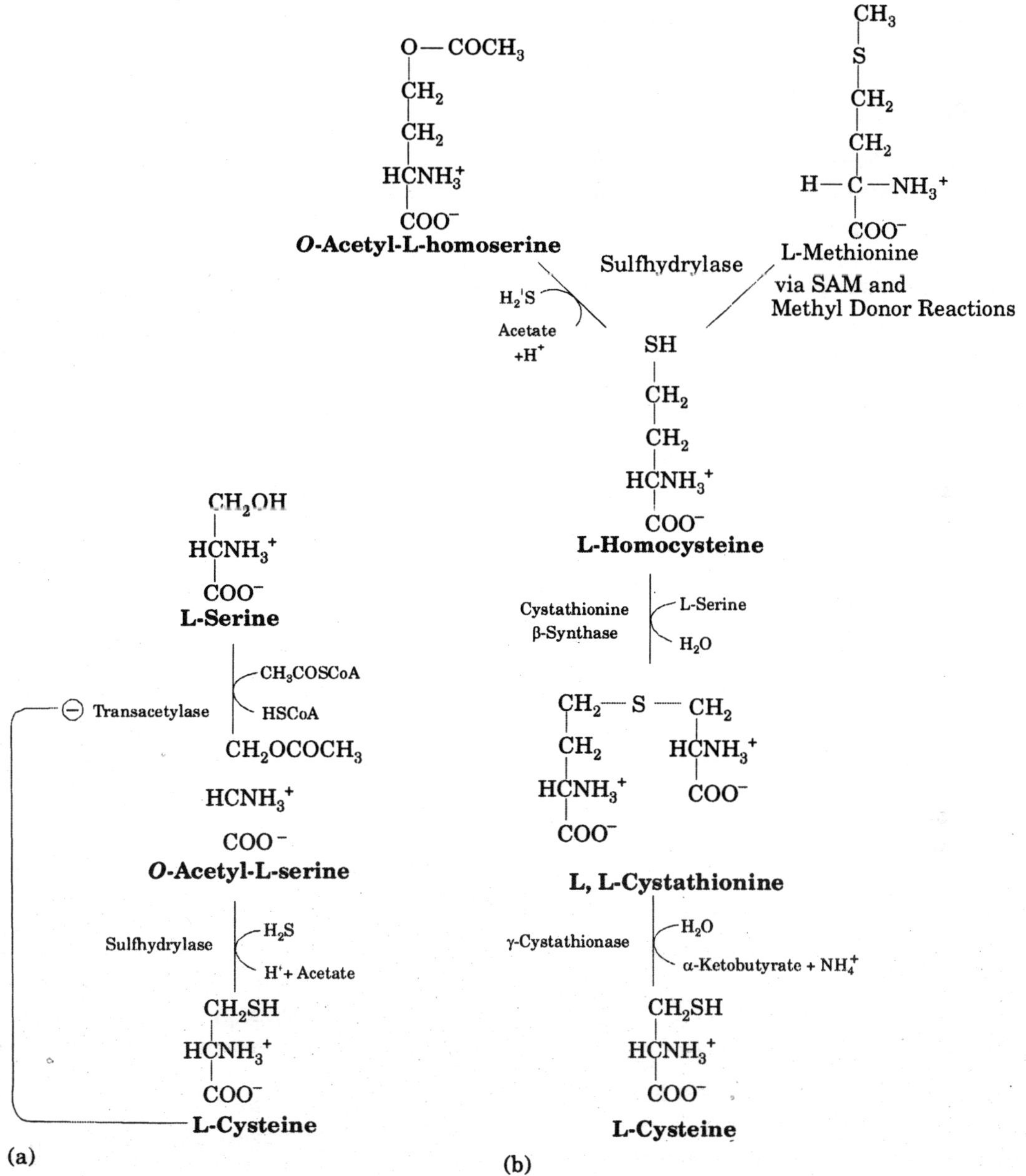


**Fig. 5.7. The biosynthesis of cysteine by direct sulfhydrylaiton and by a transsulfuration route in which the sulfur is derived from homocysteine. The direct sulfhydrylation pathway (a) is indicated as occurring with $H_2S$ as the source of sulfur. The transsulfuration pathway (b) passes through homocysteine. Two routes leading to homocysteine are shown. One starts with L-methionine. This is the sole route in animals. The second route involves the homocysteinesynthase-catalyzed conversion of *O*-acetyl-L-homoserine as shown.**

## Isoleucine and Valine Biosynthesis

The first step in valine biosynthesis is a condensation between pyruvate and "active" acetaldehyde (probably hydroxyethyl thiamine pyrophosphate) to yield α-acetolactate. The enzyme *acetohydroxy acid synthase* usually has a requirement for FAD, which, in contrast to most flavoproteins, is rather loosely bound to the protein. The very same enzyme transfers the acetaldehyde group to α-ketobutyrate to yield α-aceto-α-hydroxybutyrate, an isoleucine precursor. Unlike pyruvate, the α-ketobutyrate is not a key intermediate of the central metabolic routes; rather it is produced for a highly specific purpose by the action of a deaminase on L-threonine. Conversion of the acetohydroxy acids to the α, β-dihydroxyacid precursors of valine and isoleucine is catalyzed by *acetohydroxy acid isomeroreductase.*

The α, β-dihydroxy acids are both converted to the α-keto acid precursors of valine and isoleucine by a *dihydroxy acid dehydrase.* Finally, the two amino acids are formed in trans-amination reactions in which glutamate is the amino donor (branched-chain amino acid glutamate transaminase).

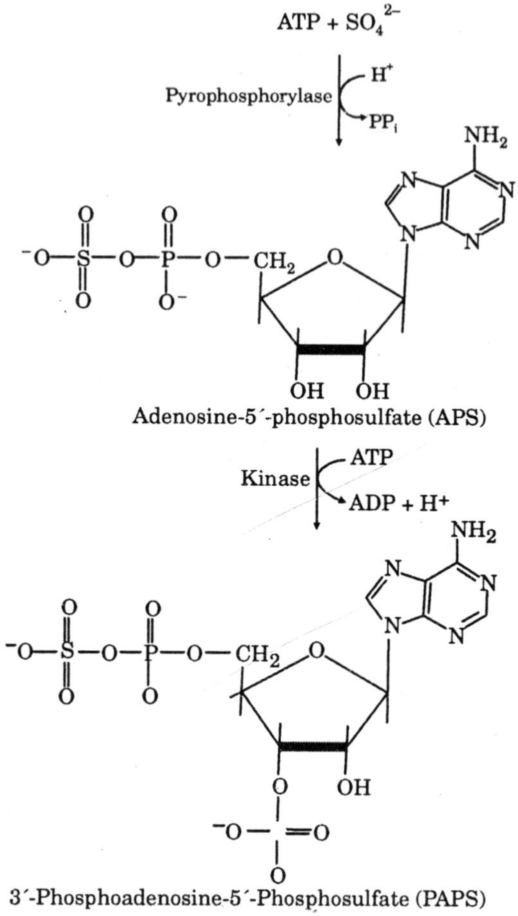

ATP + $SO_4^{2-}$

Pyrophosphorylase ($H^+$, $PP_i$)

Adenosine-5′-phosphosulfate (APS)

Kinase (ATP, ADP + $H^+$)

3′-Phosphoadenosine-5′-Phosphosulfate (PAPS)

**Fig. 5.8. Formation of 3′-phosphoadenosine-5′-phosphosulfate (PAPS), an active intermediate involved in sulfate reduction. The eight-electron reduction of $SO_4^{2-}$ to $H_2S$ is poorly understood except for the initial steps in the activation of sulfate (shown in yellow). Reduction in yeast and plants involves the APS derivative shown. In *E. coli* a PAPS derivative is used.**

## Targets for Safe Herbicides

In recent years, agribusiness firms have developed empirically several compounds that inhibit essential steps in the biosynthesis of amino acids found in plants but missing in animals. One of these compounds, glyphosate, is a highly specific inhibitor of 5-enol pyruvylshikimate-3-phosphate synthase (an enzyme needed for aromatic amino acid biosynthesis). Glyphosate is the active ingredient in the widely used herbicide Roundup.

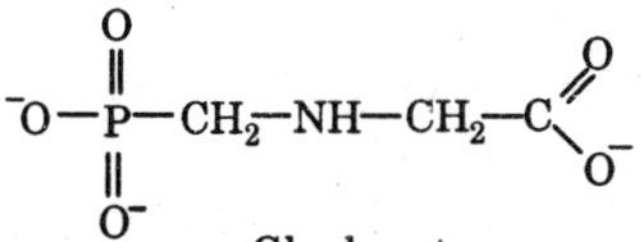

$$^{-}O-P(=O)(O^{-})-CH_2-NH-CH_2-C(=O)O^{-}$$

Glyphosate

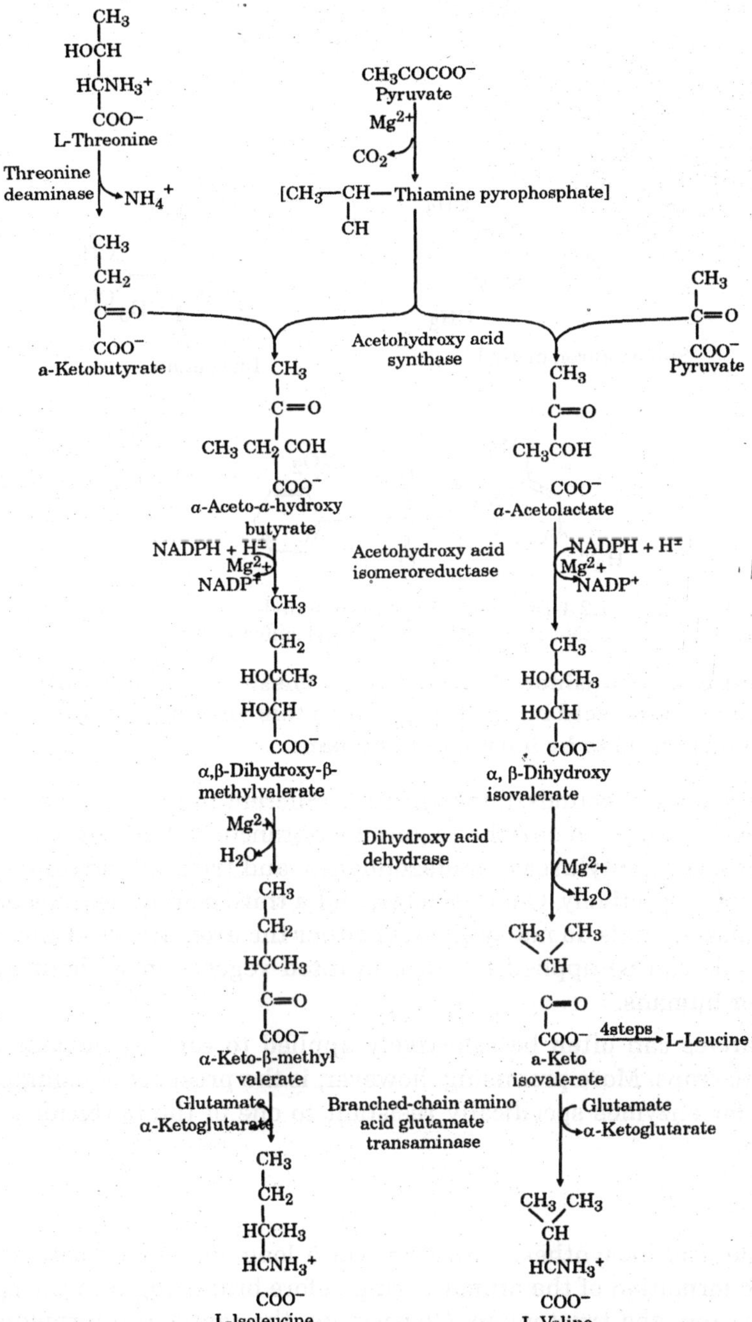


**Fig. 5.9. The biosynthesis of isoleucine and valine. The reactions leading to valine are catalyzed by the same enzymes that catalyze the corresponding reactions in isolecuine biosynthesis. Common enzymes are screened in yellow.**

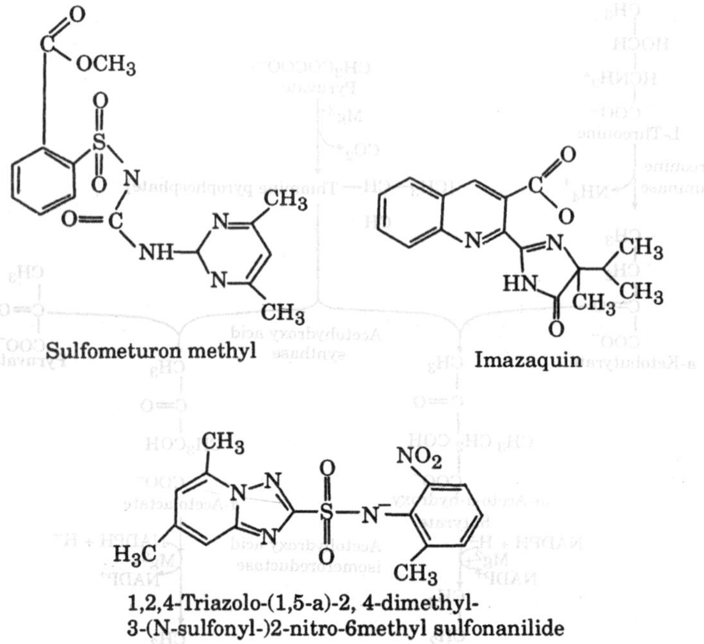

Sulfometuron methyl

Imazaquin

1,2,4-Triazolo-(1,5-a)-2, 4-dimethyl-
3-(N-sulfonyl-)2-nitro-6methyl sulfonanilide

**Fig. 5.10. Three herbicides considered to be "safe." These compounds all act on an enzyme required for branched amino acid biosynthesis. Since this enzyme is not present in mammals, the herbicides are considered to be harmless to humans.**

Three other classes of compounds, although quite different from each other, are all inhibitors of acetohydroxy acid synthase (an enzyme required for branched-chain amino acid biosynthesis. These three classes are sulfonylureas, imidazolinones, and triazolpyrimidines. which are the active ingredients in, respectively, Oust, Sceptor, and a third commercial herbicide still under development. Because animals do not synthesize either the aromatic or branched-chain amino acids, these materials can be applied to kill unwanted vegetation without causing harm to domestic animals or humans.

Certain derivatives can often be selectively applied to combat noxious plants without appreciable harm to crops. More promising, however, is the prospect of using biotechnology to incorporate genes for enzymes specifically resistant to one of the herbicides into seeds used for crop production.

## Chorismate

Aromatic amino acid biosynthesis proceeds via a long series of reactions, most of them concerned with the formation of the aromatic ring before branching into the specific routes to phenylalanine, tyrosine, and tryptophan. Chorismate, the common intermediate of the three aromatic amino acids, is derived in eight steps from erythrose-4-phosphate and phosphoenolpyruvate. We focus on the biosynthesis of tryptophan, which has been intensively studied by both geneticists and biochemists.

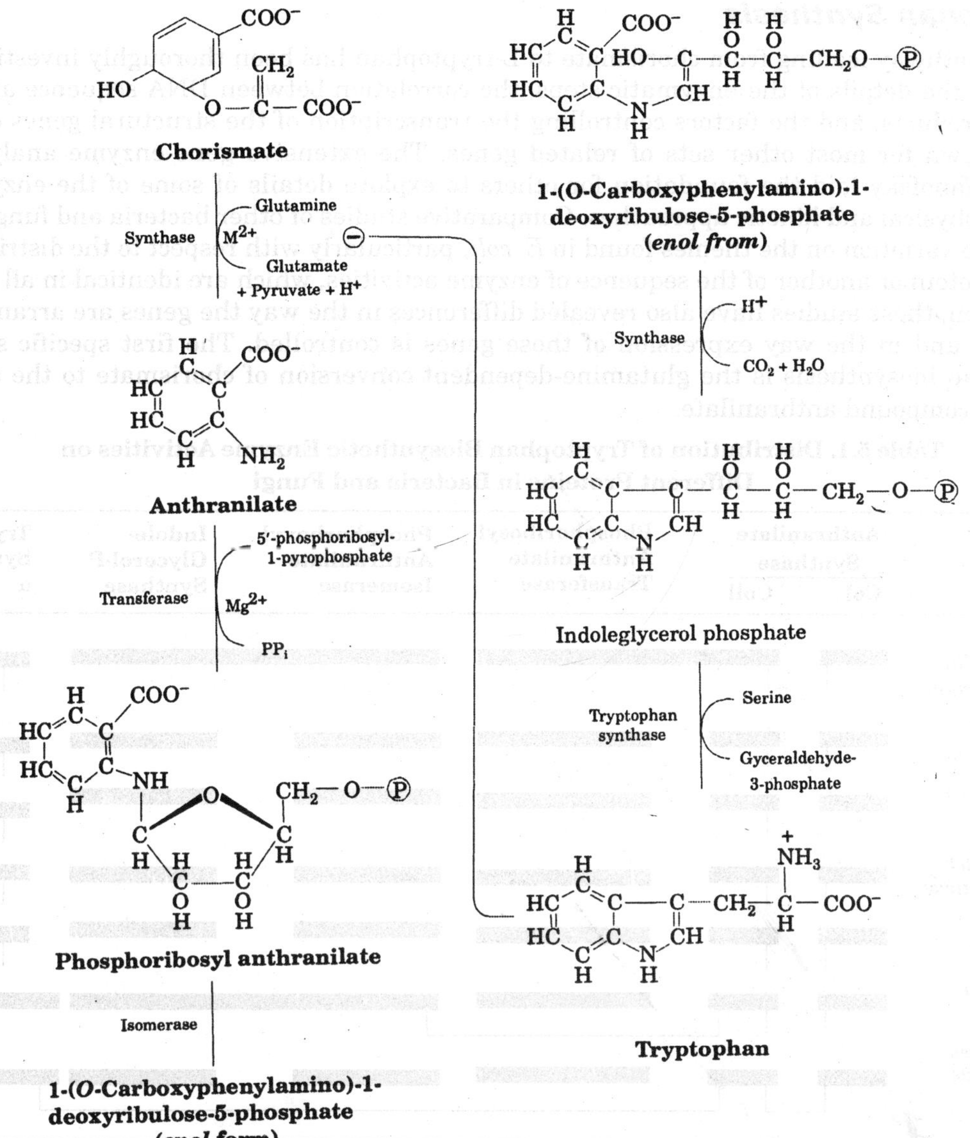


**Fig. 5.11. The biosynthesis of tryptophan from the branchpoint compound, chorismate in *E. coli*. The first step involves the conversion of chorismate to the aromatic compound anthranilate. The anthrranilate is transferred to a ribose phosphate chain. The product is cyclized to indoleglycerol phosphate by the removal of water and loss of the ring carboxyl by indoleglycerol phosphate synthase. Finally, in a replacement reaction catalyzed by tryptophan synthase, glyceraldehyde-3-phosphate is removed from indoleglycerol phosphate, and the enzyme-bound indole is condensed with serine. The structure of phosphoribosyl pyrophosphate Shown in Fig. 5.13. The red arrow indicates that the end product of this pathway inhibits the first enzyme in the pahway.**

## *Tryptophan Synthesis*

The pathway leading from chorismate to L-tryptophan has been thoroughly investigated. In *E. coli* the details of the enzymatic steps, the correlation between DNA sequence and the protein products, and the factors controlling the transcription of the structural genes exceed those known for most other sets of related genes. The extensive gene-enzyme analysis of Charles Yanofsky laid the foundation for others to explore details of some of the enzymatic steps by physical and kinetic approaches. Comparative studies of other bacteria and fungi have revealed a variation on the themes found in *E. coli,* particularly with respect to the distribution on one protein or another of the sequence of enzyme activities, which are identical in all forms. In addition, these studies have also revealed differences in the way the genes are arranged in the DNA and in the way expression of those genes is controlled. The first specific step in tryptophan biosynthesis is the glutamine-dependent conversion of chorismate to the simple aromatic compound anthranilate.

**Table 5.1. Distribution of Tryptophan Biosynthetic Enzyme Activities on Different Proteins in Bacteria and Fungi**

| Organism | Anthranilate Synthase | | Phosphoribosyl Anthranilate Transferase | Phosphorboxyl Anthranilate Isomerase | Indole-Glycerol-P Synthase | Tryptophan Synthase | |
|---|---|---|---|---|---|---|---|
| | Col | Coll | | | | α | β |
| *E. coli* *Salmonella typhimurium* | | | | | | | |
| *Serratia marcescens* | | | | | | | |
| *Pseudomonas putida* | | | | | | | |
| *Acinetobacter calcoaceticus* | | | | | | | |
| *Bacillus subtillis* | | | | | | | |
| *Neurospora crassa* | | | | | | | |
| *Sacaromyces cerevisiae* | | | | | | | |

Note ⌊___⌋ = covalent linkage; ⌊.......⌋ = obligatory association required for full activity.
A single band covering two activities indicates a single polypeptide.

Like most other glutamine-dependent reactions, the reaction can also occur with ammonia as the source of the amino group. However, high concentrations of ammonia are required. Thus far, almost all the *anthranilate synthases* examined have the glutamine amidotransferase activity (component II) and the chorismate-to-anthranilate activity (component I) on separate proteins. Anthranilate is transferred to a ribose phosphate chain in a phosphoribosyl-pyrophosphate-dependent reaction catalyzed by *anthranilate phosphoribosykransfcrase.*

Phosphoribosylanthranilate undergoes a complex reaction known as the Amadori rearrangement, in which the ribosyl moiety becomes a ribulosyl moiety. The product, 1-

(*O*-carboxyphenylamino)-1-deoxyribulose- 5′-phosphate, is cyclized to indoleglycerol phosphate by the removal of water and loss of the ring carboxyl by indole-glycerol phosphate synthase. The final step in tryptophan biosynthesis is a replacement reaction, catalyzed by tryptophan synthase, in which glyceraldehyde-3-phosphate is removed from indoleglycerol phosphate and the enzyme-bound indole undergoes a β-replacement reaction with serine.

Isotopes and mutants were instrumental in demonstrating that indole is not a true intermediate in the tryptophan biosynthetic pathway. Among different organisms the five enzyme activities required for tryptophan synthesis are distributed on different proteins. For example, in *E. coll,* indoleglycerol phosphate synthase catalyzes both the isomerization of phosphoribosylanthranilate and the cyclization step. Of particular interest is the occurrence of a single protein of the catalytic activities for nonconsecutive reactions in some cases. If in such cases the proteins were separate from each other *in* the *cell, this* arrangement, for example, in *Ncurospora,* would necessitate the product of one reaction leaving the product site of one enzyme to be acted on by another enzyme and then returning to the substrate site of a third enzyme of the same protein that exhibited the first enzyme activity.

The persistence of this arrangement during evolution makes attractive the idea that all the tryptophan biosynthetic enzymes exist in the cell as a single multienzyme (and multiprotein) complex. However, if so, the complex must be quite labile because individual gene products are so readily separated. The evolution of the gene fusion that resulted in the conversion of the α and β peptides of tryptophan synthase as found in *E. coli* to the single peptide with α and β domains as found in fungi and yeast has been studied experimentally. Fusion of the *trpB* and *trpA* genes in the order in which they occur on the chromosome to yield a single β, α chain results in a protein of only limited activity.

In contrast, when the gene fusion is done in the reverse of the genetic order (i.e., A to B), the resulting α,β chain yields a protein that is much more active. In addition, a short interdomain peptide region is required for full activity in a fusion protein.

## Negative Regulators

Metabolite flow to tryptophan is controlled by inhibition of anthranilate synthase by tryptophan. This is consistent with the principle that the end product inhibits the first enzyme which is unique to the pathway. This is a general phenomenon in amino acid biosynthetic pathways to ensure against the buildup of excess end product which is always a waste and oftentimes can cause problems. In many cases the amino acid pathway branches so that two or more amino acids are formed. Aspartate is the precursor of four other amino acids found in proteins: Iso-leucine, threonine, methionine, and lysine. The first step in this overall pathway entails the conversion of aspartate to β-aspartyl-phosphate by aspartokinase.

One might imagine that all four of the amino acid end products of this pathway would act together to inhibit this enzyme. However, in *E. coli* a different solution has been found. In this bacterium there are three aspartokinases which appear to be parts of different multienzyme complexes leading to threonine and leucine for aspartokinase I, methionine for aspartokinase II and lysine for aspartokinase III. As might be expected threonine and isoleucine inhibit aspartokinase I, methionine inhibits aspartokinase II and lysine inhibits aspartokinase III. The evolution of three different aspartokinases to carry out the same reaction in this situation underscores the tremendous importance of efficient regulation of these pathways.

## Histidine

Histidine is unusual in two respects: It is in a family by itself, and both its structure and its pathway show a strong interplay with the purine pathway. The starting point for histidine biosynthesis is phosphoribosyl pyrophosphate (PRPP) as in the purine pathway. The first specific step in histidine biosynthesis entails a condensation reaction between PRPP and ATP leading to phosphoribosyl ATP. In the fifth step most of the purine nucleotide donated in the first step is returned to the purine pathway while the histidine precursor, imidazole glycerol phosphate, which now contains a newly synthesized imidazole ring, undergoes additional conversions leading to the formation of histidine. Not only does interplay occur between the histidine pathway and the purine pathway, but the final products of the two pathways, histidine and purine nucleotide, both contain imidazole rings.

## Nonprotein Amino Acids are Derived from Protein Amino Acids

In addition to the 20 amino acids most frequently found in proteins a large group of amino acids occur in plants, bacteria, and animals that are not found in proteins. Some are found in peptide linkages in compounds that are important as cell wall or capsular structures in bacteria or as antibiotic substances produced by bacteria and fungi. Others are found as free amino acids in seeds and other plant structures. Some amino acids are never found in proteins. These non-protein amino acids, numbering in the hundreds, include precursors of normal amino acids, such as homoserine and diaminopimelate; intermediates in cata'bolic pathways, such as pipecolic acid; D enantiomers of "normal" amino acids; and amino acid analogs, such as azetidine-2-carboxylic acid and canavanine, that might be formed by unique pathways or by modification of normal amino acid biosynthetic pathways.

**Table 5.2. Some-D-Amino Acids Found in Peptide Antibiotics**

| *Antibiotic* | *D-Amino Acids Present* | *Produced by* |
|---|---|---|
| Actinomycin $C_1$(D) | D-Valine | *Streptomyces parralus* and others |
| Bacitracin A | D-Asparagine, D-glutamate, D-ornithine, D-phenylalanine | *Bacillus subtilis* |
| Circulin A | D-Leucine | *Bacillus circulans* |
| Fungisporin | D-Phenylalanine, D-valine | *Penicillium spp.* |
| Gramicidin S | D-Phenylalanine | *Bacillus brevis* |
| Malformin $A_1$, C | D-Cysteine, D-leucine | *Aspergillus niger* |
| Mycobacillin | D-Aspartate, D-glutamate | *Bacillus subtilis* |
| Polymixin $B_1$ | D-Phenylalanine | *Bacillus polymyxa* |
| Tyrocidine A, B | D-phenylalanine | *Bacillus brevis* |
| Valinomycin | D-Valine | *Streptomyces fulrissimus* |

## D-Amino Acids

Certain D-amino acids with some L enantiomers are commonly found both in microbial cell walls and in many peptide antibiotics. For example, the peptidoglycans of bacteria contain

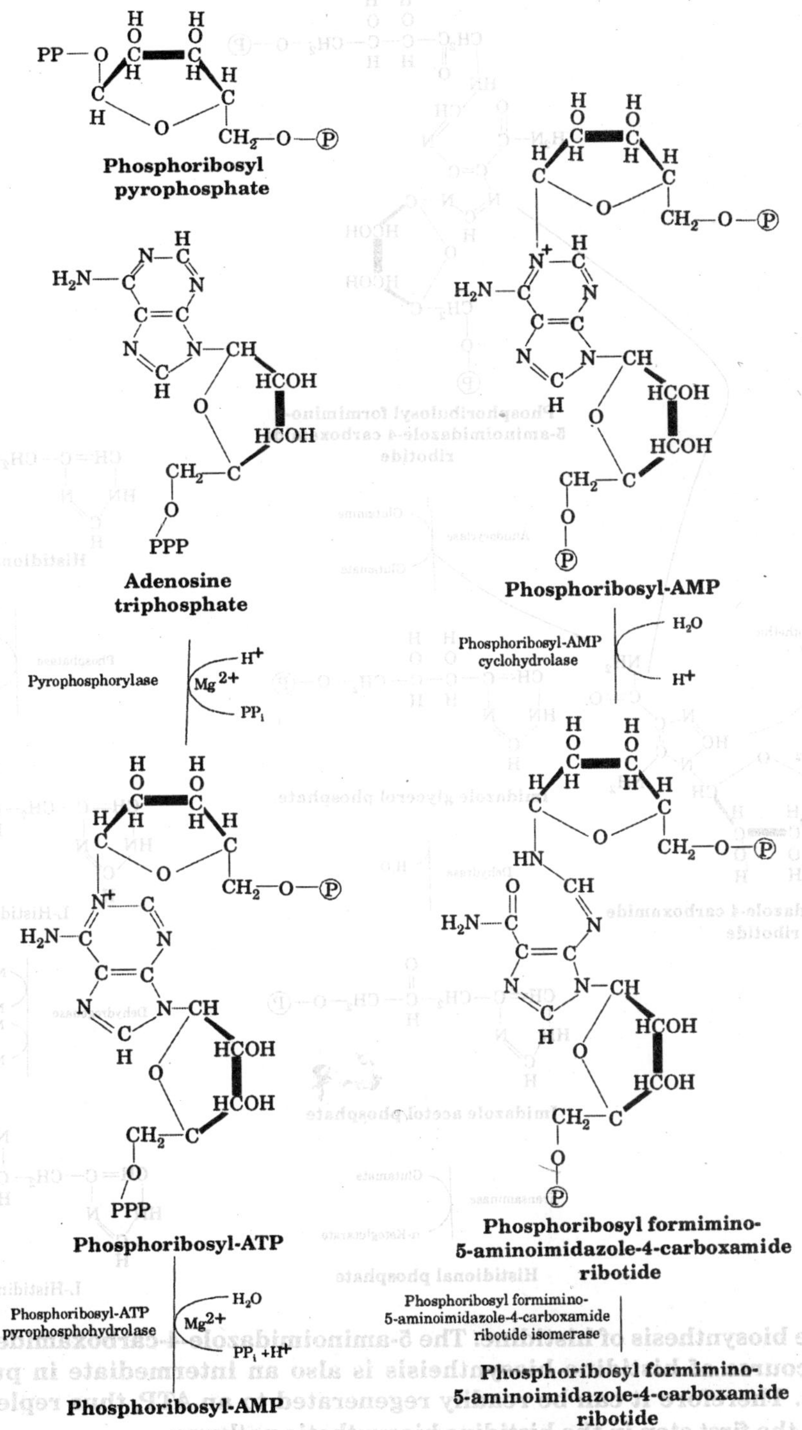

**Phosphoribosyl pyrophosphate**

**Adenosine triphosphate**

Pyrophosphorylase ($Mg^{2+}$; $H^+$, $PP_i$)

**Phosphoribosyl-ATP**

Phosphoribosyl-ATP pyrophosphohydrolase ($Mg^{2+}$; $H_2O$; $PP_i + H^+$)

**Phosphoribosyl-AMP**

**Phosphoribosyl-AMP**

Phosphoribosyl-AMP cyclohydrolase ($H_2O$; $H^+$)

**Phosphoribosyl formimino-5-aminoimidazole-4-carboxamide ribotide**

Phosphoribosyl formimino-5-aminoimidazole-4-carboxamide ribotide isomerase

**Phosphoribulosyl formimino-5-aminoimidazole-4-carboxamide ribotide**

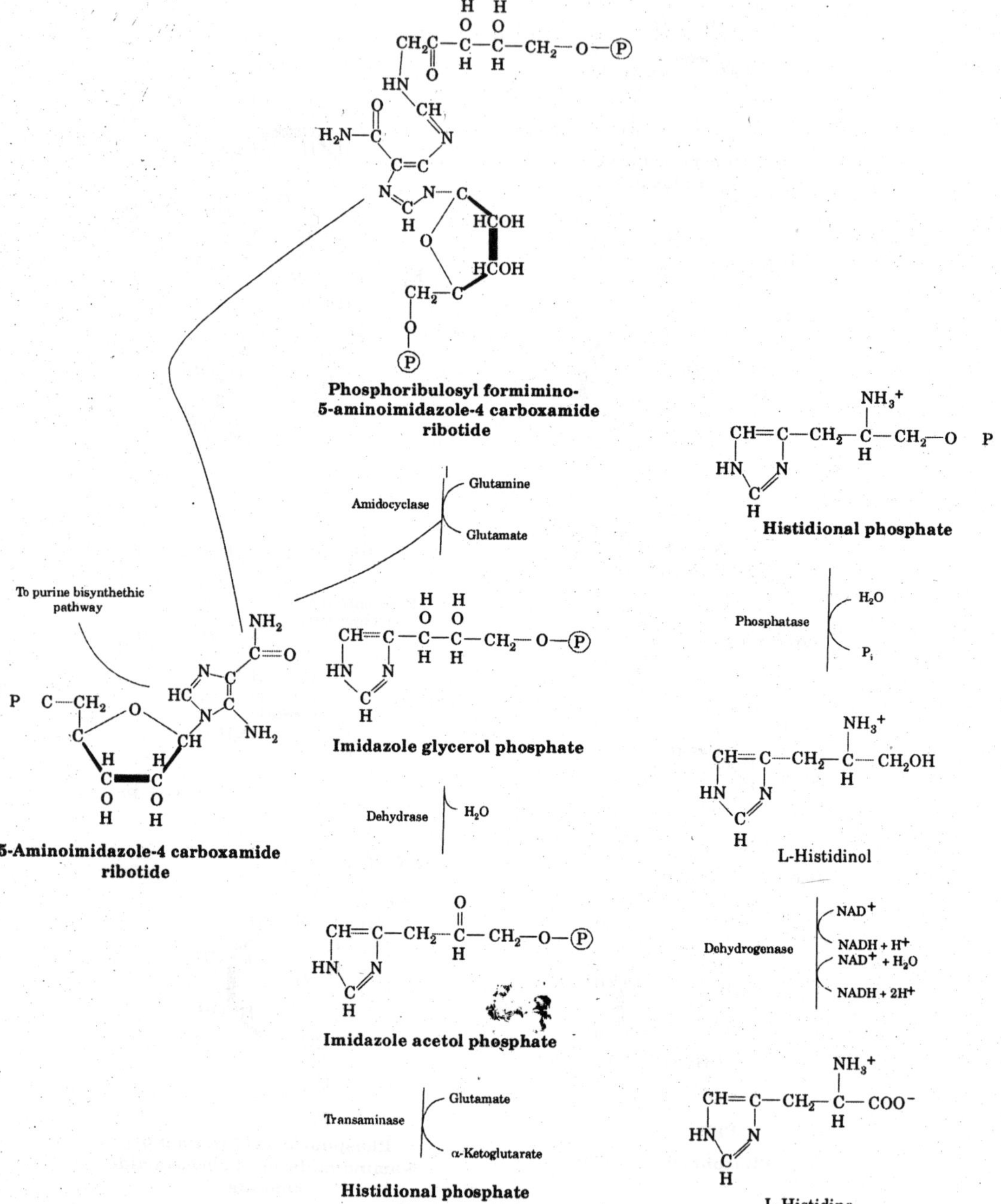


**Fig. 5.12. The biosynthesis of histidine. The 5-aminoimidazole-4-carboxamide ribotide formed during the course of histidine biosyntheisis is also an intermediate in purine nucleotide biosynthesis. Therefore it can be readily regenerated to an ATP, thus replenishing the ATP consumed in the first step in the histidine biosynthetic pathway.**

both D-alanine and n-glutamate. The latter is present in a γ-glutamyl linkage. In some forms, the α carboxyl of the D-glutamyl residue is either amidated or in peptide linkage with glycine. D-Lysine or D-ornithine is found in the glycopeptide of some Gram positive organisms. The capsule of the anthrax bacillus is composed of a nearly pure homopolymer of D-glutamate in γ linkage. Other bacilli also produce γ-linked polyglutamates, some of which form separate D-glutamate and L-glutamate chains, whereas others form a copolymer of D-and L-glutamate.

A wide variety of D-amino acids have been found in antibiotics; some of these are listed in table 4.2. D-Alanine is found in bacterial cell wall peptidoglycan. L-Alanine is converted to D-alanine by a raccmase that contains pyridoxal phosphate as a cofactor. The racemization is followed by the formation of a D-alanyl-D-alanine dipeptide, which is accompanied by the conversion of ATP to ADP. The dipeptide is subsequently incorporated into the glycopeptide. *In most cases of formation of peptides containing D-amino acid, the L form of the amino acid is the substrate for the incorporating enzyme.*

In contrast, the free D-amino acid is ordinarily a poor substrate for the incorporation reaction. Whether the racemization occurs on the enzyme or afterwards remains to be determined in most cases. In the case of the D-valyl residue formed in penicillin, a tripeptide derivative containing L-valine is an intermediate, and the conversion is thought to occur by way of an α, β-dehydro form of the valyl residue.

# CHAPTER 6

# PHARMACEUTICAL PROTEINS IN PLANTS

Molecular farming or *pharming*™is the production of pharmaceutically important and commercially valuable proteins in plants. Since the first production of a pharmaceutical protein in transgenic tobacco a broad range of proteins with potential medical applications has been expressed in plants. These range from small peptides such as enkephalins to complex, multisubunit molecules such as secreted antibodies of the IgA type. In this chapter we will address the decisions that must be made during the production and marketing of such proteins. This begins with the choice of production host, which can range from cellular green algae to a variety of crop species. For pharmaceutical applications, it is necessary to work under current good manufacturing practice (cGMP) conditions and the use of plant suspension cultures must therefore be considered as an alternative to whole plants.

Another crucial consideration is the design of the expression cassette, which mus optimized to achieve high-level accumulation of functional recombinant protein. Intrinsic factors such as how the protein is processed in the plant need to be taken into account, as well as the ability of the host species to carry out specific forms of posttranslational modification. When sufficiently high levels of expression and protein accumulation are achieved, efficient downstream processing protocols must be developed to insure product quality and the economic feasibility of production. As the demand for safe, recombinant pharmaceutical proteins continues to expand, the market potential of plant-produced recombinant proteins is considerable. Molecular farming can produce iccombinant proteins at a lower cost than traditional expression systems based on microbial or animal cell culture, and without the risk of contamination with human pathogens.

## PLANT SPECIES USED

Many species can be exploited for molecular farming with tobacco being the most widely used host plant to date. The advantages of tobacco plants include the high biomass yield (which can reach 50 – 100 tonnes $ha^{-1}$ in a high-density population depending on the cultivar), the ease of stable transformation either by co-cultivation with *Agrobacterium tumefaciens*, or transiently by infiltration with transgenic agrobacteria or transfection With viral vectors. Another benefit is that tobacco is not used as a food or feed crop, ensuring that a transformed line expressing a highly potent drug will not contaminate food resources by outcrossing or during the processing steps. A drawback of this species is its content of nicotine and other alkaloids. Although there are cultivars available with reduced alkaloid content it is necessary to remove all traces of these toxic compounds during downstream processing, especially if the recombinant protein is intended for clinical applications.

An additional disadvantage is the limited shelf life of transgenic leaf material after harvest, a problem that can be solved in other plant species by targeting the desired protein to natural storage organs like seeds or tubers. For tobacco leaves, downstream processing must commence immediately after harvest to ensure the stability of the recombinant protein, although Fiedler *et* al. reported no losses of scFv (single chain fragment variable) antibody specificity or antigen binding capacity in dried leaves three weeks after harvest. In the quest to find other plants that are suitable as bioreactors, various monocotyledonous and dicotyledonous species have been tested. These include corn, rice and wheat, alfalfa, potato, oilseed rape, pea, tomato and soybean.

The major advantage of cereal crops is that recombinant proteins can be directed to accumulate in seeds, which are evolutionar specialized for storage and thus protect proteins from proteolytic degradation. Recombinant proteins are reported to remain stable in seeds for up to five months at room temperature and for at least three years at refrigerator temperature without significant loss of activity. In addition, the seed proteome is less complex than the leaf proteome, which makes purification quicker and more economical. lt has been shown recently that the yields of recombinant protein in transgenic plants vary according to intrinsic factors such as the developmental stage and extrinsic factors such as the climate. Such factors also affect the precise nature of post-translational modification, particularly glycosylation.

Elbers and colleagues reported that the glycosylation profile, expressed as the ratio of complex to high mannose type glycans, increased with the age of tobacco leaves. Therefore, it is necessary to consider measures that ensure protein homogeneity, e.g. restricting expression to defined plant organs or specific compartments. Alternatively, it is possible to use an inducible expression system to restrict protein expression to a defined time point or period. Examples of such production methods include the postharvest system developed by Croptech and the amylase-based system for controlled expression in suspension cells.

## ALTERNATIVE EXPRESSION SYSTEM

Plant cell cultures have not been used as frequently as intact plants for molecular farming but there are several examples that demonstrate their suitability as hosts for the production of recombinant proteins. Free cell suspension is regarded as the most suitable cultivation system for large-scale biotechnology applications, but tentative experiments have also been performed with hairy roots and shooty teratomas. Examples of foreign proteins successfully expressed in suspension cells include monoclonal antibodies and their derivatives, cytokines, hormones and enzymes. The tobacco suspension cell line BY-2 appears to be a particularly suitable host cell line because of its high growth ratejtfl]. Like plant tissues, BY-2 suspension cells can be transformed easily by cocultivation with *A. tumefaciens*. In addition to this tobacco cell line, soybean cell culture has been used for the production of hepatitis B virus surface antigens and rice suspension cells have been employed for the expression of human α-1-antitrypsin and human lysozyme.

In contrast to the tobacco transformation process, gene transfer to rice is often performed by particle bombardment. Table 6.2 provides an overview of pharmaceutical proteins that have

**Table 6.1. Pharmaceutical Proteins Produced in Transgenic plant**

| *Protein* | *Regulatory elements* | *Additional elements of the construct* | | *Species/Tissue* | *Accumulation level* |
|---|---|---|---|---|---|
| α creatine kinase, MAK33 mAb (Fab) | Ca MV 35S promoter/ different termination sequence (ocs and rbcS) | LP of 2S2 storage protein of Arabidopsis *thaliana* | Fd and k chain with 3'ocs | *A. thaliana* (leaves) | 3.6% of TSP |
| | | | Fd chain with 3'ocs, κ chain with 3'rbcS | *A thalina* (leaves) | 6.5% of TSP |
| | | | Fd chain with 3'ocs, κ chain-KDEL with 3'rbcS | *A. Thaliana* (leaves) | 5.9 % of TSP |
| α human creatine kinase (fab, lgG) | CaMV 35S promoter/ ocs terminator | LP of the 2S2 storage Protein of A. *thaliana* | Fab | *Nicotiana tabacum* (callus) | 0.128% of TSP |
| | | | IgG | *N. tabacum* (callus) | 0.477% of TSP |
| | | | Fab | *A. thaliana* (callus) | 0.188% of TSP |
| | | | IgC | *A. thaliana* (callus) | 0.217% of TSP |
| α substance P (single-domain Ab) | CaMV 35S promoter/ nos terminator | *pel*B LP | | *Nicotiana benthamiana* | 1 % of TSP |
| IG of 38C13 mouse B Cell lympoma (scFv) | Viral vector TTO1A, a hybrid fusion of TMV and tomato mosaic virus, N-terminal: rice α-amylase LP | | | *N. benthamiana* (leaves) | 12.3–30.2 μg g$^{-1}$ FW |
| α Streprococcus mutans abhesin (IgA, IgG) 1 | Ca MV 35S promoter/ nos terminator | Native mouse immunoglobulin LP The γ chain constant region was replaced by an α chain constant region | | *N. tabacum* | IgG: 7.7. α g mL$^{-1}$ Plant extract IgG1/A: 1.5 μg mL$^{-}$ plant extact lgG2/A:2.1 μg mL$^{-1}$ plant extract |
| α *Streptococcus mutans* adhesin (slgA-G) | CaMV 35S promoter/ Nos terminator | Native mouse immunoglobulin LP, four transgenic lines expressing either the k, the hybrid lgA-G heavy chain, the murine J chain or the secretory component were crossed to obtain a sIgA experessing line | | *N. tabacum* | 200-500 μ g$^{-1}$ FW |
| α human (IgG)) | CaMV 35S promoter/ T7 terminator | 5'murine LP of HC and LC LC anc HC separrately transformed | | *Mddicago sativa* | 1% TSP (full-size IGG) |

**Table 6.1 Continued....**

| *Protein* | *Regulatory elements* | *Additional elements of the construct* | *Species/Tissue* | *Accumulation level* |
|---|---|---|---|---|
| α HCV-2 glycoprotein B (IgG) | CaMV 35S promoter/ nos termineator | LP of tobacco extensin | *Glycine max* | Not mentioned |
| Human hemoglobin (α + β) | CaMV enhanced 35S promoter/ CaMV 35S | Transit peptide of small subunit of Rubis CO of *Pisum sativum* | *N. tabacum* | 0.05% of seed protein |
| EGF (synathetic gene) | terminator CaMV 35S promoter/ nos terminator | Codon-optimized for high ,evel peptide expression in Escherichia coli | *N. tabcum* (leaves) | 0.001% of TSP |
| Human glutamic acid decarboxylase (hGAD) | CaMV enhanced 35S promoter/ TEV5' UTR/nos terminator | | Low alkaloid tobacco *Salanum tuberosum* | 0.4% of TSP of leave or tuber |
| Isoform of hGAD (hGAD65) | CaMV 35S promoter/ nos terminator | | *N. tabacum* (leaves) | 0.01–0.04% of TSP |
| | | | *Daucus carota* (tap root) | 0.01% of TSP |
| | | | *N. Tabacum* (leaves) | 0.04% of TSP |
| | | $GAD67_{1-87}$ (rat)/$hGAD65_{88-585}$ Fusion | | 0.19%ofTSP |
| | Expression mediated by potato virus X infection | | *N. benthamiana* (leaves) | 2.2% of TSP |
| Human collagen Pra α (1) chain | CaMV 35S promoter/ CaMV 35S terminator | Pathogenesis-related protein (PR) LP or native IP | tobacco | >100 μg $g^{-1}$ tissue |
| Hepatitis B surface antigen | CaMV enhanced 355S promoter/ TEV 5'UTR/nos terminator | | *N. tabacum* (leaves) | 66 μg $g^{-1}$ TSP |
| | Patatin promoter | | *S. tuberosum* (tuber) | 1.1 μg $g^{-1}$ FW |
| | CaMV 35S promoter/TEV 5' UTR/nos terminator | | | 0.33 μg $g^{-1}$ FW |
| | | Soybean vegetative storage protein LP (vspαS) | | 0.8 μg $g^{-1}$ FW |
| | | vspαS plus putative vacuolar targeting signal (vspαL) | | 2.4 μg $g^{-1}$ FW |
| | | Transit peptide of small subunit of RubisCO | | n.d. |
| | | KDEL | | 1.25 μg $g^{-1}$ FW |
| | CaMV 35S promoter/ΩTMV 5'UTR/nos terminator | | | 0.33 μg $g^{-1}$ FW |

**Table 6.1 Continued....**

| *Protein* | *Regulatory elements* | *Additional elements of the construct* | *Species/Tissue* | *Accumulation level* |
|---|---|---|---|---|
| | CaMV35S promoter/ Ω TMV 5'UTR/nos B terminator | | | 6.5 μg $g^{-1}$ FW |
| | CaMV/35S promoter/TEV 5'UTR/vsp B gene terminator | | | 16.0 μg $g^{-1}$ FW |
| Binding subunit of heat-labile enterotoxin B | CaMV35S promoter/TEV 5'UTR/vsp B terminator | With and without KDEL | *N. tabacum* (leaves) | 5 μg $g^{-1}$ TSP (without KDEL) 14 μg $g^{-1}$ TSP (with KDEL) |
| | | | *S. tuberosum* | 30 μg $g^{-1}$ TSP (without KDEL) 110 μg $g^{-1}$ TSP (with KDEL) |
| Cholera toxin B subunit | CaMV35S promoter nos terminator | Kozak seqence, KDEL | *Lycopersicon esculentum* | Leaves: 0.02% of TSP Fruits: 0.04% of TSP |
| IL-10 | CaMV 35S promoter/ nos terminator | Native LP, with and without KDEL | *N. tabacum* | 800 ng $g^{-1}$ TSP (without KDEL) 55 μg $g^{-1}$ TSP (with KDEL) |
| IL-2 | Patatin promoter | Native LP | *S. tuberosum* | 115 units $g^{-1}$ FW |
| Enkephalins | Promoter and terminator of 2S1 albumin gene of *A. thaliana* | Integrated in 2S1 albumin of *A. thaliana* and flanked by tryptic cleavage sites | *A thaliana* (seed) | 2.9% of TSP (200 nmol $g^{-1}$) |
| | | | *Brassica napus* (seed) | 50 nmol $g^{-1}$ |
| T84.66 α CEA (scFv) | Legumin A seed specific promoter/Ω sequence TMV 5'UTR | Light chain LP of mAb24, KDEL | *P. sativum* (seed) | 9 μg $g^{-1}$ FW |
| T84.66 α CEA (diabody) | CaMV enhanced 35S promoter/ CHS 5'UTR/Pw 3'UTR of TMV/ CaMV 35S terminator | Light chain LP of mAb24, and without KDEL | *N. tabacum* (leaves) | 1 – 5 μg $g^{-1}$ (apoplast/transient) 4 – 12 μg $g^{-1}$ (ER/ transient) 0.44 – 0.93 μg $g^{-1}$ (apoplast/stable) 3 – 9.3 μg $g^{-1}$ (ER/stable) |

**Table 6.1 Continued....**

| *Protein* | *Regulatory elements* | *Additional elements of the construct* | *Species/Tissue* | *Accumulation level* |
|---|---|---|---|---|
| Aprotinin | Maize ubiquitin promoter/ pinII terminator | Maize ubiquitin 5′intron, barley α-amylase LP | Extracellular matrix of embryo (maize seeds) | In T5: 0.35% of extractable seed proteins |
| Human serum albumin | CaMV enhanced 35S promoter/ AIMV RNA synthetic leader sequence/nos terminator | Native or PR-S leader peptide | *S. tuberosum* (leaves) | 0.02% of TSP |
| Hirudin | Oleosin promoter/ nos terminators | Fused to native oleosin | *B. napus* | 1% of seed protein |
| β-Interferon | CaMV 35S promoter/ CaMV 35S terminator | Native LP | *N. tabacum* | 170 ng $g^{-1}$ FW |
| α human Rhesus D (IgG) | CaMV 35S promoter (LC, HC and tandem construct)/E9 PolyA terminator of *P. sativum* | Murine immunoglobulin LP | *A. thaliana* | Cross of LC + HC: 0.6% of TSP, tandem : 0.3% of TSP |
| | *mas*1′2′ dual promoter (bidirectional)/nos and E9 terminator | | | Bidirectional: 0.12% of TSP |
| Dog gastric lipase | CaMV enhanced 35S promoter/ CaMV 35S terminator | LP of rabbit gastric lipase precursor | *N. tabacum* (leaves; | 7% of acid extractable proteins |
| | | LP of sweet potato sporamin precursor | *N. tabacum* (leaves) (vacuole) | 5% of acid extractable proteins |
| α Hepatitis B surface antigen (scFv) | CaMV 35S promoter/TMV Ω 5′UTR/nos terminator | | *N. tabacum* (cytosol) | n.d. |
| | | Sporamin LP | *N. tabacum* (apoplast) | 0.31 % of TSP |
| | | Sporamin pre-pro-peptide LP | *N. tabacum* (vacuole) | 0.032% of TSP |

**Table 6.1 Continued....**

| *Protein* | *Regulatory elements* | *Additional elements of the construct* | *Species/Tissue* | *Accumulation level* |
|---|---|---|---|---|
| | | Sporamin LP, KDEL | *N. tabacum* (ER) | 0.22% of TSP |
| α Hepatitis B surface antigen (scFV) | CaMV 35S promoter/TMV Ω 5'UTR/nos terminator | KDEL | *N. tabacum* (root exudates) | 630 – 760 ng $g^{-1}$ dry weight root/day |
| | | N-terminal ER LP from calreticulin | | |
| Human interferon (hIF-α2b and hIF-α8) | CaMV 35S promoter/ nos terminator) | Native LP of HIF-α2b and hIF-α8 resp. | *S. tuberosum* | 560 IU $g^{-1}$ tissue |

Abbreviations used in table 6.1: AIMV: alfalfa mosaic virus; CaMV: cauliflower mosaic virus; CEA: carcinoembryogenic antigen; CHS: chalcone synthase; Fab: antigen binding fragment; Fd: variable domain and first constant domain of the heavy chain; FW: freshweight; HC: heavy chain; hIF: human interferon; HSV: herpes simplex virus; IL: interleukin; LC: light chain; LP: leader peptide; mAb: monoclonal antibody; mas: mannopine synthase; n.d.: not detectable; nos: nopaline synthase; ocs: octopine synthase; *pel*B: pectate lyase; pin II: potato proteinase inhibitor II; PR-S: pathogenesis related protein S; Pw: pseudoknot wildtype of TMV; rbcS: ribulose-1, 5-bisphosphate carboxylase small subunit; RubisCO: Ribulose-1, 5-bisphosphate Carboxylase Oxigenase; scFv: single chain Fragment variable; TEV: tobacco etch virus; TMV: tobacco mosaic virus; TSP: total soluble protein; UTR: untranslated region; vspB: vegetative storage protein B.

been produced in plant cell cultures so far and the expression cassette elements used in each case. Possible contamination by chemical or biological substances is one of the most important concerns when producing pharmaceutical proteins. Plant cell cultures ensure the production of the desired protein in a controlled, sterile and sealed environment and can be adapted to cGMP conditions. Therefore, the risk of contamination is minimized and the production conditions can be modified more easily in a contained reactor than in the field.

Another advantage is the ability to freeze plant suspension cells in liquid nitrogen. So that master and working cell banks can be established, aprerequisitefor cGMP procedures. Although plant cell culture is not as cost effective as plant cultivation in the open field, it will become an economical process if higher protein yields can be achieved. The cultivation medium of plants is chemically defined, consisting of a carbon source, minerals, vitamins and phytohormones. Furthermore, it is protein-free and relatively inexpensive. In contrast, animal cells often require complex supplements such as fetal calf serum and/or expensive growth factors, although serum-free cultivation is possible in case of Chinese hamster ovary (CHO) cells. When trying to increase the yields of recombinant protein inplant suspension cultures, one should consider optimizing the nutrient supply and including product stabilizing agents.

Sharp and Doran showed that antibody accumulation in hairy root cultures was improved by increasing the dissolved oxygen tension to 150% air saturation and that loss of the antibody could be minimized by inhibiting protein transport in the secretory pathway with the antibiotic brefeldin A. The beneficial effects of stabilizing solutes were demonstrated among others by Magnuson *et al.* who achieved a 35-fold increase in scFv yield by adding polyvinylpyrrolidone (PVP) to the culture medium. Other proteins were stabilized to a lesser degree by supplementing the medium with the protease inhibitor bacitracin, bovine serum albumin (BSA), gelatin or dimethylsulfoxide (DMSO).

Fischer *et al.* reported a transient threefold increase in recombinant antibody levels after supplementing cell cultures with amino acids prior to harvest. All these studies show, however, that the benefits obtained by certain expression cassette elements or defined culture conditions cannot be generalized for all types of pharmaceutical proteins. These factors have to be investigated and optimized individually for every product. In order to facilitate purification of a recombinant protein from plant cell culture, the recombinant protein can be directed to the extracellular space (apoplast and culture medium) by attaching an appropriate signal sequence or using a native signal sequence if one is present. Only about 100-1000 endogenous proteins are thought to be secreted from cultured plant cells, so the purification of a recombinant protein from the culture medium is much easier than purification from whole cell extracts.

The plant cell wall is a natural barrier to protein secretion. Proteins >20 kDa tend to be retained although higher-molecular-weight proteins such as monoclonal antibodies are secreted. Secretion can be further enhanced by the addition of DMSO. Although cell cultures produce a much lower amount of biomass than plants cultivated in the open field, tobacco suspension cells have been cultivated at volumes of up to 20 $m^3$ and references therein).

## PROCESSING STEPS IN PLANTS

In order to make molecular farming commercially profitable, recombinant proteins must be produced at a sufficiently high yield and in an active form. It has become clear that, for

**Table 6.2. Pharmaceutical Proteins Produced in Plant Suspension Cultures**

| *Protein* | *Regulatory elements* | *Additional elements of the construct* | *Species* | *Accumulation level* |
|---|---|---|---|---|
| Human $\alpha_1$ antitrypsin | Promoter of rice a amylase gene *RAmy3D*/terminator of *RAmy3D* | LP of rice a amylase gene *RAmy3D* | *Orza sativa* | 4.6–5.7 mg $g^{-1}$ dry cell weight |
| | | Additional first intron of *RAmy3D* | | No significant increase |
| | | LP of rice α amylase gen *RAmy3D* | | 18.2–24.2 mg $g^{-1}$ dry cell weight |
| Human erythropoietin | CA MV 35S Promoter/CaMV35S terminator | Native LP | *N. tabacum* cv By-2 | 26ng$^{-1}$ TSP |
| α hepatitis B surface antigen-(scFv) | CaMV 35S Promoter/ nos terminator | Sporamin LP | *N. Tabacum* cv Havana SR | 1 mg $L^{-1}$ in culture supernatant and 5 mg $kg^{-1}$ wet weight from cell extracts |
| hIL-12 | CaMV enhaced 35S promoter | Native LP | *N tabcum* cv Havana SR | 175μ $L^{-1}$ |
| hIL-2, hIL-4 | CaMV 35S promoter/ T7 terminator | Native LP | *N. Tabacum* NT-1 | 0.1 mg $L^{-1}$ (hIL-2) and 0.18 mg $L^{-1}$ (hIL4) |
| Human Serum albumin | CaMV enhyanced 35S promoter/ AIMV RNA synthetic leader sequence/nos terminator | Native or PR-s LP | *N. tabacum* | 250 μg $g^{-1}$ TSP |
| Bryodin 1 | CaMV 35S promoter/ nos terminator | Extensin LP | *N. tabacum* NT-1 | 30mg$L^{-1}$ |
| hG-CSF | CaMV enhanced 35S promoter/ TMV Ω 5'UTR/nos terminator | | *N. tabacum* | 105 μg $L^{-1}$ extracellular |
| GM-CSF | CaMV35Spromoter/TEV5' UTR/T7 terminator | | *N. tabacum* cv BY-2 | 150 μg $L^{-1}$ intracellular and 250 ug $L^{-1}$ extracellular |
| | CaMV enhanced 35 S promoter/ TMV Ω 5'UTR/nos terminator | Native GM-CSF LP | *N. tabacum* cv Havana SR | 180 ug $L^{-1}$ extracellular, after adding of gelatin increase to 738 μg$L^{-1}$ |

Abbreviations used in table 2: AIMV: alfalfa mosaic virus; BY-2: *Nkotiana tabacum* cv Bright Yellow 2; CaMV: cauliflower mosaic virus; FW: freshweight; GM-CSF: granulocyte macrophage colony stimulating factor; hG-CSF: human granulocyte colony stimulating factor; hIL: human interleukin; LP: leader peptide; nos: nopaline synthase; PR-S: pathogenesis related protein S; RAmy 3D: rice amylase gene promoter 3D; scFv: single chain Fragment variable; TEV: tobacco etch virus; TMV: tobacco mosaic virus; TSP: total soluble protein; UTR: untranslated region.

high-level protein accumulation, the stability of transgene expression can be as important as the expression level itself. The quantity of protein is determined by the rate of protein synthesis, assembly as well as proteolytic degradation. Processes that influence the production of a stable and functional end product in plants and plant cell suspension cultures include transgene integration, transcriptional and translational activity, posttranslational modification. The first hurdle is stable transgene integration and expression in the genome of the transgenic plants or transformed cells, since transgenes are often silenced either immediately after integration or in subsequent generations.

A novel and very interesting method for dealing with this problem in transient gene expression is the use of tomato bushy stunt virus P19 protein. This is a known suppressor of posttranscriptional gene silencing (PTGS) and transient expression of a range of proteins was increased 50-fold in the presence of p19. Another approach is the inclusion of scaffold attachment regions (SARs) or matrix attachment regions (MARs) within the expression construct. Such elements, which attach to the nuclear matrix and may help divide interphase chromatin into functional domains, have been shown to enhance transformation efficiency to achieve stable gene expression and to increase transgene expression levels up to 650-fold although the level of enhancement appears to be dependent on the promoter. Increased transcription levels are assumed to result in increased protein synthesis. One approach to reach this goal is to raise the transgene copy number by the use of amplification-promoting sequences derived from a spacer sequence of tobacco ribosomal DNA. Posttranscriptional processes such as capping, splicing and polyadenylation are important for high protein yields, and it is also important to maximize mRNA stability. The synthesis of a protein may be hampered by suboptimal translation initiation or termination sites. Lukaszewicz *et al.* studied the context sequence of the initiation codon and Sawant *et al.* were able to increase the expression level of a reporter gene by inserting a sequence commonly found in highly expressed plant proteins.

Other problems can include the presence of cryptic splice or polyadenylation sites, and unfavourable codon bias (i.e. clusters of rare codons). If problems of this type are encountered, the construction of a synthetic gene can be considered. The artificial synthesis of a whole gene is expensive, but codon optimization at the 5′ end of the open reading frame can be sufficient to enhance gene expression. Batard *et al.* demonstrated an increase in wheat P450 enzyme activity in transgenic tobacco after optimization of the first 111 bp of the wheat open reading frame. This corresponds to 7% of the entire open reading frame. An useful alternative to nuclear transformation is the direct transformation of plastids, since plastid genomes are not affected by epigenetic silencing and they have a codon bias different to that of nuclear genes.

By creating transplastomic plants, very high levels of recombinant proteins have been obtained. In the case of pharmaceutical proteins, 7% total soluble protein (TSP) has been achieved for human somatotropin and 10% TSP has been achieved for human serum albumin (HSA). This exceeds the levels obtained through nuclear transformation, although the production of homoplastic plants is unfortunately time consuming and has only been achieved for a few species. Posttranslational modifications can have dramatic effects on the accumulation of a recombinant protein. If proteins are not properly cleaved, pro-proteins will accumulate instead of mature proteins.

For example, it has been shown that one of the causes for improper processing of soybean glycinins in potato tubers is the absence of the necessary proteases in the host plant. Some proteins need specific modifications to become active, but these are not carried out in plants. These include the addition of certain sugar molecules or the hydroxylation of proline for the production of native-like collagen in tobacco plants. Differences in glycan structures between plants and mammals (e.g. the presence of xylose and fucose residues in plants) have raised concerns regarding the potential immunogenicity of recombinant proteins, especially those used as human therapeutics. In order to obtain an antibody with a glycan profile similar to the mammalian type, Bakker *et al.* expressed human β-1, 4 galactosyltransferase in tobacco plants and crossed them with plants expressing a murine antibody. The result was a plantibody with partially galactosylated N-glycans. The folding of polypeptide chains and the assembly of multiple subunits are critical requirements when complex and multimeric proteins such as full size antibodies are expressed in plants.

Correct folding of the protein may be favoured by retention in the endoplasmic reticulum (ER), which can be achieved by adding a C-terminal K/HDEL ER-retention signal. This prolongs the exposure of the protein to endogenous chaperones like BiP (binding protein). Bouquin *et al.* expressed the heavy and light chains of an antibody individually in *Arabidopsis thaliana* plants. Strongly expressing parental lines were selected and crossed to obtain a hybrid line producing the full-size antibody. The multimeric antibody accumulated to higher levels than either of the individual chains in the parental lines, indicating that assembly of the mature protein had a stabilizing effect in comparison to the individual subunits. Expressing recombinant proteins as N-terminal fusions with ubiquitin is another strategy that can help to achieve proper folding.

Additionally, this is an elegant way to obtain proteins that do not start with a methionine residue, since the ubiquitin moiety is cleaved off by endogenous ubiquitin-specific proteases. If the desired protein is prone to proteolytic degradation, the use of appropriate targeting signals to direct the protein to different cellular compartments (e.g. the ER, chloroplast, vacuole or apoplast) or the use of tissue specific promoters (e. g. the hordein promoter for protein expression in barley grains) could be considered. Targeting is especially important if the recombinant protein is toxic to the production host. For example, targeting the non-pharmaceutical protein avidin to the cytosol in transgenic tobacco plants was toxic, but plants were regenerated successfully when this molecule was targeted to the vacuole. The expression of a secreted version of avidin in maize *(Zea mays)* resulted in partial or complete male sterility

## IMPROVED PROTEIN YIELDS

We will focus on recombinant antibodies (rAbs) and their derivatives (e.g scFvs, dia-bodies) in this section because these classic plant-produced pharmaceuticals have been expressed in many diverse production systems using a variety of regulatory elements. However, even in the case of this extensively studied class of proteins, direct comparisons between systems are difficult to make, since different genes, regulatory elements and production hosts have been used. Some investigators have compared the expression of a particular rAb in different plant species, or have targeted the antibody to different tissues or subcellular compartments in one particular host. Generally, such comparisons have revealed significant variation in expression and accumulation levels. Stoger *et al.* expressed the scFv antibody T 84.66, which recognizes the carcinoembryogenic antigen (CEA), and systematically compared several plant systems, regulatory elements and subcellular targeting signals.They concluded that the major factor determining overall accumulation levels was the stability of the antibody, which could be controlled in part by subcellular targetig.

In agreement with others, the highest levels of active antibody were obtained by targeting the protein to the secretory pathway using its native signal peptide, and then retaining the antibody in the ER lumen using a KDEL tag. Presumably because of the favourable biochemical environment, the lack of proteases and the presence of chaperones, rAbs accumulate to higher levels in the ER than in any other compartment. Furthermore, this strategy can limit the extent to which core glycans are replaced by plant-specific carbohydrates like α(1,3)-fucose and β(1, 2)-xylose.

Without ER retention, antibodies are secreted to the apoplast where accumulation *i* levels are typically 10 – 100-fold lower than in the ER. The cytosol appears to be an inappropriate compartment for antibodies, resulting in very low accumulation levels. An exception is described by De Jaeger *et al.* who targeted a phage display-derived scFv to the cytosol and achieved a level of 1% TSP. Stoger *et al.* also observed that between species, the amounts of scFv per unit fresh weight were in the same range, and did not correlate with the total protein content in the plant. For example, even though pea is a much more proteinaceous crop than rice, the amounts of antibody measured as a percentage of TSP were considerably leach species, individual promoters resulted in distinct, tissue-dependent accumulation patterns. The cauliflower mosaic virus (CaMV) 35S promoter, for example, led to high-level accumulation in callus and leaves whereas the maize ubiquitin-1 promoter was the best choice for producing recombinant proteins in cereal seeds even though it is not in itself seed-specific.

The lack of such comparative studies for proteins other than rAbs makes_it_ difficult to generalize an optimal expression strategy for all proteins. List recombinant proteins expressed in plants and provide details of the production system, promoters and other regulatory elements used in each case.

## DOWNSTREAM PROCESSING

When choosing a plant expression system, downstream processing should be taken into account since this will contribute a substantial proportion of the overall costs. A cost distribution for the production of β-glucuronidase in maize has been made by Evangelista *et al.* This is not a pharmaceutical protein, but the principles of cost analysis should hold true anyway. It was stated that protein extraction and purification would account for 40% and 48%, respectively, of the total annual operating costs, assuming a product purity of 83%. When higher purity is needed, as for pharmaceutical proteins, downstream processing could account for an even larger proportion of the total costs. Various strategies have been proposed to simplify downstream processing. We have already mentioned the use of tissue-specific promoters or targeting signals that allow protein secretion or accumulation in a particular organelle.

A useful example is protein targeting to oil bodies. which is used in combination with the oleosin-partitioning technique for protein isolation from oilseed crops. For new products that have yet to undergo clinical testing, the addition of an affinity tag can be considered to simplify protein isolation. Such tags have unique properties, which allow fusion proteins to be isolated from cellular extracts or media by affinity to particular ligands.

Alternatively, the tag may confer a novel physical property on the protein. An example is the synthetic elastin-like polypeptide (ELP) tag. When a temperature transition occurs, the tags fold into spirals that aggregate. These aggregates can easily be isolated by centrifugation. However, if such tags are used, they must not adversely affect the function of the protein. Importantly, the presence of affinity tags on pharmaceutical proteins may result in non-compliance with regulatory issues. A minimalist extraction buffer is often recommended for protein purification since most additives provide only a marginal improvement in yields, but will increase costs significantly.

The removal of phenolics, for example by tangential-flow ultrafiltration/diafiltration is an important step that should be carried out as early as possible in the purification procedure, since these molecules can become covalently linked to amino acid side chains and can oxidize certain residual groups. When using plant cell cultures for the production of biopharmaceuticals, targeting the protein to the apoplast (and from there to the culture medium) is beneficial for later purification. Plant cell culture media are protein-free, which makes the purification of the product relatively easy. Unfortunately, secreted recombinant proteins are often unstable in the cell culture medium. One strategy to overcome this problem was presented by James *et al* in form of an affinity chromatography bioreactor. By initiating protein purification while the cells are still growing, the exposure of the protein to destabilizing agents in the culture medium is minimized.

## PLANT-DERIVED PHARMACEUTICALS

The production of pharmaceutical proteins in plants has clear advantages over traditional systems in terms of cost-efficiency and product safety, since there is no risk of contamination with human pathogens. Furthermore plants are much less likely than mammalian cells to be affected by the expression of certain human proteins, such as growth factors and cell cycle inhibitors. Therefore, plants provide a strategic complement to existing microbial and animal production systems. Antibodies account for more than 20% of all biotechnology-derived molecules, and the market is expected to reach $5 billion by 2005. More than 250 companies are working on over 700 therapeutic antibodies, of which 220 are in clinicaltrials. These numbers demonstrate the ongoing commercial expansion of pharmaceutical protein production and indicate a growing demand for production capacity.

It has been estimated that plant-derived antibodies can be produced in a 250 $m^2$ greenhouse at about one tenth of the cost of hybridoma-derived antibodies. Planet Biotechnology Inc. presents an even rosier picture, estimating that the cost of plant-derived antibodies will be 5 % of the cost of antibodies produced in animal cells. This calculation, however, was based on yields of 500 $\mu g\ g^{-1}$ leaf fresh weight, a rather ambitious level for antibody production in plants. Once again, the greatest proportion of the cost reflects downstream processing and purification. When purification is needed, it is estimated that a recombinant protein must be expressed at greater than 0.01% of fresh weight or 1% of TSP for molecular farming to be commercially feasible. To date, several antibodies with therapeutic potential have been produced in plants. Among them are the Guy's 13 antibody that recognizes *Streptococcus mutans* (the causative agent of dental caries), humanized antibodies that recognize viral antigens (herpes simplex and respiratory syncitial virus), an antibody against CEA and a tumor-specific vaccine for the treatment of lymphoma.

Although none of these antibodies is produced commercially at the current time, the lymphoma scFv has completed phase I clinical trials and the Guy's 13 antibody has reached phase II clinical trials in the USA *(www.planetbiotechnology.com/products.html).* Regarding other pharmaceutical proteins, enterotoxin B and aprotinin have both reached phase I trials, while gastric lipase is undergoing clinical phase IIa trials in France and in Germany *(www.meristem-therapeutics.com/GB/intro.htm).* Although no plant-derived pharmaceutical protein is yet available on the market, two technical proteins (avidin and β-glucuronidase) are successfully produced in corn for commercial purposes.

## CONTAINMENT STRATEGIES

The issue that concerns people the most, with respect to the safety of plant biotechnology, is the possibility that transgenes could spread in the environment by out-crossing. Methods to contain foreign genes within transgenic plants are reviewed by Daniell and include maternal inheritance, male or seed sterility and induced transgene excision before flowering. Chloroplast genetic engineering takes advantage of the fact that the chloroplasts of most crops are maternally inherited. Male sterility can be achieved by crossing plants that produce the pharmaceutical protein of interest with plants that are male sterile or through the use of genetic methods to prevent the production of functional pollen.

Seed sterility can be achieved using Monsanto's Terminator Technology™ or the recoverable block of function systern. The Cre-loxP svstein has been used for transgene excision. Another strategy is the cultivation of transgenic plants in greenhouses. Although this increases production costs considerably, it could still be the cheapest solution overall. Khoudi *et al.* calculated the cost of 1 g of purified antibody produced in greenhouses, and found that it was only 10% of the cost of hybridoma technology. A further advantage of molecular farming in greenhouses is that the product itself would be contained, a desirable outcome in the case of biopharmaceuticals that are potentially harmful, capable of persisting in the environment or of accumulating in non-target organisms. Such safety issues came to the fore after the ProdiGene fiasco in late 2002, prompting the US Department of Agriculture in March 2003 to tighten its permit conditions relating to the field-testing of plants engineered to produce pharmaceutical and industrial compounds.

Another means to improve safety standards is the use of inducible promoters, which could be used to delay protein expression until the crop has reached the factory. This would also eliminate protein degradation problems caused by the delay between harvesting and processing. Selectable markers are an essential part of the transformation construct, since they allow the selective propagation of transformed cells.

Antibiotic resistance is the most widely used selectable phenotype, but there are concerns that resistance genes could be transferred into bacteria and might spread in the environment. For example Mercer *et al.* reported that oral bacteria can be transformed with naked DNA in human saliva. The persistence of DNA in the field was reviewed recently by Dale *et al.* who stated that there is no evidence for the incorporation of functional plant (trans)genes into the meat products of animals fed with genetically-modified feed. Nevertheless, the use of such markers has been legally challenged in Europe, and the commercial use of antibiotic resistance markers will be prohibited from 2005 (EU Directive 2001/18/EC released 21/03/2001). Other types of marker gene may be used, such as metabolic markers that allow the use of an alternative carbon source or growth-regulation markers that express enzymes involved in the biosynthesis of plant hormones. Another approach is the use of removable markers. For a review on these selection procedures see. Also, a recent publication proves the feasibility of marker-free transformation, which must be regarded as the best possible technique to prevent an unintended spread of antibiotic resistance genes.

## CONCLUDING REMARKS

Numerous issues must be taken into account in order to make molecular farming a commercial success. These issues include the yield and quality of the target protein, downstream processing methods, storage and purification, transgene containment and the cost of production. Currently, there are few examples of plant-derived pharmaceutical proteins that have been produced in a manner that addresses such concerns. We are facing a growing demand for therapeutics and diagnostics, but almost the entire capacity for cell-based production is already in use, leaving little room for the estimated 50-60 new protein pharmaceuticals that will be approved over the next 6-7 years. A shift to plant-bioreactors could therefore become a necessity. Although molecular farming still faces a number of challenges, one can clearly see the benefits and possibilities of the technology. Further research is needed to meet these challenges and facilitate the cost-effective and safe production of pharmaceuticals in plants.

CHAPTER 7

# RESPIRATION AND LIPID METABOLISM

Whereas photosynthesis provides the carbohydrate substrate upon which the plant (and all life) depends, glycolysis and respiration are the processes whereby the energy stored in carbohydrates is released in a controlled manner. Since many plants store their photosynthate in a more highly reduced form as lipids, a phenomenon common in seeds, we will also consider pathways of lipid biosynthesis and degradation in relation to the energy metabolism of seeds. Aerobic respiration is common to all eukaryotic organisms, and in its broad outlines, the respiratory process in higher plants is similar to that found in animals and lower eukaryotes. However, a number of specific aspects of plant respiration distinguish it from its animal counterpart.

*Aerobic respiration* refers to the biological process by which reduced organic compounds are mobilized and subsequently oxidized in a controlled manner. During respiration, free energy is released and incorporated into a form (ATP) that can be readily utilized for the maintenance and development of the plant. From a chemical standpoint, respiration is more commonly expressed in terms of the oxidation of the six-carbon sugar glucose, as outlined below.

$$C_6H_{12}O_6 + 6O_2 + 6H_2O \rightarrow 6CO_2 + 12H_2O$$

This equation is, of course, the reverse of that used to describe the photosynthetic process, and it represents a coupled redox reaction in which glucose is completely oxidized to $CO_2$ while oxygen serves as the ultimate electron acceptor, being reduced to water. The standard free energy change for the reaction as written involves the release of roughly 2880 kj (686 kcal) per mole (180 g) of glucose oxidized, and it is the controlled release of this free energy and its coupling to the synthesis of ATP that serves as the primary, though by no means the only, role of respiratory metabolism.

While glucose is most commonly cited as the substrate for respiration, it should be recognized that in a functioning plant cell the reduced carbon is actually derived frori such sources as the glucose polymer starch, the disaccharide sucrose, fructose-containing polymers (fructosans), and other sugars, as well as lipids (primarily triacylglycerols), organic acids, and, on occasion, proteins. Considering the large amount of free energy released in the oxidation of glucose, it is not surprising that respiration is a multistep process in which glucose is oxidized through a series of reactions. These reactions can be subdivided into three stages: (1) glycolysis, (2) the tricarboxylic acid (TCA) cycle, and (3) the electron transport chain. *Glycolysis is* carried out by a group of soluble enzymes located in the cytosol.

Chemically, glucose undergoes a limited amount of oxidation to produce two molecules of pyruvate (a three-carbon compound), a little ATP, and stored reducing power in the form of a reduced pyridine nucleotide, NADH. The *TCA cycle* and the electron transport chain are both located within the confines of the membrane-bounded organelle known as the mitochondrion. The *TCA cycle* brings about the complete oxidation of pyruvate to $CO_2$ and, in so doing, generates a considerable amount of reducing power (about 10 NADH equivalents per glucose). With one exception, these reactions involve a series of soluble enzymes located in the internal aqueous compartment, or matrix, of the mitochondrion.

The *electron transport chain* consists of a collection of electron transport proteins bound to the inner of the two mitochondrial membranes. This system transfers electrons from NADH (and related species), produced during glycolysis and the TCA cycle, to oxygen. The electron transfer releases a large amount of free energy, much of which is conserved through the conversion of ADP and $P_i$ to ATP. This final stage completes the oxidation of glucose. We can now formulate a more complete picture of respiration as related to its role in cellular energy metabolism:

$$C_6H_{12}O_6 + 6O_2 + 6H_2O + 32ADP + 32P_i \rightarrow 6CO_2 + 12H_2O + 32ATP$$

As will become apparent, not all the carbon that enters the respiratory pathway ends up as $CO_2$. Many important metabolic intermediates appear in the reactions of the glycolytic and TCA cycle pathways, allowing these pathways to serve as starting points for many other cellular pathways in addition to their central role in energy metabolism. Examples of metabolites derived from respiratory intermediates inelude several amino acids, the pentoses used in cell wall and nucleotide biosynthesis, precursors of porphyrin biosynthesis, and the primary compound (acetyl CoA) needed to synthesize fatty acids and isoprenoid compounds (carotenoids, gibberellins, and abscisic acid, among others).

## Energy Is Stored in NADH and ATP

Glycolysis is a process that occurs in all living organisms (prokaryotes and eukaryotes). From an evolutionary standpoint, it is the oldest of the three stages of respiration. No oxygen is required to convert glucose to pyruvate, and glycolytic metabolism can become the primary mode of energy production in plant tissues where oxygen levels are low—for example, in roots in a flooded soil. As noted, glycolysis involves a series of soluble enzymes located in the cytosol, and until recently it was not associated with any particular organelle or subcellular organizational network.

However, there is now evidence that in animal cells the enzymes of the glycolytic pathway do not exist independently in the cytosol but are associated in a supramolecular complex that may be bound to the surface of the outer mitochondrial membrane. This complex is thought to facilitate the conversion of substrates to products in the multistep glycolytic process. The principal reactions associated with the classical glycolytic and fermentative pathways. The two monosaccharides, glucose and fructose, can readily enter the pathway. In vivo, these two sugars are most commonly derived from the breakdown of either sucrose or starch. However, in higher plants, starch synthesis and catabolism take place in plastids, so carbon obtained from starch probably enters the glycolytic sequence in the cytosol at the level of the three-carbon sugars glyceraldehyde 3-phosphate and dihydroxyacetone phosphate.

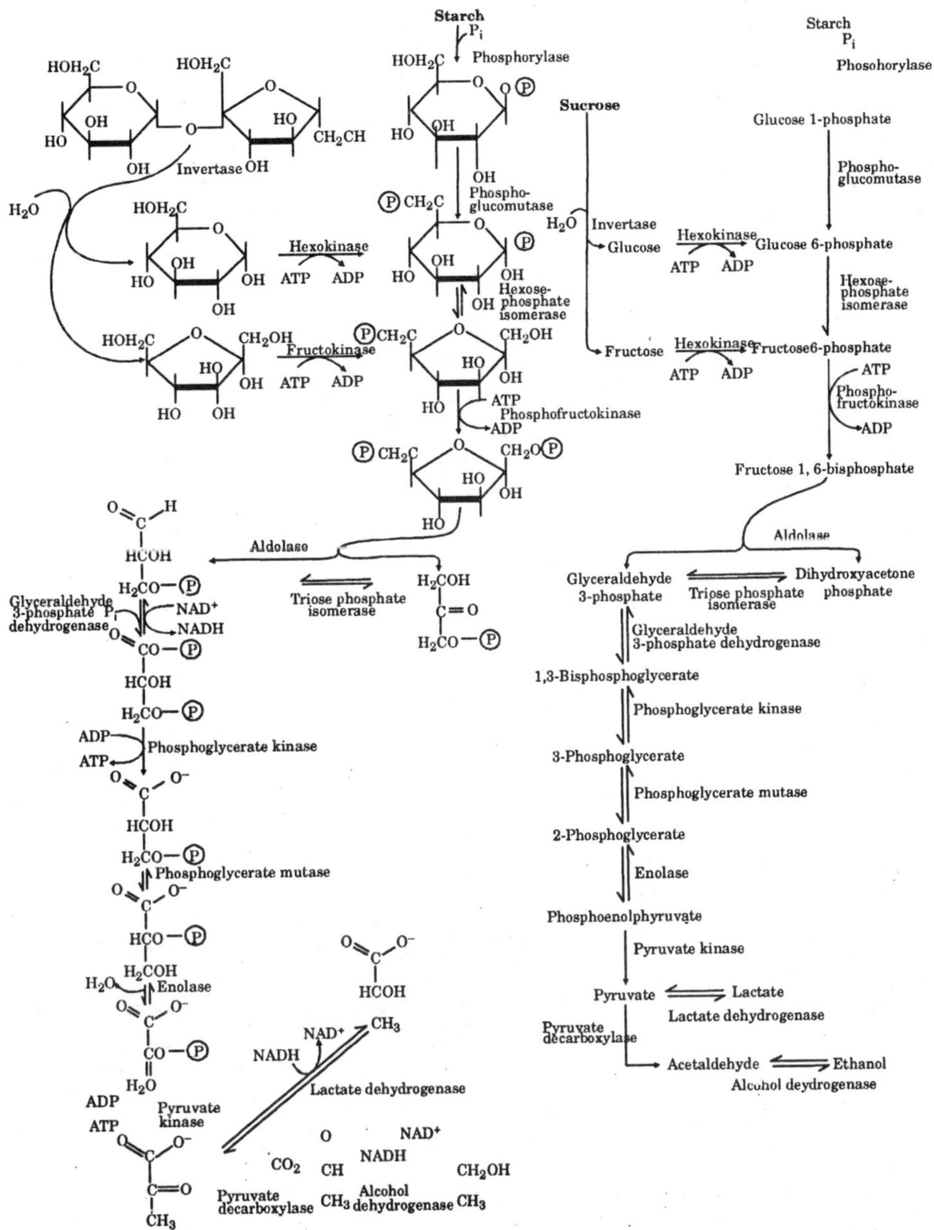


**Fig. 7.1. (left) The reactions of glycolysis and anaerobic fermentation. For each molecule of glucose that is metabolized, two molecules of glyceraldehyde 3-phosphate are formed, which ultimately produce two molecules of pyruvate. NAD+ is reduced to NADH by glyceraldehyde-3-phosphate dehydrogenase, and the resulting NADH is reoxidized during fermentative metabolism (indicated by the shaded areas) by either lactate dehydrogenase or alcohol dehydrogenase. ATP input is required at the level of the hexo-and phosphofructokinase reactions, while ATP is synthesized by the phosphoglycerate and pyruvate kinases. The double arrows (⇌) denote the reversible reactions of glycolysis and the single arrows (→) denote irreversible reactions.**

Because sucrose is the major translocated sugar in higher plants and is therefore the form of carbon that most nonphotosynthetic tissues import, sucrose *(not* glucose) should be thought of as the true "sugar substrate" for plant respiration. All the enzymes shown in Fig. 7.1 have been measured in plant tissues at levels sufficient to support the respiration rates observed in intact tissue. In the initial series of glycolytic reactions, the entering hexose (glucose or fructose) is phosphorylated twice and then split, producing two molecules of the three-carbon sugar, glyceraldehyde 3-phosphate. This requires the input of two molecules of ATP per glucose and involves two of the three essentially irreversible (i.e., $K_{eq} > 500$) reactions of the glycolytic pathway catalyzed by hexokinase and phosphofructokinase.

The phosphofructokinase reaction serves as one of the control points of glycolysis in both plants and animals. Recent work on plants has indicated that this part of the pathway is even more complex than shown in Fig. 7.1 and this will be discussed in a later section. With the appearance of glyceraldehyde 3-phosphate, the glycolytic pathway can begin to extract usable energy. The enzyme glyceraldehyde-3-phosphate dehydrogenase catalyzes the oxidation of the aldehyde to a carboxylic acid, releasing sufficient free energy to allow the concomitant reduction of $NAD^+$ to NADH and the phosphorylation (using inorganic phosphate) of glyceraldehyde 3-phosphate to produce 1, 3-bisphosphoglycerate. $NAD^+$ (nicotinamide-adenine dinucleotide) is an organic cofactor associated with many enzymes that catalyze cellular redox reactions. $NAD^+$ is the oxidized form of the cofactor, and it undergoes a reversible two-electron reaction yielding NADH ($NAD^+ + 2e^- + H^+ \rightleftharpoons NADH$).

The standard reduction potential for this redox couple is about –320 mV, which, biologically speaking, makes it a relatively strong reductant (i.e., electron donor). NADH is thus a good species in which to store free energy released during the stepwise oxidations of glycolysis and the TCA cycle. The subsequent oxidation of NADH by oxygen via the electron transport chain then releases sufficient free energy (220 kj $mol^{-1}$ or 52 kcal $mol^{-1}$) to drive the synthesis of ATP. A related compound, NADP (nicotinamide-adenine dinucleotide phosphate), performs a redox function in photosynthesis and the oxidative pentose phosphate pathway to be discussed later. The phosphorylated carboxylic acid on C-l of 1, 3-bisphosphoglycerate represents a mixed acid anhydride that has a large free energy of hydrolysis (–54.5 kj $mol^{-1}$ or –11.8 kcal $mol^{-1}$). This makes 1,3-bisphosphogly-cerate a strong donor of phosphate groups. In the next step of glycolysis, catalyzed by phosphoglycerate kinase, the phosphate of carbon 1 is transferred to a molecule of ADP to give ATP and 3-phosphoglycerate.

For each glucose entering the pathway, two ATPs are generated by this reaction: one for each 1, 3-bisphosphoglycerate. This type of ATP synthesis is referred to as a substrate-level *phosphorylation* because it involves the direct transfer of a phosphate moiety from a substrate molecule to ATP. It is distinct from the ATP synthesis by oxidative phosphorylation that is utilized by the electron transport chain in the final stage of respiration. In the last series of glycolytic reactions, the phosphate on 3-phosphoglycerate is transferred to carbon 2 and a molecule of water is removed to give the compound phosphoenolpyruvate (PEP).

The phosphate group on PEP also has a high free energy of hydrolysis (– 62 kj $mol^{-1}$ or – 14.8 kcal $mol^{-1}$) by virtue of its locking the parent compound, pyrurate, into an energetically unfavoured enol (—C=C—OH) configuration, and this makes PEP an extremely good phosphate donor as well. In the final step of glycolysis, the enzyme pyruvate kinase catalyzes a second substrate-level phosphorylation to yield ATP and pyruvate. This final step, which is the third irreversible step in glycolysis, yields two additional molecules of ATP for each glucose that enters the pathway.

## Regeneration of $NAD^+$

In the absence of oxygen, the TCA cycle and electron transport chain cannot function. This creates a problem for the continued operation of glycolysis because the cell's supply of $NAD^+$ is limited, and once all the $NAD^+$ becomes tied up in the reduced state (NADH) the glyceraldehyde-3-phosphate dehydrogenase reaction will not be able to take place. To overcome this problem,' plants and other organisms can further metabolize pyruvate by carrying out one or more forms of fermentation metabolism. In lactic acid fermentation (common to mammalian muscle as well as plants), the enzyme lactate dehydrogenase uses NADH to reduce pyruvate to lactate, thus regenerating $NAD^+$.

In alcoholic fermentation (also common in plants but more widely known in brewer's yeast), the two enzymes pyruvate decarboxylase and alcohol dehydrogenase act on pyruvate, ultimately producing ethanol and $CO_2$ and oxidizing NADH in the process. Under some circumstances, plant tissues may subjected to low (hypoxic) or zero (ancxic) concentrations of ambient oxygen, which forces them to carry out fermentative metabolism. The best-known example of this involves flooded or waterlogged soils in which the diffusion of oxygen to the roots is sufficiently reduced to cause root tissues to become hypoxic.

In corn, the initial response involves lactic acid fermentation, but subsquent induction of the enzyme alcohol dehydrogenase promotes a switch to ethanolic fermentation. Ethanol is thought to be a less deleterious end product of fermentation than lactate, because accumulation of lactate promotes acidification of the cytosol. There are numerous other examples of plants functioning under near-anaerobic conditions by carrying out some form of fermentation.

## Glycolytic Pathway In Plants

The reactions described above occur in all organisms that carry out glycolysis. In addition, organisms can utilize this pathway operating in the opposite direction to synthesize glucose from other organic molecules. The operation of the glycolytic pathway to synthesize glucose is known as *gluconeogenesis*. Gluconeogenesis operates in higher plants in the germination of seeds, such as castor bean and sunflower, that store a significant quantity of their carbon reserves in the form of fat.

During germination, much of the fat is converted via gluconeogenesis to sucrose, which is then used to support the growing seedling. Because the reaction catalyzed by phosphofructokinase is not readily reversible, an additional enzyme, fructose bisphosphatase, acts at this step to convert fructose 1, 6-bisphosphate to fructose 6-phosphate during gluconeogenesis. Fructose bisphosphatase is found in higher plants. Further, these two enzymes appear, in both plants and animals, to represent a major control point of carbon flux through the glycolytic and gluco-neogenic pathways. However, in higher plants the interconversion of fructose 6-phosphate and fructose 1,6-bisphosphate is made more complex by the presence in plant tissues of an additional enzyme, a pyrophosphate ($PP_i$)-dependent phosphofructokinase (pyrophosphate-fructose-6-phosphate 1-phosphotransferase), which catalyzes the reversible interconversion of fructose 6-phosphate and fructose 1, 6-bisphosphate:

$$\text{Fructose 6-P} + PP_i \rightleftharpoons \text{fructose 1,6-}P_2 + P_i$$

This enzyme is found in the cytosol of many plant tissues at levels that can be considerably higher than those of phosphofructokinase. Since the $PP_i$-phosphofructokinase reaction is readily reversible, this enzyme could operate during either glycolytic or gluconeogenic carbon fluxes, although its operation in glycolysis would require a source of pyrophosphate. More important, this enzyme, like phosphofructokinase and fructose bisphosphatase, appears to be under metabolic control, leading to the conclusion that, under some circumstances, the operation of the glycolytic (or gluconeogenic) pathway in higher plants may not conform exactly to that commonly associated with other organisms.

## Anaerobic Fermentation

Before leaving glycolysis, some consideration should be given to the efficiency of fermentation. The standard free energy change ($\Delta G^{\circ\prime}$) for the complete oxidation of one mole of glucose is -2880 kJ $mol^{-1}$ (–686 kcal $mol^{-1}$). The $\Delta G^{\circ\prime}$ for the synthesis of ATP is –31.8 kJ $mol^{-1}$ (–7.6 kcal $mol^{-1}$). However, under the nonstandard conditions that normally exist in plant cells, the synthesis of ATP requires an input of free energy of approximately –50.2 kJ $mol^{-1}$ (–12 kcal $mol^{-1}$). Given the net synthesis of two molecules of ATP for each glucose that is converted to lactate (or ethanol), the efficiency of anaerobic fermentation—that is, the energy stored as ATP relative to the energy potentially available in a molecule of glucose—is only about 4%.

Most of the energy available in the glucose remains in the reduced by-product of fermentation, lactate (or ethanol). During aerobic respiration, the pyruvate produced by glycolysis is transported into the mitochondria, where it is further oxidized, resulting in much more efficient conversion of the free energy originally available in the glucose.

## Semiautonomous Organelles

The next two stages of respiration take place within a double membrane-bound organelle, the *mitochondrion*. Plant mitochondria were originally identified by light microscopy as particles that stained with the dye Janus green B. With the advent of electron microscopy, mitochondrial morphology became clearly defined. Plant mitochondria generally appear as spherical or rodlike entities ranging from 0.5 to 1.0 μm in diameter and up to 3 μm in length. Highly reticulate networks representing a single complex mitochondrion are commonly observed in animal cells and have been reported in unicellular algae, but their occurrence is not common in higher plant tissues. The number of mitochondria observed per plant cell is usually directly related to the metabolic activity of the tissue, reflecting the mitochondrial role in energy metabolism. The ultrastructural features of plant mitochondria are similar to those of mitochondria in nonplant tissues.

Plant mitochondria have two membranes, a smooth outer membrane that completely surrounds a highly invaginated *inner membrane*. The aqueous phase contained within the inner membrane is referred to as the mitochondrial matrix, while the region between the two mitochondrial membranes is known as the intermembrane space. Invaginations of the inner membrane give rise to structures known as *cristae*. Intact mitochondria are osmotically active; that is, they take up water and swell when placed in a hypoosmotic medium. Most inorganic ions and charged organic molecules are not able to diffuse freely into the matrix space.

The site of the osmotic barrier is the innermembrane, the outer membrane being permeable to solutes having molecular masses less than approximately 10,000 Da (i.e., most cellular metabolites and ions). The lipid fraction of both membranes is made up primarily of phospholipids, 80% of which are either phosphatidylcholine or phosphatidylethanolamine. Mitochondria, like chloroplasts are semiautonomous organelles, since they contain ribosomes, RNA, and DNA, which encodes a limited number (15 – 25) of mitochondrial proteins. Plant mitochondria are thus able to carry out the various steps of protein synthesis and to transmit genetic information from one generation to the next. The phenomenon of cytoplasmic male sterility, which is commonly used by plant breeders to develop hybrid lines, in most instances seems to involve a mitochondrially linked trait.

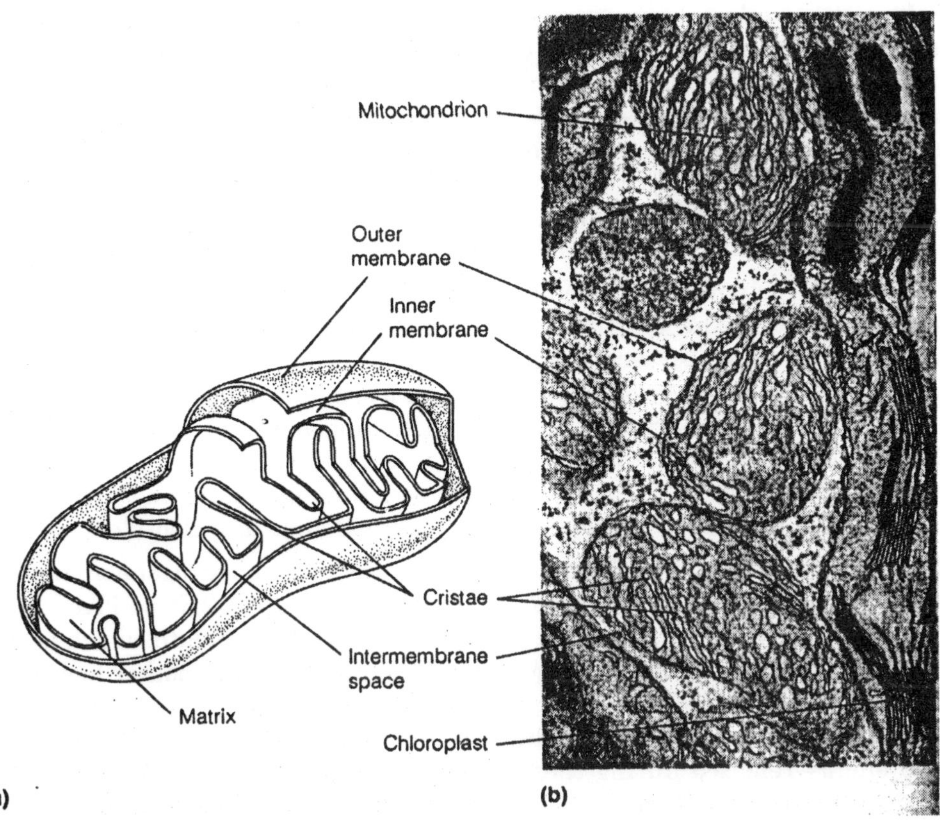


**Fig. 7.2. Three-dimensional representation of a mitochondrion showing the invaginations of the inner membrane that give rise to the cristae and the location of the matrix and intermembrane spaces. The outer membrane is permeable to molecules smaller than 10,000 Da; the inner membrane serves as the major osmotic barrier of the organelle. (b) Electron micrograph of corn mesocotyl mitochondria (31,600 x).**

## TCA Cycle

The TCA cycle is also known as the Krebs cycle in honour of Hans A. Krebs, who first proposed a cyclic series of reactions in the aerobic breakdown of pyruvate, and as the *citric acid cycle*, because of the importance of citrate as an early intermediate. This cycle represents the second stage in respiration and takes place in the mitochondrial matrix. Its operation necessitates transport of the pyruvate generated in the cytosol during glycolysis through the rather impermeable inner mitochondria membrane. This transport involves a pyruvate (monocarboxylate) translocator that catalyzes an electroneutral exchange of pyruvate and $OH^-$ across the inner membreane. Once inside the mitochondrial matrix, pyruvate is oxidatively decarboxylated by the enzyme pyruvate dehydrogenase to produce NADH (from $NAD^+$), $CO_2$, and acetic acid, which is linked via a thioester bond to a sulfur-containing cofactor, coenzyme A (Co A), to form acetyl CoA.

Pyruvate dehydrogenase exists as a complex of several enzymatic activities that catalyze the overall reaction in three separate steps. Following this reaction, the enzyme citrate synthase combines acetyl CoA with a four-carbon dicarboxylic acid, oxaloacetic acid (OAA), to give a six-carbon tricarboxylic acid, citrate. Following isomerization of citrate to isocitrate by the enzyme aconitase, the next two reactions involve successive oxidative decarboxylations, each of which produces one NADH and releases a molecule of $CO_2$. At this point, three molecules of $CO_2$ have been produced for each pyruvate that enters the mitochondrion, so the complete oxidation of glucose has actually been accomplished. The remainder of the TCA cycle involves the conversion of succinyl CoA to OAA to allow continued operation of the cycle.

Initially, the large amount of free energy available in the thioester .bond of succinyl *Co*A is conserved through the synthesis of ATP from ADP and $P_i$ via a substrate-level phosphorylation catalyzed by succinyl-CoA synthetase. (In the citrate synthase step, the free energy available in acetyl CoA was used to form a carbon-carbon bond in citrate.) The resulting succinate is oxidized to fumarate by succinate dehydrogenase, which is the only membrane-associated enzyme of the TCA cycle and is also considered to be a component of complex II of the electron transport chain. The electrons removed from succinate end up not on $NAD^+$ but on another electron-transferring cofactor, FAD (flavin-adenine dinucleotide).

FAD is covalently bound to the active site of succinate dehydrogenase and undergoes a reversible two-electron reduction to produce $FADH_2$ ($FAD + 2e^- + 2H^+ \rightleftharpoons FADH_2$). In the final two reactions of the TCA cycle, fumarate is hydrated to produce malate, which is subsequently oxidized by malate dehydrogenase to regenerate OAA and produce another molecule of NADH. The OAA produced is now able to react with another acetyl CoA and continue the cycle. The stepwise oxidation of pyruvate in the TCA cycle gives rise to three molecules of $CO_2$, and much of the free energy released during these oxidations is stored in the form of reduced $NAD^+$ (four NADH) and FAD one $FADH_2$).

In addition, one molecule of ATP is produced by a substrate-level phosphorylation. All the enzyme activities associated with the TCA cycle have been measured in mitochondria of higher plants, and it is likely that the enzymes are associated together in a multienzyme complex.

## Unique Features of TCA Cycle

The TCA cycle reactions outlined in Figure 7.3 are not all indentical to those carried out by animal mitochondria. For example, the succinyl-CoA synthetase step produces ATP in plants and GTP in animals. In animal mitochondria, ATP is formed in a second enzyme reaction that involves the transfer of a phosphate from GTP to ADP. Anotner feature of the TCA cycle that

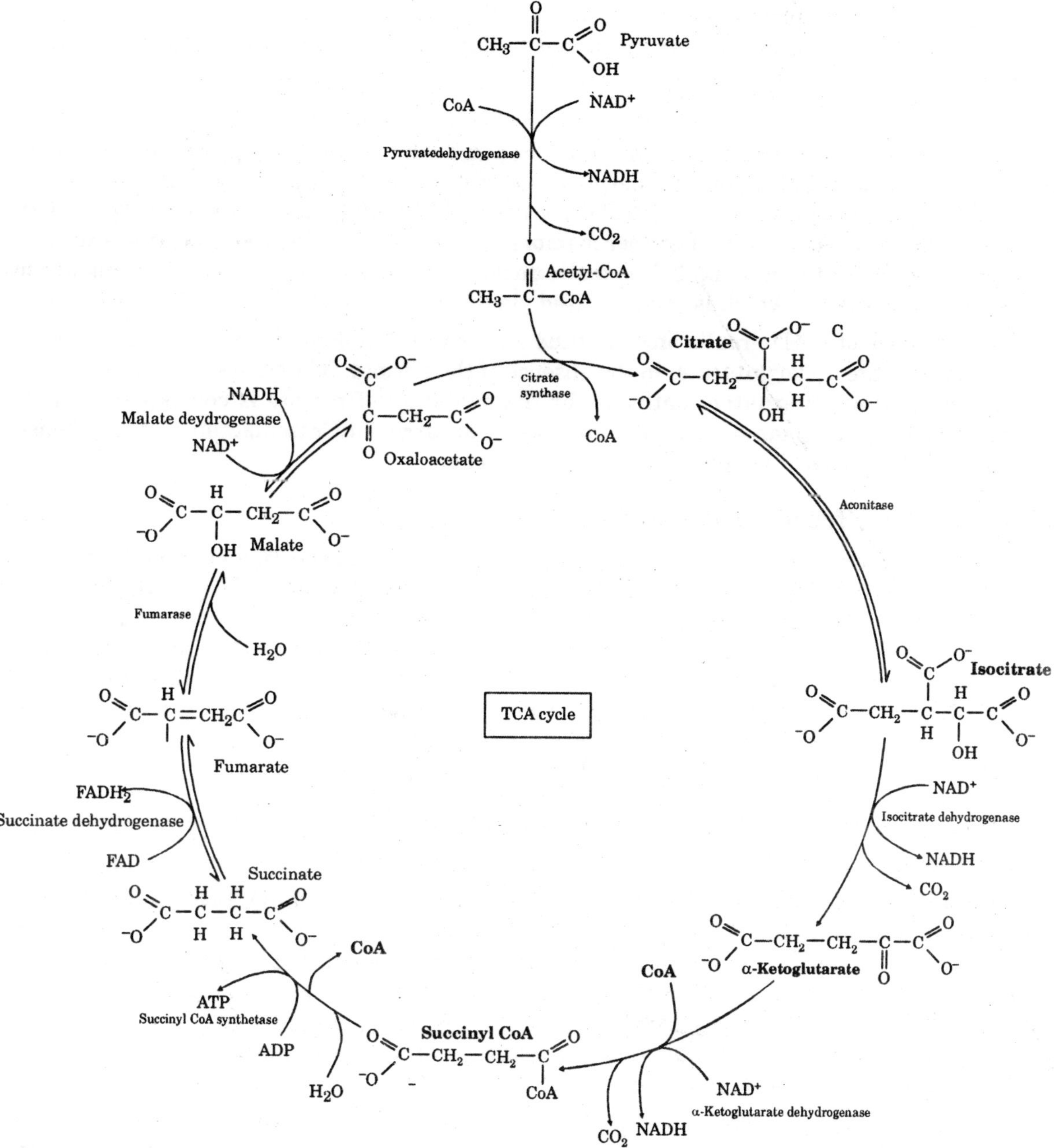


**Fig. 7.3. The reactions and enzymes of the TCA cycle. In the cycle, pyruvate is completely oxidized to three molecules of $CO_2$. The electrons generated during these oxidations are used to reduce four molecules of $NAD^+$ to NADH and one molecule of FAD to $FADH_2$. One molecule of ATP is also synthesized by a substrate-level phosphorylation during the succinyl-CoA synthetase reaction.**

is generally unique to plants is the significant $NAD^+$ malic enzyme activity, which has been observed in all plant mitochondria measured to date. This enzyme catalyzes the oxidative decarboxylation of malate.

$$\text{Malate} + NAD^+ \rightarrow \text{pyruvate} + CO_2 + NADH$$

The presence of $NAD^+$ malic enzyme enables plant mitochondria to operate an alternative pathway for the metabolism of PEP derived from glycolysis. Malate can be synthesized from PEP in the cytosol via the enzymes PEP carboxylase and malate dehydrogenase. The malate is then readily transported into the inner membrane that catalyzes the electroneutral exchange of malate and $P_i$. In the matrix, $NAD^+$ malic enzyme promotes the metabolism of malate by oxidizing it to pyruvate, which is then oxidized via the TCA cycle as outlined previously.

The presence of NAD+ malic enzyme thus allows the complete oxidation of organic acids (e.g., malate citrate α-oxoglutarate) in the absence of the normal TCA cycle substrate, pyruvate. The presence of an alternative pathway for the oxidation of malate is consistent with the observation that many plants, in addition to those that carry out Crassulacean acid metabolism store significant levels of malate in their central vacuole.

## Electron Flow from NADH to $O_2$

For each molecule of glucose oxidized through glycolysis and the TCA cycle pathways, two molecules of NADH are generated in the cytosol while eight molecules NADH plus two molecules of $FADH_2$ (on succinate dehydrogenase) appear in the mitochondria matrix. As discussed previously, these reduced compounds must be reoxidized or the entire respiratory process will come to a halt. The electron transport chain catalyzes an electron flow from NADH (and $FADH_2$) to oxygen, the final electron acceptor of the respiratory process. For the oxidation of NADH, the overall two-electron transfer is represented as

$$NADH + H^+ + \tfrac{1}{2}\,O_2 \rightarrow NAD^+ + H_2O$$

From the midpoint reduction potentials for the NADH/ $NAD^+$ couple, –320 mV, and the $H_2O/\tfrac{1}{2}O_2$ couple, + 810 mV, the standard free energy released during this overall reaction ($= -nF\,\Delta E°'$) is about 220 kJ $mol^-$ (52 kcal $mol^{-1}$) (per two electrons). Because the $FADH_2$/FAD reduction potential (–45 mV) is somewhat higher than that of NADH/$NAD^+$, only 167.5 kJ $mol^{-1}$ (40 kcal $mol^{-1}$) is released for each two electrons generated during the oxidation of succinate. The role of the electron transport chain is to bring about the oxidation of NADH (and $FADH_2$) and, in the process, utilize some of the free energy released to drive the synthesis of ATP. The electron transport chain of higher plants contains approximately the same general set of electron carriers found in nonplant mitochondria.

The individual electron transport proteins are organized into a series of four multiprotein complexes (I–IV), each localized on the inner mitochondrial membrane. Electrons from the NADH generated in the mitochondrial matrix during the TCA cycle are oxidized by complex I (an NADH dehydrogenase), which in turn transfers these electrons to ubiquinone, a *p*-benzoquinone species that exists as a pool within the inner membrane and is not tightly associated with any of the membrane proteins. Chemically and functionally, ubiquinone is similar to plastoquinone in the photosynthetic electron transport chain. The electron carriers in complex I include a tightly bound FMN (flavin mononucleotide, chemically similar to FAD) and three or four iron-sulfu proteins.

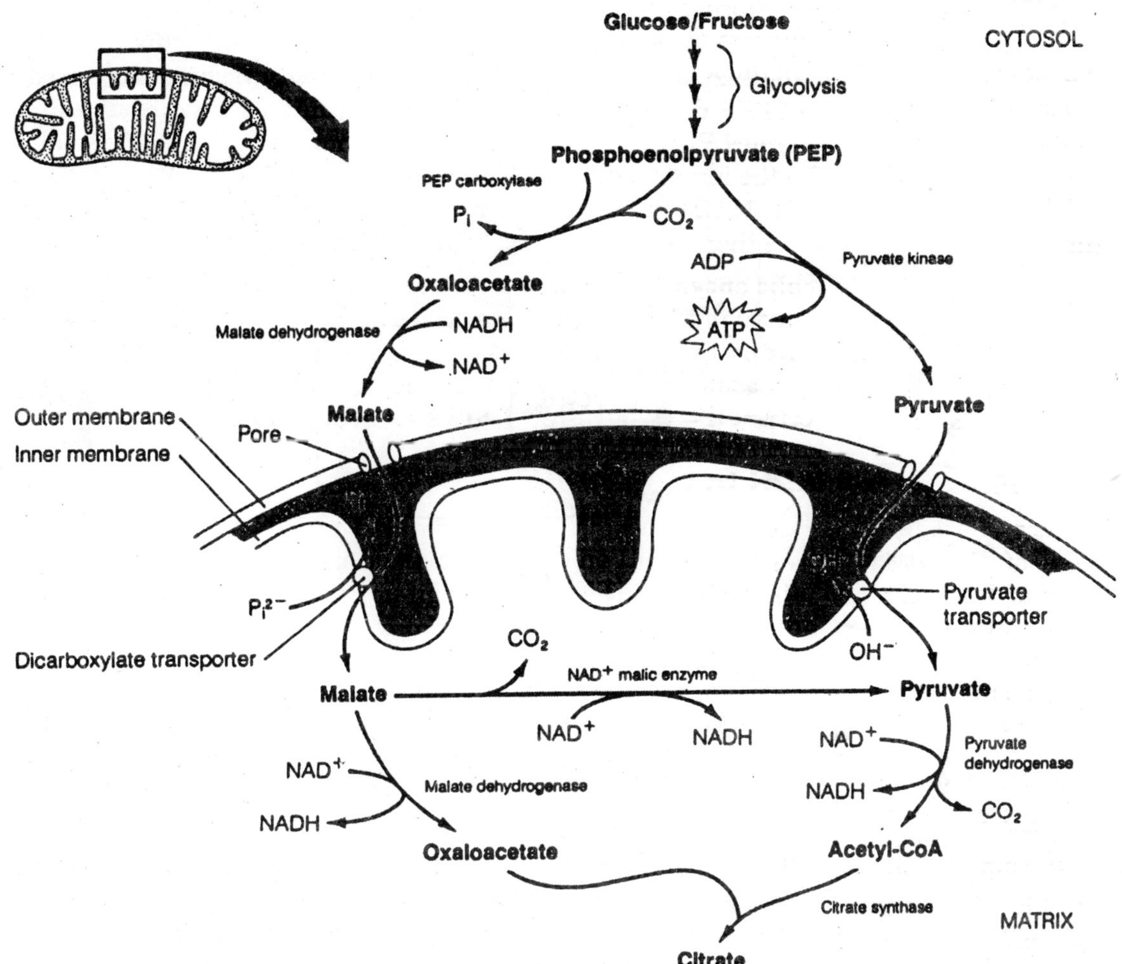


**Fig. 7.4. Pathways for the movement of malate andpyruvate into the TCA cycle from glycolysis in higher plants. Pyruvate generated in the cytosol during glycolysis is transported across the inner mitochondrial membrane in an electroneutral exchange with $OH^-$ involving a pyruvate-specific transporter. Because plant mitochondria contain $NAD^+$ malic enzyme, they are also capable of completely oxidizing malate (in the absence of added pyruvate). This pathway involves the conversion of PEP, generated in the cytosol during glycolysis, to malate and the movement of malate into the mitochondria through the dicarboxylate transporter. The latter transporter catalyzes the electroneutral exchange of malate (and other dicarboxylates) and $P_i^{2-}$ across the inner mitochondrial membrane.**

The iron-sulfur proteins can be distinguished on the basis of their redox potentials and their electron spin resonance (ESR) spectra. The TCA cycle enzyme succinate dehydrogenase is part of complex II, so that electrons derived from the oxidation of succinate are transferred via the $FADH_2$ and a group of three ironsulfur proteins into the ubiquinone pool. Complex III acts as a ubiquinone :cytochrome-c oxidoreductase, oxidizing reduced ubiquinone (ubi-quinol) and transferring the electrons through an iron-'sulfur center, two *b*-type cytochromes, and a membrane-bound cytochrome $c_1$ to cytochrome *c*. The electron flow through this part of the electron transport chain is complex and does not proceed in the linear fashion suggested by Fig. 7.5

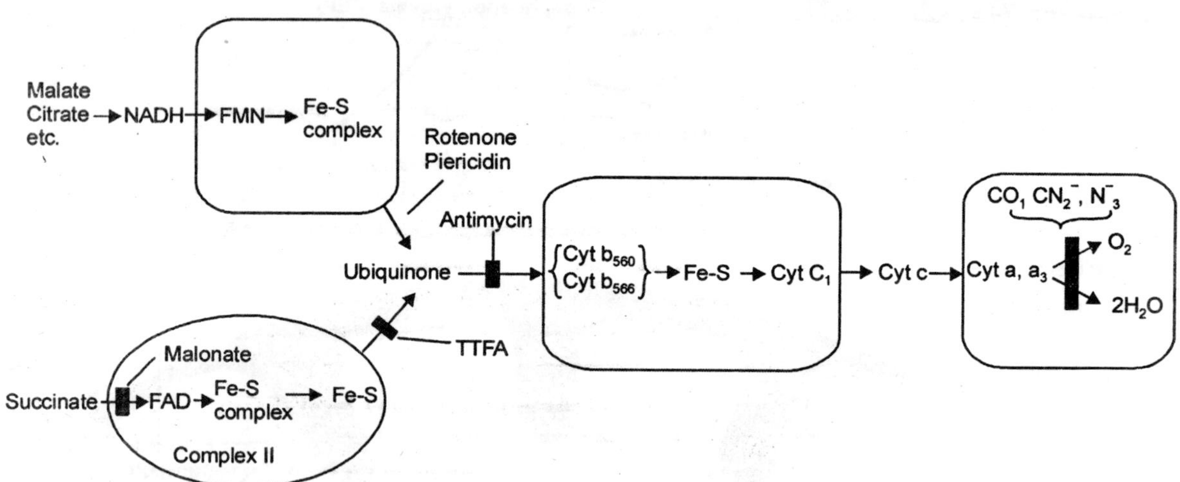


**Fig. 7.5. A linear representation of the cyanide-sensitive (main) pathway of plant mitochondrial electron transport, showing complexes I-IV and the sites of action of several inhibitors. Complex I, NADH : ubiquinone oxidoreductase; complex II, succinate:ubiquinone oxidoreductase; complex III, ubiquinone:cytochrome-c oxidoreductase; complex IV, cytochrome-c oxidase. Specific inhibitors include rotenone and piericidin, which inhibit electron transfer between the complex I iron-sulfur (Fe-S) complex and ubiquinone pool; malonate, which acts as a competitive inhibitor of succinate in complex II; TTFA (thenoyltrifluoroacetone), which blocks electron transfer from complex II into the ubiquinone pool; antimycin, which inhibits electron transfer in the cytochrome *b* region of complex III; and CO, cyanide ($CN^-$), and azide ($N_3^-$), all of which are competitive inhibitors of oxygen at cytochrome-c oxidase (complex IV).**

Cytochrome *c* is the only protein in the electron transport chain that is not an integral membrane protein, and it serves as a mobile carrier to transfer electrons between complexes III and IV (cytochrome-c oxidase). Cytochrome oxidase contains two a-type cytochromes and two atoms of copper, which bring about the four-electron reduction of $O_2$ to two molecules of $H_2O$. The organization of these complexes on the inner membrane is quite specific. The oxidation of NADH and succinate takes place on the matrix or M side of the membrane, as does the reduction of oxygen.

Cytochrome *c* is located in the intermembrane space, on the cytoplasmic or C side of the membrane. It appears that the complex III iron-sulfur center and cytochrome $C_1$ though both membrane-bound, are oriented facing the C side of the inner membrane, while the two cytochromes *b* of complex III and cytochromes *a* and $a_3$ of cytochrome oxidase are localized toward opposite sides of the inner membrane. This sort of organization necessitates transmembrane movement of electrons as they traverse the electron transport chain, and its functional significance will be dealt with more fully when we consider the mechanism of ATP synthesis later in this chapter. Finally, these individual electron transfer complexes apparently are not permanently associated into larger "supercomplexes"; rather, electron transfer occurs during random collisions between reacting species (e.g., between complex I and ubiquinone) as they diffuse llaterally within the fluid phospholipid bilayer.

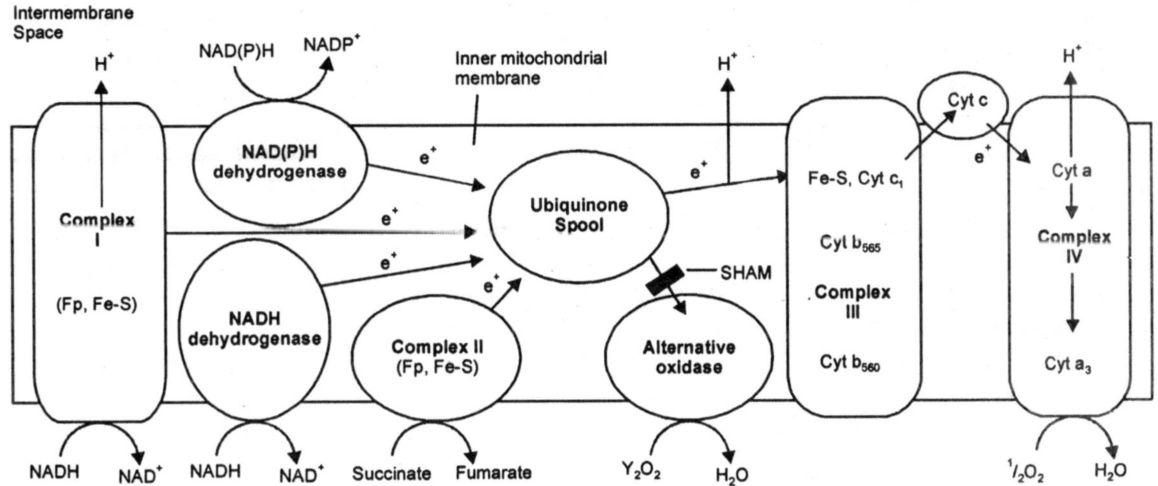


**Fig. 7.6. A two-dimensional representation of the electron carriers on the plant inner mitochondrial membrane (complexes I-IV, cytochrome c, two additional NADH dehydrogenases, and the cyanide-resistant [alternative] oxidase), the three sites of energy conservation (vertical arrows indicating $H^+$ extrusion), and the site of action of the specific inhibitor of the alternative pathway, salicylhydroxamic acid (SHAM).**

## Electron Carriers of Plant Mitochondria

In addition to the collection of electron carriers described above, plant mitochondria contain some components not commonly found in other mitochondria. These include an NADH dehydrogenase complex that faces the intermembrane space or C side of the inner membrane and apparently facilitates the oxidation of cytoplasmic NADH and, possibly, NADPH (Moller and Lin, 1986). Electrons from this *external* NADH dehydrogenase enter the main electron transport chain at the level of the ubiquinone pool. A second unusual feature of plant mitochondria is the presence of two pathways for oxidizing matrix NADH. Electron flow through the classical complex I described above is sensitive to inhibition by a number of compounds, including rotenone and piericidin.

Isolated plant mitochondria consistently display the presence of a rotenone-resistant pathway for the oxidation of NADH derived from TCA cycle substrates. Although much has been learned about the properties of this rotenone-resistant bypass, its exact role in plant metabolism is still a matter of speculation Finally, most, if not all, plant mitochondria have varying levels of an "alternative" pathway for the reduction of oxygen. This pathway involves an oxidase that, unlike cytochrome-c oxidase, is insensitive to inhibition by cyanide, azide, or carbon monoxide. The nature and physiological significance of this cyanide-resistant pathway will be considered more fully below.

## ATP Synthesis in the Mitochondrion

The transfer of electrons to oxygen via complexes I–IV is coupled to the synthesis of ATP, with the number of ATPs synthesized depending on the nature of the electron donor. With isolated mitochondria, electrons derived from internal (matrix) NADH give ADP:O ratios (number of ATPs synthesized per two electrons transferred to oxygen) of 2.4-2.7. Succinate and externally added NADH each give values in the range 1.6–1.8, while ascorbic acid, which serves as an artificial electron donor to cytochrome *c*, gives values of 0.8-0.9. Results such as these (for both plant and animal mitochondria) have led to the general concept that three sites of energy conservation exist along the electron transport chain, at complexes I, III, and IV, respectively.

Because of the observed connection between mitochondrial electron flow and ATP synthesis, this conversion of ADP and $P_i$ to ATP became known as oxidative phosphorylation long before any of its mechanistic features were understood. However, in recent years a remarkable amount of progress has been achieved in understanding the mechanism of this energy transformation. The currently accepted mechanism of mitochondrial ATP synthesis is based on the chemiosmotic hypothesis, proposed in the early 1960s by Peter Mitchell as a general mechanism of energy conservation across biological membranes. In this theory, the asymmetric orientation of electron carriers within the mitochondrial inner membrane allows for the transfer of protons ($H^+$) across the inner membrane during electron flow.

Numerous studies have confirmed that mitochondrial electron transport is associated with a net transfer of protons from the mitochondrial matrix to the intermembrane space. The free energy associated with the formation of this proton electrochemical gradient ($\Delta\bar{\mu}_{H+}$, also referred to as a *proton motive force* when expressed in units of volts) is made up of an electrical (membrane) potential component, $\Delta E_m$, and a chemical potential component, ΔpH. The $\Delta E_m$ results from the asymmetric distribution of a charged species ($H^+$) across the membrane, and the ΔpH is due to the proton concentration difference across the membrane. Both contribute to the proton motive force in plant mitochondria, although $\Delta E_m$ is consistently found to be of greater magnitude. The input of free energy required to generate $\Delta\bar{\mu}_{H+}$, comes from the free energy released during electron transport.

Exactly how electron transport is coupled to proton translocation is not well understood in all cases. Mitchell originally proposed a series of alternating hydrogen ($H^+ + e^-$) and electron ($e^-$) carriers suitably situated within the membrane to give rise to a series of redox loops. This idea holds, in somewhat modified form, for proton translocation at complex III but seems unlikely to explain the coupling of proton and electron transfer at complexes I and IV. Once generated, the $\Delta\bar{\mu}_{H+}$, is reasonably stable because of the low permeability (conductance) of the inner membrane to protons, and the free energy stored in the proton electrochemical gradient can be utilized to carry out chemical work (*i.e.*, ATP synthesis). The $\Delta\bar{\mu}_{H+}$, is coupled to the synthesis of ATP by an additional protein complex associated with the inner membrane, the $F_0F_1$-*ATP synthase* (also called complex V). This complex consists of two major components, $F_1$ and $F_0F_1$ is a peripheral membrane protein complex that is composed of at least five different subunits and contains the catalytic site for converting ADP and $P_i$ to ATP (or hydrolyzing ATP to ADP and $P_i$ when the ATPase activity is being measured). This complex is attached to the matrix side of the inner membrane. $F_0$ is an integral membrane protein complex that consists of at least three different polypeptides that form the channel through which protons are able to cross the inner membrane.

The passage of protons through the channel activates the ATP synthase to synthesize ATP and simultaneously dissipates the $\Delta\bar{\mu}_{H+}$. The structure and function of the mitochondrial ATP synthase are similar to those of the $CF_0$-$CF_1$ ATPase in photosynthetic photophosphorylation. Acceptance of the general features of a chemiosmotic mechanism of ATP synthesis has several implications. First, the true site of ATP formation on the mitochondrial inner membrane is the ATP synthase complex, not complex I, III, or IV. The latter serve as sites of *energy conservation* whereby electron transport is coupled to the generation of a $\Delta\bar{\mu}_{H+}$. In fact, electron transport is not obligatorily linked to ATP synthesis if an alternative method of generating a proton gradient across the inner membrane exists. This can be demonstrated experimentally when electron transport is inhibited and ATP synthesis is made to proceed through an artificially generated pH gradient. Second, it is not necessary for an integral number of ATPs to be produced at each of the three sites of energy conservation.

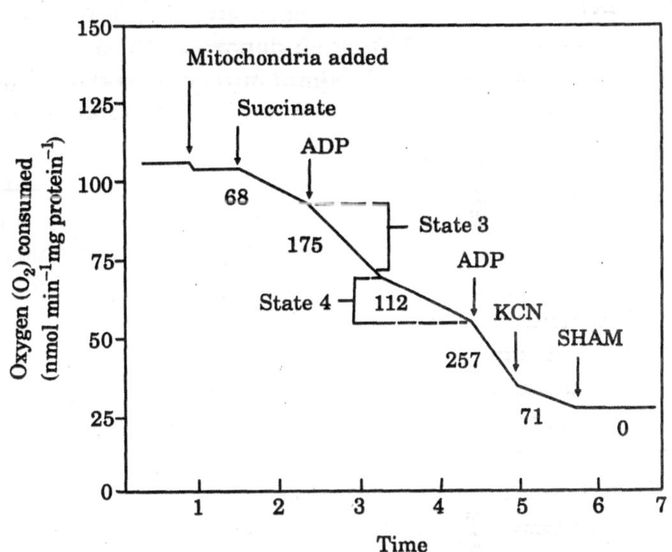


**Fig. 7.7. Regulation of respiratory rate by ADP in isolated mitochondria during succinate oxidation. The vertical arrows indicate successive additions of purified mitochondria, 10 mM succinate, 150 μM ADP, 450 μ M ADP, 0.5 mM KCN, and 1.0 mM SHAM. Addition of succinate initiates mitochondrial electron transfer, which is measured with an oxygen electrode as the rate 'of oxygen reduction (to $H_2O$). Addition of ADP stimulates electronansfer (state 3) by facilitating the dissipation of the proton electrochemical gradient. Once all the ADP has been converted to ATP, electron transfer reverts to a lower (state 4) rate. Addition of cyanide ($CN^-$) inhibits electron flow through the main pathway and diverts it to the cyanide-resistant pathway, which is subsequently inhibited by the addition of SHAM. The numbers below the traces refer to the rate of oxygen uptake expressed as $O_2$ consumed.**

The overall ADP:O ratio will simply be a function of the number of protons translocated per two electrons times the number of

ATPs synthesized per proton translocated. This need not be an integral value, so the ADP:O values of 2.4-2.7 measured for internal NADH oxidation are probably correct and do not have to be rounded off to 3.0 when attempting to calculate the total number of ATPs generated during aerobic respiration. Finally, the chemiosmotic theory readily explains the mechanism of action of *uncouplers,* a wide range of chemically unrelated compounds (including dinitrophenol, FCCP [*p*-trifluoromethoxycarbonylcyanide phenylhydrazone], and many detergents] that inhibit mitochondrial ATP synthesis but are often observed to stimulate the rate of electron transport. These compound all make the inner membrane leaky to protons, which prevents the buildup of a sufficiently large $\Delta\bar{\mu}_{H+}$, to drive ATP synthesis.

The stimulation of electron transport upon addition of uncouplers is related to an effect seen with isolated mitochondria. Experimentally, lower rates of electron flow (measured as the rate of oxygen uptake in the presence of an appropriate TCA cycle substrate (Refferred Fig. 7.8) are observed in the absence of ADP (referred to as state 4) than upon addition of ADP (state 3). ADP provides a substrate that stimulates the dissipation of the $\alpha\bar{\mu}_{H}^{+}$ through the $F_0F_1$-ATPase during ATP synthesis. Once all the ADP has been converted to ATP, the $\Delta\bar{\mu}_{H+}$, builds up again and exerts a restraint on the rate of electron flow. The ratio of the state 3 and state 4 'rates is referred to as the respiratory control ratio (RCR) and, although some caveats apply it and the ADP:O ratio are used as characteristic measures of the quality of a mitochondrial preparation. The proton electrochemical gradient also plays a role in the movement of the substrates and products of the TCA cycle and oxidative phosphorylation into and out of mitochondria.

Whereas mitochondria synthesize ATP in the mitochondrial matrix, most of the ATP is used outside the organelle, so an efficient mechanism is needed for moving ADP into and ATP out of the mitochondrion. This mechanism involves another inner membrane protein, the ADP/ATP (adenine nucleotide) transporter which catalyzes an exchange of ADP and ATP across the inner membrane. The electrical potential gradient ($\Delta E_m$) generated during electron transfer (negative inside) is such that there will be a net movement of the more negatively charged $ATP^{4-}$ out of the mitochondria in exchange for $ADP^{3-}$. Likewise, the uptake of $P_s$ involves an active phosphate·transporter that uses the ΔpH component of the proton motive force to catalyze the electroneutral exchange of $P_i^-$ (in) for $OH^-$ (out). As long as a ΔpH is maintained across the inner membrane, the $P_i$ content within the matrix will remain high.

Similar reasoning applies to the uptake of pyruvate, which is driven by the electroneutral exchange of pyruvate for $OH^-$, leading to continued uptake of pyruvate from the cytoplasm. Most of the proton electrochemical potential (~ – 200 mV) across the mitochondrial inner membrane is due to the $\Delta E_m$.

## Aerobic Respiration

The complete oxidation of glucose leads to the net formation of four molecules of ATP by substrate-level phosphorylation (two during glycolysis and two in the TCA cycle), two molecules of NADH in the cytosol, and eight molecules of NADH and two molecules of $FADH_2$ (via succinate dehydrogenase) in the mitochondrial matrix. Based on measured ADP:O values a total of approximately 28 molecules of ATP will be

**Table 7.1. ADP:O Ratios in Isolated Plant Mitochondria**

| *Substrate* | *ADP:0 ratio* |
|---|---|
| Malate | 2.4–2.7 |
| Succinate | 1.6–1.8 |
| $NADH_{ext}$ | 1.6–1.8 |
| Ascorbate | 0.8–0.9 |

generated per glucose by oxidative phosphorylation. This results in a total of 32 ATPs per glucose. (Note the difference between this number and the value of 36 commonly seen in textbooks; the latter value is based on rounded-off integral values for the number of ATPs synthesized during the oxidation of each molecule of matrix or external NADH and succinate rather than the actual measured values given in Table. Using 50.2 kj $mol^{-1}$ (12 kcal $mol^{-1}$) for the actual free energy of formation of ATP in vivo, we find that about 1606 kj $mol^{-1}$ (384 kcal $mol^{-1}$) of free energy is conserved per mole of glucose oxidized during aerobic respiration. This represents about 56% of the total free energy available; the rest is lost as heat. This is a marked improvement over the 3-5% efficiency associated with fermentative metabolism.

## Cyanide-Resistant Respiration Pathway

If cyanide (1 mM)is added to actively respiring animal tissues, cytochrome oxidase is inhibited and the respiration rate quickly drops to less than 1% of its initial level. However, most plant tissues display a level of *cyanide-resistant respiration* that represents 10-25%, but in some tissues can be up to 100% , of the uninhibited control rate. The enzyme responsible for this oxygen uptake has been identified as a cyanide-resistant oxidase of the plant mitochondrial electron transport chain. Although the detailed mechanism of the unusual oxidase associated with this alternative pathway has yet to be characterized, current evidence indicates that electrons feed off the main electron transport chain into the alternative pathway at the level of the ubiquinone pool.

The cyanide-resistant terminal oxidase associated with the alternative pathway catalyzes a four-electron reduction of oxygen to water and is specifically inhibited by several compounds, most notably salicylhydroxamic acid (SHAM) and *n*-propyl gallate. Because electrons branch to the alternative pathway from the ubiquinone pool, two sites of energy conservation (at complexes III and IV) are bypassed, and there is no evidence for an energy conservation site on the alternative pathway between ubiquinone and oxygen. Therefore, the free energy that would normally be stored as ATP is lost as heat when electrons are shunted through the alternative pathway. How can such an energetically wasteful process as the cyanide-resistant pathway contribute to plant metabolism? One example of its utility appears during floral development in certain members of the Araceae (skunk cabbage, *Symplocarpus foetidus)*.

Just prior to pollination, tissue within a clublike organ on the developing floral apex (the spadix) undergoes high rates of respiration through the alternative pathway, causing the spadix to heat up by as much as 14°C above ambient temperature and to volatilize odiferous compounds that attract pollinating insects. Salicylic acid, a compound related to aspirin, has been identified as the chemical signal responsible for initiating this thermogenic event. In most plants, however, both the respiratory rates and the level of cyanide-resistant respiration are too low to generate significant levels of heat.

It has been suggested that the alternative pathway might serve as an energy overflow, oxidizing respiratory substrates that accumulate in excess of those needed for growth, storage, or synthesis of ATP. In this view, electrons "spill" into the alternative pathway only after the capacity of the main pathway is saturated. This occurs experimentally when cyanide is added, but it could take place in vivo if a state 4 situation developed and the respiration rate exceeded the cell's demands for ATP. There is evidence from research that the alternative pathway serves an energy overflow function in higher plants.

## Regulation of Respiration

The control of carbohydrate metabolism in plants is poorly understood, but it is clear that it operates at numerous levels, starting with the rate at which reduced carbon is synthesized and/or imported into a plant tissue. In addition, because one of the major functions of respiration is to generate ATP, the rate of ATP utilization by a cell or tissue will be of primary importance in controlling the level of ADP present, and this will markedly influence the rates of glycolysis in the cytosol oxidative phosphorylation in the mitochondria througe mechanisms described below. Control points appear to exist at all three stages of respiration, and only a brief overview of some obvious features will be given here.

In vivo, glycolysis appears to be regulated at the level of the enzymes phosphofructokinase and pyruvate kinase, just as it is in animal respiration. Pyruvate kinase is markedly inhibited by its product, ATP. As a result, high cytosolic levels of the substrate ADP promote the pyruvate kinase reaction, whereas high levels of ATP in the cytosol inhibit this reaction. Further PEP, which builds up in the cytosol in response to ATP inhibition of pyruvate kinase, is a potent inhibitor of the ATP-dependent phosphofructokinase activity. This in hibitory effect of PEP on phosphofructokinase is enhanced in the presence of ATP but strongly attenuated by inorganic phosphate. These regulatory controls allow the flow of carbon through glycolysis to respond, to a first approximation, to the cell's energy demands by monitoring changes in the cytoplasmic [ATP]/[ADP] ratio. However, the regulation of glycolysis is more complex than outlined above.

For example, the ability of plant mitochondria to metabolize malate in addition to pyruvate makes it possible for plant cells to bypass the regulatory features of the pyruvate kinase reaction. With phosphofructokinase, the situation is made more complex by the fact that plants can contain appreciable levels of the $PP_i$-dependent phosphofructokinase activity as well as the metabolic effector fructose 2,6-bisphosphate. Fructose 2,6-bisphosphate (fructose 2,6-$P_2$) plays an important regulatory role in glycolysis in animals, stimulating phosphofructokinase activity and inhibiting that of fructose 1,6-bisphosphatase. Fructose 2,6-$P_2$ is found in the cytosol of plant tissues, as are the enzymes fructose 6-phosphate 2-kinase and fructose 2,6-bisphosphatase, which act to synthesize and break down fructose 2,6-$P_2$, respectively.

Although the role of fructose 2,6-$P_2$ in the regulation of plant carbohydrate metabolism is not fully understood, several points have been established. Fructose 2,6-$P_2$ is present at varying levels in the cytoplasm. It markedly inhibits the activity of fructose 1,6-bisphosphatase, stimulates the activity of $PP_i$-dependent phosphofructokinase, and has little or no effect on the activity of the ATP-dependent phosphofructokinase. Further information is needed before we will fully understand the regulation at this point of the glycolytic/gluconeogenic pathways. However, it is clear that the $PP_i$-dependent phosphofructokinase and fructose 2,6-$P_2$ play a role in plant carbohydrate metabolism.The primary control of plant mitochondrial oxidations seems to take place at the level of cellular adenine nucleotides. Exactly what feature is most important is less certain.

Both the *adenylate energy charge*—([ATP] + $\frac{1}{2}$[ADP])/([ATP] + [ADP] + [AMP])—and the [ATP]/ [ADP] *ratio* have been cited as regulatory indicators of .the cell's energy status. From the limited number of studies of the mechanisms of regulatory control in plant mitochondria, it appears that the absolute concentration of ADP available (i.e., in the cytosol) is the most ' important factor in the regulation of mitochondrial respiratory rates. According to this view, as the cell's idemand for ATP in the cytosol decreases relative to the rate of synthesis of ATP in the mitochondria, less ADP will be available, and the electron transfer chain will Eoperate in more of a state 4 situation. The resulting slowdown of electron transport could be communuunicated to TCA cycle enzymes through increases in 1(1) the matrix NADH:$NAD^+$ ratio, which affects several TCA cycle dehydrogenase activities, and (2) the matrix concentration of ATP, which has an allosteric influence on the pyruvate dehydrogenase and citrate synthase reactions.

Decreased electron flow through the main electron transport chain would also lead to use of the cyanide-resistant pathway as the steady-state level of reduction of the ubiquinone pool increased in response to decreased cellular energy demands (consistent with its role in energy overflow). Clearly, much more to be learned about the various controls that modulate plant respiration rates.

## The Pentose Phosphate Pathway

The glycolytic pathway is not the only route available for oxidation of glucose in the cytosol of plant cells. The *oxidative pentose phosphate pathway* (also known as the hexose monophosphate shunt) can also accomplish this. The first two reactions of this series represent the oxidative events of the pathway, converting the six-carbon glucose 6-phosphate to a five-carbon sugar, ribulose 5-phosphate, with loss of a $CO_2$ and generation of two molecules of NADPH (not NADH). The remaining reactions of the pathway bring about the conversion of ribulose 5-phosphate to the glycolytic intermediates glyceraldehyde 3-phosphate and fructose 6-phosphate. Studies of the release of $^{14}CO_2$ from isotopically labeled glucose indicate that glycolysis is the more dominant pathway, accounting for 80-95% of the total carbon flux in most plant tissues.

However, the pentose phosphate pathway does contribute, and developmental studies indicate that its contribution increases as plant cells go from a meristematic to a more differentiated state. The pentose phosphate pathway plays several roles in plant metabolism. The products of the two oxidative steps are NADPH, and it is thought that this NADPH is used to drive reductive steps associated with various biosynthetic reactions that occur in the cytosol. Such a role has commonly been accepted for the operation of this pathway in animal tissues.

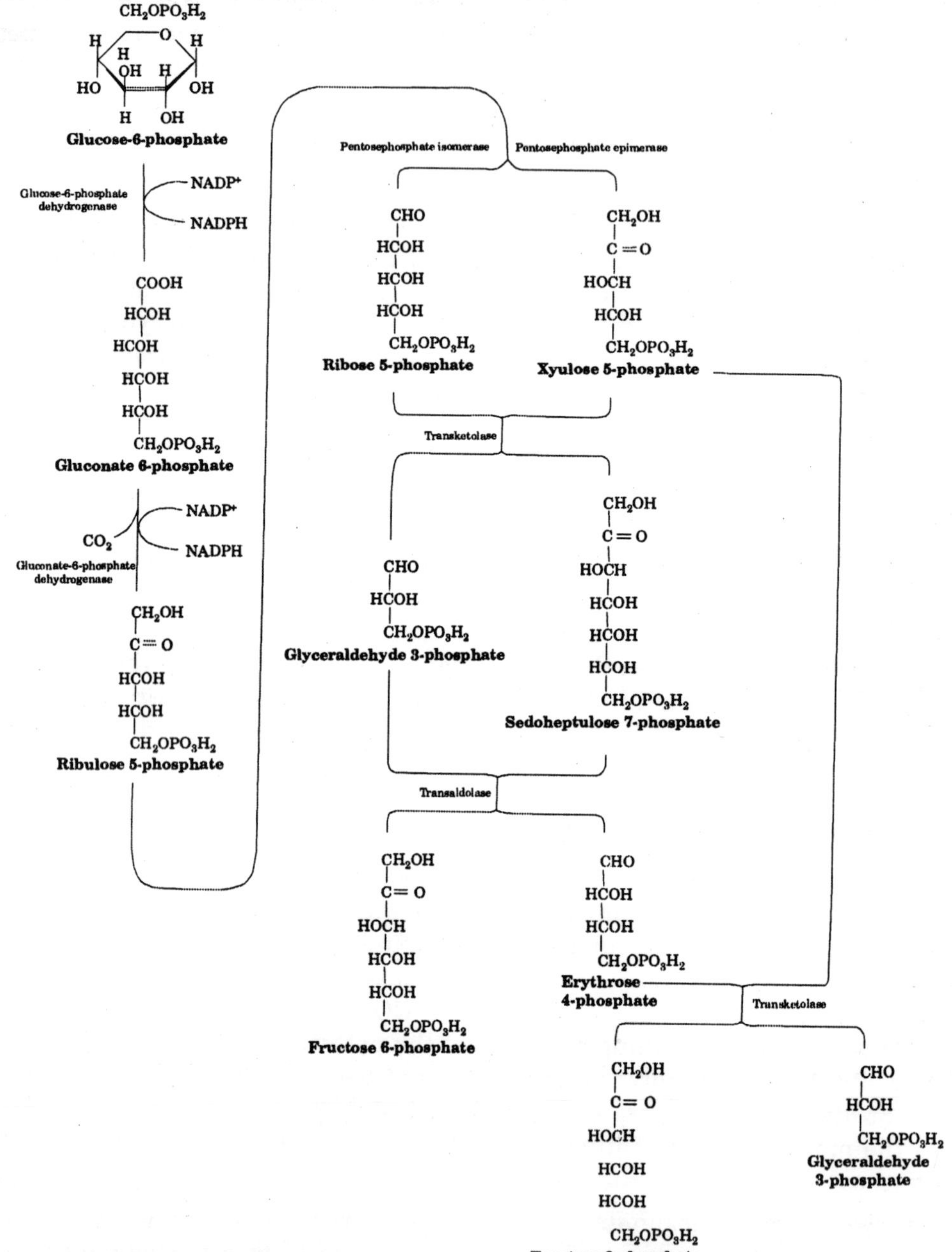


**Fig. 7.8. Reactions of the oxidative pentose phosphate pathway in higher plants. NADPH is generated in the first two reactions of the pathway, and the resulting ribulose 5-phosphate is converted to the glycolytic intermediated fructose 6-phosphate and glyceraldehyde 3-phosphate through a series of metabolic interconversion. The first two reactions are essentially irreversible, while the interconversions between ribulose 5-phosphate and the glycolytic compounds are freely reversible.**

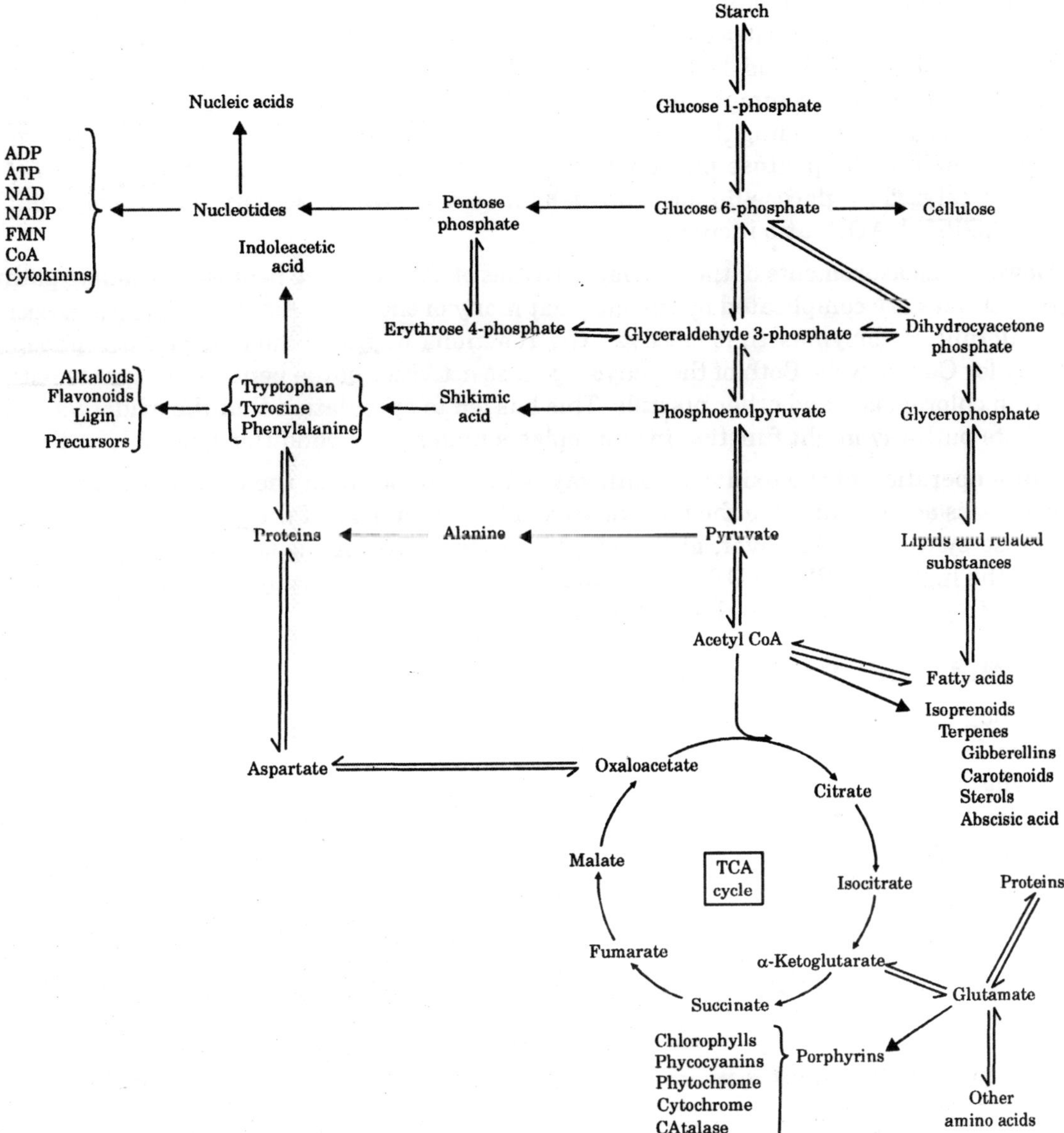


**Fig. 7.9. Glycolysis and the TCA cycle contribute intermediates to a variety of other metabolic pathways in higher plants. This figure is not intended to be complete. Rather, it gives some indication of the extent to which plant metabolism depends on the flux of carbon through these two pathways and emphasizes the fact that not all the carbon that enters the glycolytic pathway is oxidized to $CO_2$.**

However, because the NADH dehydrogenase facing the cytosol on the mitochondrial inner membrane is also capable of oxidizing NADPH, it is possible that in plant cells some of the reducing power generated by the pentose phosphate pathway contributes to cellular energy metabolism—that is, electrons from NADPH may end up reducing $O_2$ and generating ATP. The pathway also produces ribose 5-phosphate, a precursor of the ribose and deoxyribose needed in the synthesis of RNA and DNA. Another intermediate on this pathway, the four-carbon erythrose 4-phosphate, combines with PEP in the initial reaction in the production of plant phenolic compounds, including the aromatic amino acids and the precursors of lignin, flavonoids, and phytoalexins. The pentose phosphate pathway is controlled by the initial reaction of the pathway catalyzed by glucose-6-phosphate dehydrogenase, whose activity is markedly inhibited as the NADPH.-NADP ratio increases.

However, measurements of the enzyme activities of the oxidative pentose phosphate pathway in green tissues are complicated by the fact that many of the same Activities are also associated with chloroplast enzymes that catalyze the reactions of the reductive pentose phosphate pathway for Calvin cycle. Both of the dehydrogenase activities of the pentose phosphate pathway appear in chloroplasts and other plastids. This has led to speculation that the oxidative pentose phosphate pathway might function in chloroplasts under some conditions, notably in the dark.

Little operation of the oxidative pathway is likely to occur in the chloroplast in the light, because mass action will drive the nonoxidative interconversions of the patpway in the direction of pentose synthesis. Moreover, glucose-6-phosphate dehydrogenase will be inhibited in the light by the high NADPH:NADP ratio in the chloroplast as well as by a reductive inactivation involving the ferredoxin:thioredoxin system.

## Other Metabolic Pathways in the Cell

Although much of this chapter has focused on the role of respiration in energy metabolism, it should be pointed out that glycolysis and the TCA cycle are both linked to a number of important metabolic pathways. Glycolysis and the TCA cycle are central to the production of a wide variety of plant metabolites, including amino acids, lipids and related compounds, isoprenoids, and porphyrins. Not shown in Fig. 7.10 are two additional reactions associated with plant mitochondria. First, two enzyme activities found in the matrix of mitochondria from chlorophyll-containing tissues, glycine decarboxylase and serine hydroxymethyltransferase, catalyze the $CO_2$-releasing step in photorespiration. Second, plant mitochondria provide some of the reactions associated with the breakdown of stored fats in developing seeds, which is considered more extensively later in this chapter.

## Whole-Plant Respiration

Many factors can affect the respiration rates of a plant as well as the individual organs of the plant. These include the species and growth habit of the plant, the type and age of the specific organ, and environmental variables such as the external oxygen concentration, temperature, and plant water status. Whole-plant respiration rates, particularly when considered on a fresh-weight basis, are generally lower than respiration rates reported for animal tissues. This is due in large part to the considerable fraction of a mature plant cell that is metabolically inactive in respiration (i.e., the central vacuole and the cell wall.

Nonetheless, respiration rates in some plant tissues equal the rates observed in actively respiring animal tissues so the plant respiratory process is not inherently slower than respiration in animals. Even though plants generally have low respiration rates, the contribution of respiration to the overall carbon economy of the plant can be substantial. A survey of several herbaceous species indicated that 30-60% of the daily gain in photosynthetic carbon was lost to dark respiration, although these values tended to decrease with age of the plants.

Young trees lose roughly a third of their daily photosynthate to respiration, and this loss can double in older trees as the ratio of photosynthetic to nonphotosynthetic tissue decreases. In tropical areas, 70-80% of the daily photosynthetic gain can be lost to respiration owing to the high nighttime respiration rates associated with elevated night temperatures.

## Rate of Respiration

A useful rule of thumb is that the greater the overall metabolic activity of a given tissue, the higher is its repiration rate. Developing buds usually show very high rates of respiration (on a dry-weight basis), and respiration rates of vegetative tissues usually decrease from the growing tip to more differentiated regions. In mature vegetative tissues, stems generally have the lowest respiration rates, and leaf and root respiration varies with the plant species and the conditions under which the plants are growing. When a plant tissue has reached maturity, its respiration rate will either remain roughly constant or decrease, slowly as the tissue ages and ultimately undergoes senescence.

An exception to this behaviour is the marked rise in respiration known as the *climacteric* that accompanies the onset of ripening in many fruits (avocado, apple, banana) and senescence in detached leaves and flowers. Both ripening and the climacteric respiratory rise are triggered by the endogenous production of ethylene, and both processes can be stimulated to occur prematurely by exogenous application of ethylene. Although ripening in these fruits is associated with increased rates of protein synthesis, which require cellular energy, the exact role of the respiratory rise is not clear.

Some fruits (citrus, pineapple, grapes) do not show a respiratory climacteric, and these fruits are also not generally stimulated to ripen by addition of ethylene. Many studies have shown that fruits that undergo a climacteric respiratory rise display a high level of the cyanide-resistant pathway prior to ripening, but apparently not during the respiratory burst itself.

## Environmental Factors Affecting Respiration Rates

The environmental factors that can affect plant respiration clearly include oxygen because of its role as a substrate in the overall process. At 25°C, the equilibrium concentration of $O_2$ in an air-saturated (20% $O_2$) aqueous solution is about 250 μM. The $K_m$ for oxygen in the cytochrome oxidase reaction is difficult to measure exactly but is well below 1 μM, so there should be no apparent dependence of the respiration rate on external $O_2$ concentrations. In fact, little effect on measured respiration rates is observed until the atmospheric oxygen concentration drops below 5% for whole tissues or below 2-3% for tissue slices. Isolated mitochondria are unaffected by oxygen depletion approaching 0%. That there is any effect of oxygen concentration at all is due to limitations on the diffusion for oxygen through the aqueous phase between the atmosphere and the mitochondria.

The diffusion coefficient of oxygen dissolved in an aqueous medium is $10^4$ less than that of oxygen diffusing through air. The fact that atmospheric oxygen tensions as high as 5% can reduce the observed respiration rate in whole tissues indicates that the diffusive movement of oxygen represents a true limitation on plant respiration. Further, the existence of a diffusion-limited, aqueous pathway of oxygen movement implies that intercellular air spaces are important in facilitating oxygen movement within plant tissues. If there were no gaseous diffusion pathway throughout the plant, the cellular respiration rates of many plants would be limited by an insufficient supply of oxygen. This diffusion limitation is even more significant iwhen the medium in which the tissue is growing is a liquid.

When plants are grown hydroponically it is necessary to aerate the solutions vigorously to keep oxygen levels high in the vicinity of the roots, and the probleml of oxygen supply also arises with plants growing in predominantly wet or flooded soils. Some plants, particularly trees, are restricted in geographic distribution by the need to maintain a supply of oxygen to their roots. For instance, dogwood and tulip tree poplar can survive only in well-drained, aerated soils because their roots are unable to tolerate more than a limited exposure to a flooded condition. Many other plants, however, can adapt to growth in flooded soils.

Herbaceous species such as rice and sunflower often rely on a network of intercellular air spaces running from the leaves to the roots to provide a continuous, gaseous pathway for the movement of oxygen to the flooded roots. The problem can be even more acute for trees that have very deep roots growing in wet soils. Such roots must survive on anaerobic (fermentative) metabolism and develop structures that facilitate the movement of oxygen to the roots. Good examples of the latter are outgrowths of the roots, called *pneumatophores,* which protrude out of the water and provide a gaseous pathway for oxygen diffusion into the roots. Pneumatophores are found in *Avicennia* and *Rhizophora* trees that grow in mangrove swamps under continuously flooded conditions. While superficially similar in appearance to mangrove pneumatophores, the "knees" of cypress *(Taxodium)* do not perform such a function. Although lowering the external oxygen concentration ultimately reduces the rate of aerobic respiration, it does not necessarily decrease the rate of carbohydrate utilization.

In 1861, Louis Pasteur discovered that yeast cells consume less glucose in the presence of air than under anaerobic conditions. This response, known as the Pasteur effect, occurs in higher plant tissues and manifests itself as an increased rate of fermentative metabolism under anaerobic conditions. In aerobic respiration, cytosolic levels of ATP are high, producing an inhibitory effect on the regulatory enzymes of the glycolytic pathway, ATP-dependent phosphofructokinase and pyruvate kinase. In tissues maintained under anaerobic conditions, the cytosolic ATP level and its attendant inhibitory effects are decreased, leading to higher rates of carbohydrate metabolism. The effect of atmospheric oxygen (and temperature) on respiration is utilized in fruit storage, where it is common to maintain fruits in the cold under conditions of 2-3% oxygen and 3-5% $CO_2$. The reduced temperature serves to lower the respiration rate (see below), as does the reduced oxygen.

Low levels of oxygen are used instead of anoxic conditions to avoid lowering tissue oxygen tensions enough to stimulate fermentative metabolism. Carbondioxide has little direct effect on the respiration rate even at 3-5% , well in excess of the 0.035% normally found in the atmosphere. High concentrations of $CO_2$ are used in fruit storage to block the effect of the hormone ethylene on fruit ripening. In the physiological temperature range, respiration is temperature-dependent. The increase in respiration rate for every 10°C increase in temperature, commonly referred to as the $Q_{10}$, is slightly greater than two, although in some plants an unexplained decrease in the $Q_{10}$ has been observed at temperatures above 30°C.

High night temperatures are thought to account for the large respiratory component in tropical plants, and lowered temperatures are utilized to retard post harvest respiration rates during the storage of fruits and vegetables. However, complications may arise from this practice. For instance, when potato tubers are stored at temperatures above 10°C, respiration and ancillary metabolic activities are sufficient to allow sprouting to proceed. Below 5°C, respiration rates and sprouting are reduced in most tissues, but the breakdown of stored starch and its conversion to sucrose impart an unwanted sweetness to the tubers. As a compromise, potatoes are stored at 7-9°C, which prevents the breakdown of starch while minimizing respiration and germination. Other factors that can affect plant respiration include inorganic ions and injury.

Addition of ions to plants previously grown in distilled water stimulates respiration—a phenomenon called salt respiration. The simplest explanation for this effect is that respiratory-linked ATP is required to support the enhanced ion uptake, but other factors also seem to be involved. Physical injury to higher plant tissues often stimulates oxygen uptake because of increases in respiration (mitochondria-linked oxygen uptake) and in nonmitochondrial activities (e.g., lipoxygenase, polyphenoli oxidase, peroxidase).

## Lipid Metabolism

Starch and sucrose are not the only substrates available for energy production in the plant. Fats and oils are important storage forms of reduced carbon in many seeds, including those of agriculturally important species such as soybean, sunflower, peanut, and cotton. Oils often serve a major storage function in nondomesticated plants that produce small seeds. Some fruits, such as olives and avocados, also store fats and oils. Fats and oils belong to the general class of *lipids,* a structurally diverse group of hydrophobic compounds' that are soluble in organic solvents and highly insoluble in water. They represent a more reduced form of carbon than carbohydrates, so the complete oxidation of one gram of fat or oil (~40 kJ or 9.3 kcal) can produce considerably more ATP than the oxidation of one gram of starch (~15.9 kJ or 3.79 kcal). Conversely, the biosynthesis of fats, oils, and related molecules such as the phospholipids of membranes requires a correspondingly large investment of metabolic energy.

Other lipids are important for plant structure and function but are not used for energy storage. These include waxes, which make up the protective cuticle that reduces water loss from exposed plant tissues, and terpenoids (also known as isoprenoids), a family of compounds derived from the condensation of successive five-carbon isoprene units. Important terpenoids include the carotenoids involved in photosynthesis and the sterols, which are present in many plant membranes.

## Triglycerides

Fats and oils exist mainly in the form of triglycerides, or triacylglycerols, in which fatty acid molecules are linked by ester bonds to the three hydroxyl groups of glycerol. The fatty acids in plants are usually straightchain carboxylic acids having an even number of carbon atoms. The carbon chain lengths can be as short as 12 and as long as 20 but are more commonly 16 and 18. Oils ; liquid at room temperature, primarily due to the presence of a number of unsaturated bonds in their component fatty acids, whereas fats are solid, having a higher* proportion of saturated fatty acids. The major fatty acids appearing in plant lipids are shown in Table 7.2.

The percent composition of fatty acids in plant lipids varies. With the speciesy. For example, peanut oil is about 9% palmitate, 59% oleate, and 21% linoleic acid, while cottonseed oil is 20% palmitate, 30% oleate, and 45%, linoleic acid Triacylglycerols in most seeds are stored in the cytoplasm of either cotyledon or endosperm cells in organelles known as *spherosomes* (also called lipid bodies and oleosomes). Spherosomes have an unusual membrane barrier that separates the triglycerides from the aqueous cytoplasm.

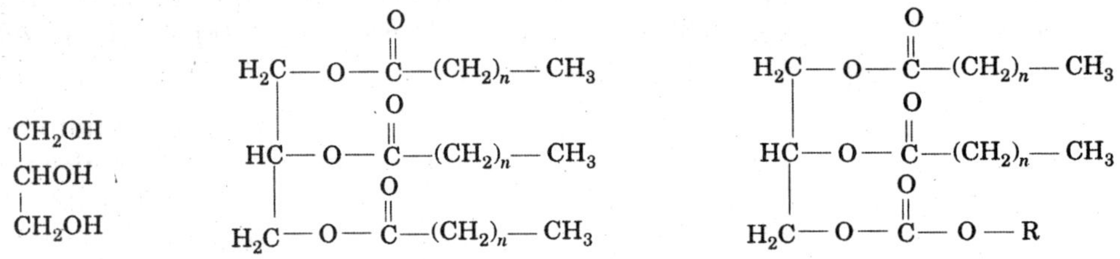

$CH_2OH$ — $CHOH$ — $CH_2OH$

**Glycerol**

$H_2C—O—C(=O)—(CH_2)_n—CH_3$
$HC—O—C(=O)—(CH_2)_n—CH_3$
$H_2C—O—C(=O)—(CH_2)_n—CH_3$

**Triglyceride**

$H_2C—O—C(=O)—(CH_2)_n—CH_3$
$HC—O—C(=O)—(CH_2)_n—CH_3$
$H_2C—O—C(=O)—O—R$

**Diglycerides**

| R = | |
|---|---|
| $R = —H$ | **Dlacylglycerol (DAG)** |
| $= —HPO_3^{-2}$ | **Phosphatidic acid (PA)** |
| $= —PO_3—CH_2—CH_2—N(CH_3)_2$ | **Phosphatidylcholine (lecithin)** |
| $= —PO_3—CH_2—CH_2—NH_2$ | **Phosphatidylethanolamine** |
| = —galactose(s) | **Galactolipids** |

**Fig. 7.10. The structural features of triacylglycerols and amphipathic lipids in higher plants. The carbon chain lengths of the fatty acids, which always have even numbers of carbons, range from 12-20 but are typically 16-18. Thus, the value of *n* is usually 14-16.**

**Table 7.2. Common Fatty Acids in Higher Plant Tissues.**

| *Name* | *Structure* |
|---|---|
| ***Saturated fatty acids*** | |
| Lauric acid (12:0)* | $CH_3(CH_2)_{10}CO_2H$ |
| Myristic acid (14:0) | $CH_3(CH_2)_{12}CO_2H$ |
| Palmitic acid (16:0) | $CH_3(CH_2)_{14}CO_2H$ |
| Stearic acid (18:0) | $CH_3(CH_2)_{16}.CO_2H$ |
| ***Unsaturated fatty acids*** | |
| Oleic acid (18:1) | $CH_3(CH_2)_7CH=CH(CH_2)_7CO_2H$ |
| Linoleic acid (18:2) | $CH_3(CH_2)_4CH=CH—CH_2—CH=CH(CH_2)_7CO_2H$ |
| Linolenic acid (18:3) | $CH_3CH_2CH=CH—CH_2—CH=CH—CH_2—CH=CH—(CH_2)_7CO_2H$ |
| Arachidonic acid (20:4) | $CH_3(CH_2)_3(CH_2—CH=CH)_4(CH_2)_3CO_2H$ |

* Each fatty acid has a numerical abbreviation. The number before the colon represents the total number of carbons, while the number after the colon is the number of double bonds.

A single layer of phospholipids (i.e., a half belayes) surrounds the spherosome with the hydrophilic ends of the phospholipids exposed to the cytosol and the hydrophobic acyl side chains facing the triacylglycerol interior. Several proteins are also located within the phospholipid half bilayer. This unique structure appears to result from the pattern of triglyceride biosynthesis. Triglyceride synthesis is completed by enzymes in the endoplasmic reticulum (ER), and the resulting fats accumulate between the two monolayers of the ER membrane bilayer.

The bilayer swells apart as more fats are put into the growing structure, and ultimately a mature spherosome buds off from the ER.

## Triacylglycerol Biosynthesis

Triacylglycerol biosynthesis is a multicompartment process, and the biochemical pathway in plants is similar to that in animals. Saturated fatty acids are synthesized by the repeated addition of two-carbon units derived from acetyl CoA. However, the acetyl CoA is first carboxylated to malonyl CoA, which serves as the two-carbon donor. The reaction series involves several enzymes, including acetyl-CoA carboxylase and a fatty acyl synthetase complex. Throughout the biosynthetic process, the growing fatty acid remains covalently attached to a low-molecular-weight protein known as an *acyl carrier protein* (ACP). The biosynthesis of fatty acids is not cheap energetically; one molecule of ATP and two molecules of NADPH are required for the addition of each two-carbon unit. In animals, fatty acid biosynthesis takes place in the cytosol, but in plants it is localized in chloroplasts in leaves and in proplastids in nongreen tissue.

The initial product of this series of reactions is the $C_{16}$ saturated fatty acid palmitate (16:0) A two-carbon elongation to stearate (18:0) also takes place in the plastid (by a mechanism that does not involve the fatty acid synthetase complex), as does the initial desaturation of stearate to oleate (18:1). Further modifications of the fatty acids (e.g., desaturation of oleate to linoleate) take place in the ER following transfer of the fatty acids from the plastid to the ER as a fatty acyl CoA intermediate. The glycerol moiety of triacylglycerols is derived from the reduction of dihydroxyacetone phosphate to glycerol 3-phosphate in the cytosol. Three fatty acyl CoAs are subsequently esterified to the glycerol by enzymes in the ER membrane, leading to the buildup of fats in the growing spherosome.

## Phospholipid Biosynthesis

The lipids that make up plant membranes are generally *amphipathic* molecules composed of a hydrophobic diacylglycerol moiety connected to one of several types of hydrophilic head groups. The nature of the head group varies from one membrane to the next. In mitochondria and most other plant membranes, phospholipids Predominate, whereas in chloroplasts the thylakoid membranes are largely composed of galactolipids. The biosynthesis of membrane lipids is complex, involves several different cell compartments, and is not Icompletely understood in most cases (Moore, 1982). Here we will give a brief overview of the facts that seem most well established. Two routes of phospholipid biosynthesis are recognized within the ER, both of which share a common pathway initially. These are presented in Fig. 7.13.

At one time, it was thought that the addition of head groups in phospholipid biosynthesis took place exclusively in the ER. Now there is evidence that some phospholipid biosynthesis can also occur in other cell compartments. dost notably, mitochondria contain the enzymes necesary for converting diacylglycerol to phosphatidylcholine I and phosphatidylethanokimine, and plant mitochondria are capable of synthesizing phosphatidic acid, CDP-diacylglycerol, and phosphatidylglycerol. Further, mitochondria appear to be the sole site of synthesis of a unique mitochondrial inner membrane lipid, cardiolipin (diphosphatidylglycerol).

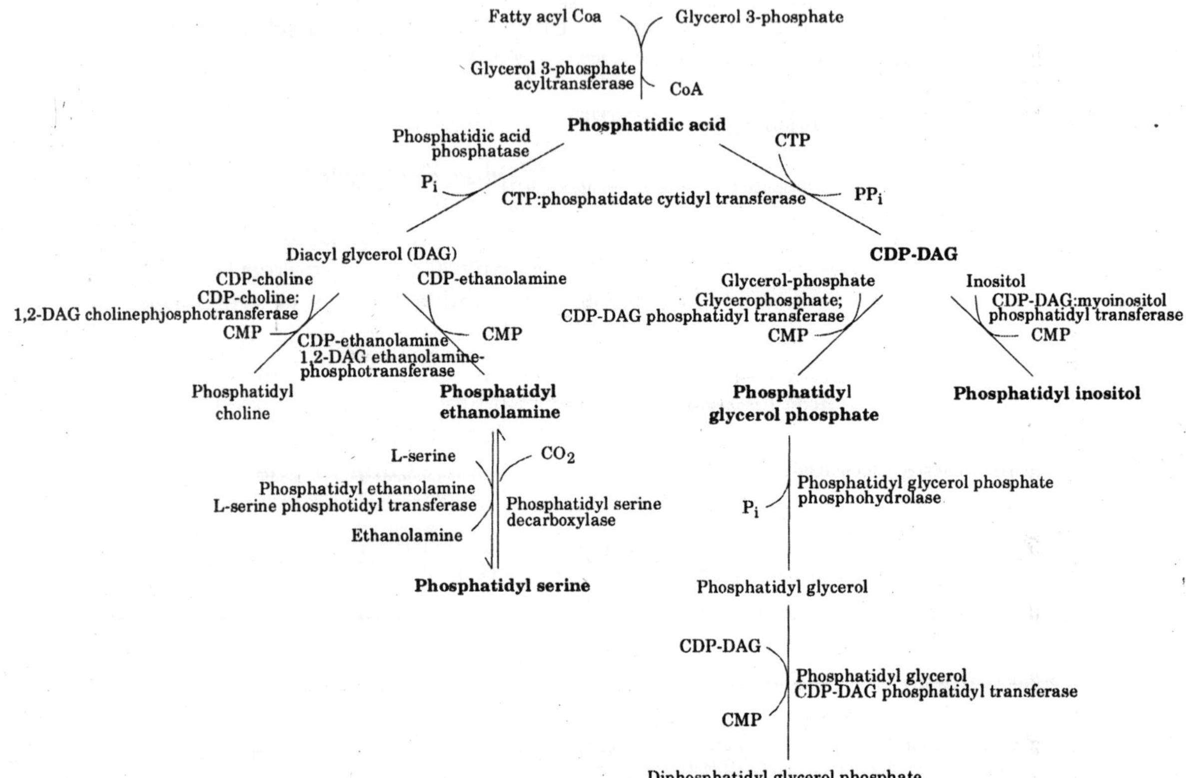


**Fig. 7.11. Phosphatidic acid is synthesized from fatty acyl CoA and glycerol 3-phosphate in the endoplasmic reticulum membrane as shown in Figure 7.12. Subsequent metabolism involves either the hydrolysis of phosphatidic acid to diacylglycerol (DAG) or reaction with CTP to give CDP-diacylglycerol. Diacylglycerol leads to the formation of phosphatidylcholine and phosphatidylethanolamine by reacting with the appropriate CDP adduct, and CDP-diacylglycerol reacts with inositol to produce phosphatidylinositol. Diphosphatidylglycerol (cardiolipin) is also synthesized from CDP-diacylglycerol, and phosphatidylserine is produced from phosphatidylethanolamine.**

The extent to which mitochondrial membranes import lipids from the ER during organelle biogenesis is not known, nor is the mechanism of exchange between the ER and the outer and inner mitochondrial membranes. Phospholipids in the Golgi complex, plasma membrane, and nuclear envelope (all components of the endomembrane system that includes the ER) are probably derived from phospholipids initially synthesized in the ER and transferred to the appropriate membrane by vesiculation of specific regions of the ER. This ER-vesiculation model also explains most of the observed features of microbody (peroxisome, glyoxysome) membrane biogenesis. The synthesis of diacylglycerol and galactolipids for the chloroplast thylakoid membranes has been localized in spinach to a group of enzymes within the inner membrane of the chloroplast envelope.

Synthesis of galactolipids is carried out by an enzyme that transfers the galactosyl moiety from UDP-galactose to diacylglycerol to form monogalactosyldiacylglycerol. Formation of digalactosyldiacylglycerol involves reaction with a second molecule of UDP-galactose.

## Conversion of Lipids into Carbohydrates

Metabolism of stored fats by oil-containing seeds involves the conversion of lipids to carbohydrates following germination by a process involving several different cellular sites Plants are not able to transport fats from the endosperm to the root and shoot tissues of the germinating seedling, so they must convert stored lipids to a more mobile form of carbon, generally sucrose. The process of converting lipid to sugar in oil seeds is triggered by germination and begins with the hydrolysis of triglycerides stored in the spherosomes to free fatty acids, followed by oxidative breakdown of the fatty acids to produce acetyl CoA.

The latter reaction takes place in a type microbody called glyoxysomes, organelles bounded by a single membrane that are found in the oil-rich storage tissues of seeds. The discovery of glyoxysomes in castor beans and much of our understanding of glyoxysomal function are due to the work of Harry Beevers and his colleagues at Purdue University and the University of California, Santa Cruz. Acetyl CoA is metabolized in the glyoxysome via the glyoxylate cycle to produce succinate, which is transported from the glyoxysome to the mitochondria, where it is further metabolized to oxaloacetate.

The process ends in the cytosol with the conversion of oxaloacetate to glucose via gluconeogenesis. Although some of this carbon is diverted to other metabolic reactions in certain oil seeds, in castor bean the process is so efficient that each gram of lipid metabolized results in the formation of one gram of carbohydrate, equivalent to 40% recovery of free energy in the form of carbon bonds. The initial step in the conversion of lipids to carbohydrate involves the breakdown of triglycerides stored in the spherosomes by the enzyme lipase, which, at least in castor bean, is located on the half membrane that serves as the spherosome outer boundary.

The lipase hydrolyzes triglycerides to three molecules of fatty acid and glycerol. Corn and cotton also contain a lipase activity in the spherosome, but peanut, soybean, and cucumber show lipase activity in the glyoxysome instead. During the breakdown of lipids, there is generally a close physical association between spherosomes and glyoxysomes. Following hydrolysis of the triglycerides, the resulting fatty acids enter the glyoxysome, where they are activated to fatty acyl CoA by the enzyme fatty-acyl-CoA synthase. Fatty acyl CoA is the initial substrate for the β-oxidation series of reactions in which $C_n$ fatty acids are sequentially broken down to $n/2$ molecules of acetyl CoA. This reaction sequence is similar to that associated with the breakdown of fatty acids in animal tissues and involves the reduction of $\frac{1}{2}O_2$ to $H_2O$ and the formation of one NADH for each acetyl CoA producecd.

In mammalian tissues, the four enzyme activities associated with β-oxidation appear in the mitochondrion, whereas in plant seed storage tissues they are exclusively localized in the glyoxysome. Interesting, in plant vegetative tissues (e.g., mung bean hypocotyl, potato tuber), the β-oxidation reactions are localized in a related organelle, the peroxisome, rather than in the mitochondria suggesting that the limited breakdown of lipids that does occur in plant nonstorage tissues takes place in peroxisomes and not mitochondria. The acetyl CoA produced by β-oxidation is further metabolized in the glyoxysomes through a series of reactions that make up the *glyoxylate cycle*.

Initially, the acetyl CoA reacts with oxaloacetate to give citrate, which in turn undergoes isomerization to isocitrate. Both of the enzymes involved here (citrate synthase and aconitase) catalyze the same reactions as in the TCA cycle in the mitochondria. The next two reactions are unique to the glyoxylate pathway. First, isocitrate ($C_6$) is cleaved by the enzyme isocitrate lyaseto give succinate ($C_4$) and glyoxylate ($C_2$). Next, malate synthase combines a second molecule of acetyl CoA with glyoxylate to produce malate. According to the classical view of the operation of the glyoxylate cycle, the malate is then oxidized by malate dehydrogenase to oxaloacetate, which can combine with another acetyl CoA to continue the cycle. The function of the glyoxylate cycle is, in essene, to convert two molecules of acetyl CoA to succinate.

The succinate then moves to the mitochondria, where it is converted to malate by the normal TCA cycle reactions. The resulting malate can be moved out of the mitochondria in exchange for the incoming succinate via the dicarboxylate transporter. The malate is then oxidized to oxaloacetate by cytosolic malate dehydrogenase, and the resulting oxaloacetate is converted to carbohydrate. This conversion requires circumventing the irreversibility of the pyruvate kinase reaction and is facilitated by the enzyme PEP carboxykinase, which utilizes the phosphorylating ability of ATP to convert oxaloacetate to PEP and $CO_2$. (Fig. 7.14) From PEP, gluconeogendsis can proceed to the production of glucose. Sucrose is the ultimate product of this process, being the primary form in which reduced carbon is translocated from the cotyledons to the growing seedling tissaes. Although the pathway for the mobilization of triglycrides has been best characterized in castor bean (Ricinus *communis*), it seems to be similar in the storage tissues of other oil seeds.

However, not all seeds carry out a quantitative conversion of fat to sugar. In castor bean, the endosperm degenerates after the fat and protein reserves are fully utilized. In many oil seeds, such as sunflower (*Helianthus annus*), cotton (*Gossipium hirsutum*), and members of the squash family (*Cucurbits*), the cotyledons differentiate into actively photosynthesizing organs after the food reserves are used up. In these tissues only part of the stored lipid is converted to exported carbohydrate. Much of the lipid-derived carbon remains in the cotyledons, where it contributes to the synthesis of chloroplasts and other cellular structures. As the greening process takes place, there is a transition in the microbody population of these cells that shows fewer of the characteristics of glyoxysomes and more of those of leaftype peroxisomes. Such a transition is in keeping with the decreased requirement for the breakdown of stored lipids and the increased need to metabolize the products of photorespiration as the tissue goes from a heterotrophic to a more autotrophic mode of metabolism.

## Concluding Remarks

Higher plant respiration couples the complete oxidation of reduced cellular carbon generated during photosynthesis to the synthesis of ATP. Respiration takes place in three stages: glycolysis, the tricarboxylic acid cycle, and the electron transport chain. In glycolysis, carbohydrate is converted in the cytosol to pyruvate with the synthesis of a small amount of ATP via substrate-level phosphorylation. Pyruvate is subsequently oxidized within the mitochondrial matrix through the TCA cycle, generating a large number of reducing equivalents in the form of NADH and $FADH_2$. In the third stage, NADH and $FADH_2$ are oxidized by the electron transport chain which is associated with the inner mitochondrial membrane. The free energy released durirg electron transport is used to synthesize a large amount of ATP in a process known as oxidative phosphorylation. ATP synthesis is accomplished by the generation of a proton gradients across the inner mitochondrial membrane during electron transfer.

The $F_oF_1$-ATP synthase complex then couples the break-down of the proton gradient to the conversion. of ADP and $P_i$ to ATP. The regulation of substrate oxidation during respiration involves control points at glycolysis, the TCA cycle, and the electron transport chain. Aerobic respiration in higher plants has several unique features, including the presence of a cyanide-resistant oxidative pathway. Carbohydrates can also be oxidized via the pentose phosphate pathway, in which the reducing power generated produces NADPH for biosynthetic purposes.

Numerous glycolytic and TCA cycle reactions also provide the starting points for different biosynthetic pathways. Many factors can affect the respiration rate observe at the whole-plant level. These include the nature and age of the plant tissue and such environmental factors as the external oxygen concentration and temperature. Because respiration rates contribute to the overall net carbon' balance of a plant, variations in whole-plant respiration rates can affect final agronomic yields. Lipids play a major role in higher plants; amphipathic lipids serve as the primary nonprotein components of plant membranes, while triglycerides (fats and oils) are an efficient storage form of reduced carbon, particularly in seeds.

Triglycerides are stored in subcellular organelles known as spherosomes. Synthesis of the long-chain fatty acids that make up triglycerides takes place in the plastids of higher plants, and the conversion of the fatty acids to triglycerides occurs on the membranes of the endoplasmic reticulum. During germination seeds that store triglycerides, the stored lipids are metabolized to carbohydrates in a series of reactions the involve a metabolic sequence known as the glyoxylate cycle. This cycle takes place in organelles called glyoxysomes. The reduced carbon generated during lipid breakdown in the glyoxysomes is ultimately converted 1 carbohydrate in the cytosol by the process known gluconeogenesis.

## Does Respiration Reduce Crop Yields?

Plant respiration can consume an appreciable amount of the carbon fixed each day during photosynthesis over and above the losses due to photorespiration. To what extent can changes in a plant's respiratory metabolsim affect crop yields? Attempts to establish a quantitative relation between respiratory energy metabolism and the various processes going on in the cell have led to a splitting of respiration into two components (Lambers, 1985). *Growth respiration* involves the processing of reduced carbon to bring about the growth of new plant matter. *Maintenace respiration* is the component of respiration needed to keep existing, mature cells in a viable state. Utilization of energy by the latter is not well understood, but estimates indicate that it can represent more than 50% of the total respiratory flux. on top of all this is the specter of the cyanide-resistant alternative pathway, with its potential for utilizing considerable amounts of the cell's reduced carbon to no apparent useful end. Estimates of the alternative pathway in wheat roots alone suggest a loss of carbon equivalent to 6% of the final grain yield.

Although numerous questions remain regarding the issues cited above, there are several examples of empirical relations between plant respiration rates and crop yield. In the forage crop parennial ryegrass (*Lolium perenne*), yield increases of 10 – 20% were correlated with a 20% decrease in the leaf respiration have been found for other plants, including corn and tall fescue (Lambers, 1985). Although there is a potential for increasing crop yield through reduction of respiration rates, a better understanding of the sites and mechanisms that control plant respiration is needed before such changes can be exploited in a logical fashion by plant physiologists, geneticists, and molecular biologists. Much remains to be established regarding the general applicability of such observations and the conditions under which slower-respiring lines might actually be at a disadvantage.

# CHAPTER 8

# PURIFICATION OF PROTEINS

*Proteins can be isolated and characterized according to their size, shape, and charge.*

In the previous three chapters we described the structures of amino acids and proteins, and in two cases we examined how these structures relate to their function. Some of the methods for structure determination were also discussed. To analyze the structure of a protein we must isolate it from the complex mixture in which it exists in whole cells. The primary object of this chapter is to acquaint you with techniques used for protein purification. Because these procedures arc often used for protein characterization as well, they will add to your repertoire of methods for protein characterization. In the first part of this chapter methods for protein fractionation are discussed in isolation.

Success in protein purification depends on picking a number of procedures and combining them in an effective order. Two examples are given in which trains of procedures are combined to purify specific proteins from crude whole-cell extracts.

## Methods of Protein Fractionation

Many more techniques For protein analysis exist than can be covered in a single chapter. The emphasis here is on some of the more popular procedures and the principles involved in their use.

### *Differential Centrifugation*

A typical crude broken-cell preparation contains disrupted cell membranes, cellular organelles, and a large number of soluble proteins, all dispersed in an aqueous buffered solution. The membranes and the organelles can usually be separated from one another and from the soluble proteins by *differential centrifugation*. Differential centrifugation divides a sample into two fractions: the pelleted fraction, or sediment, and the supernatant fraction, that is, the fraction that is not sedimented. The two fractions may then be separated by decantation. According to its purpose, differential centrifugation involves the use of different speeds and different times of centrifugation.

For example,if the protein of interest is in the mitochondrial fraction, the crude cell lysate is centrifuged first at 4,000 × *g* for 10 min to remove cell membranes, nuclei, and (in the case of plant material) chloroplasts. The supernatant from this step contains, among other elements, the mitochondria, and is decanted and recentrifuged at 15,000 × *g* for 20 min to obtain a sediment primarily containing mitochondria. If ribosomes instead of mitochondria are the goal, then the crude lysate is centrifuged at 30,000 × *g* for 30 min, and the resulting supernatant is decanted and centrifuged at 100,000 × *g* for 180 min to obtain a ribosomal sediment. If obtaining the soluble protein fraction is the goal, then the entire lysate is centrifuged at 100,000 × *g* for 180 min, and the resulting supernatant, containing the soluble protein, is carefully decanted for further processing.

## Differential Precipitation

Once a crude extract of protein has been made, it is common to separate this mixture into different fractions by a precipitation step. The solubility of a protein reflects a delicate balance between different energetic interactions—both internally within the protein and between the protein and the surrounding solvent. Consequently, the choice of solvent can affect both the solubility and the structure of a protein.

**Table 8.1. Sedimentation Conditions for Different Cellular Fractions.**

| *Fraction Sedimented* | *Centrifugal Force (x, g)* | *Time (min)* |
|---|---|---|
| Cells (eukaryotic) | 1,000 | 5 |
| Chloroplasts; cell membranes; nuclei | 4,000 | 10 |
| Mitochondria; bacteria cells | 15,000 | 20 |
| Lysosomes; bacterial membranes | 30,000 | 30 |
| Ribosomes | 100,000 | 180 |

## A Minimum of Solubility Occurs at the Isoelectric Point

Proteins typically have charged amino acid side chains on their surfaces that undergo energetically favourable polar interactions with the surrounding water. The total charge on the protein is the sum of the side-chain charges. However, the actual charge on the weakly acidic and basic side-chain groups also depends on the solution pH. In fact, the acidic and basic groups within the protein can be titrated just like free amino acids to determine their number and their pK values.

Proteins tend to show a minimum solubility at their isoelectric pH — a fact that is apparent for β-lactoglobulin. The decrease in sloubility at the isoelectric pH reflects the fact that the individual protein molecules, which all have similar charges at pH values away from their isoelectirc points, cease to repel each other. Instead, they coalesce into insoluble aggregates.

## Salting In and Salting Out

Proteins also show a variation in solubility that depends on the concentration of salts in the solution. These frequently complex effects. may involve specific interactions between charged side chains and solution ions or, particularly at high salt concentrations, may reflect more comprehensive changes in the solvent properties. For the effect of salt concentration on the solubility of β-lactoglobulin, again see figure 8.1 Most globulins are sparingly soluble in pure water. The increase in solubility that occurs after adding salts such as sodium chloride is often referred to as salting in. The effects of four different salts on the solubility of hemoglobin at pH 7 can be seen in figure 8.2.

All four salts produce the salting-in effect with this protein, two of them, sodium sulfate and ammonium sulfate, also produce a greatly decreased solubility of the protein at high salt concentrations. This result is called *salting out* and occurs with salts that effectively compete with the protein for available water molecules. At high salt concentrations, the protein molecules tend to associate with each other because protein-protein interactions become energetically more favourable than protein-solvent interactions. Each protein has a characteristic salting-out point, a fact we can exploit to make protein separations in crude extracts.

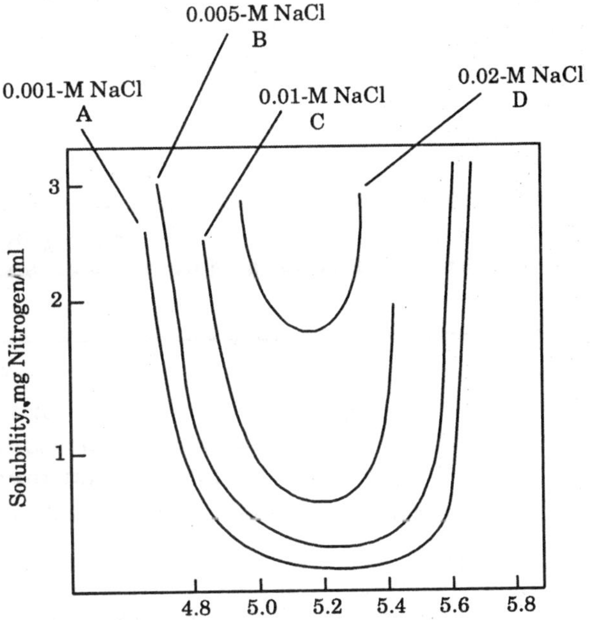

**Fig. 8.1. Solubility of β-lactoglobulin as a function of pH and ionic strength. The isoelectric pH (pI) for this protein is about 5.2, which corresponds to the point of minimum solubility.**

For this purpose $(NH_4)_2SO_4$ is the most commonly used salt because it is very soluble and is generally effective at lower concentrations than many other salts. Typically, the desired protein precipitates over a range of salt concentrations. If it precipitates in the range of 20-30% by weight, we first add ammonium sulfate to a concentration slightly below 20% and then centrifuge to remove by sedimentation any proteins that precipitate in the 0-20% range. To the supernatant from this centrifugation we then add more ammonium sulfate, to 30%. Centrifugation at this point brings down the desired protein, as well as other proteins that precipitate in this range of salt concentrations. The supernatant is discarded and the sediment saved for further purification.

## Column Procedures

Column procedures are the most effective and most varied of purification methods. Common to all column procedures is the use of a glass cylinder with an opening at the top and bottom. The cylinder is filled with a column of hydrated material, and the protein sample is applied to the top of the column. Then further buffer is passed through the column. In most column procedures, proteins bind differentially to the column material, and a change in the elution buffer causes them to be eluted differentially according to their degree of affinity for the column material.

The eluant exiting from the bottom of the column is collected in equal-sized fractions with the help of an automatic fraction collector. Each fraction is analyzed, and fractions containing an appreciable amount of the desired protein are pooled for further purification. Many variations on this basic procedure are in common use:

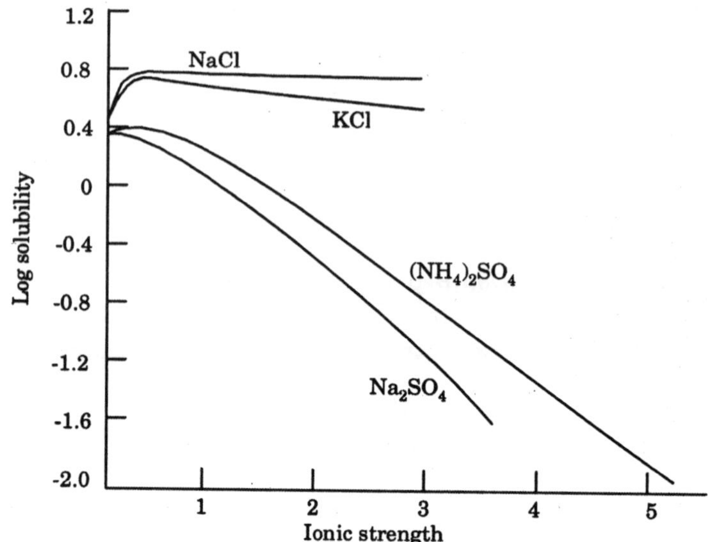


**Fig. 8.2. Solubility of horse carbon monoxide hemoglobin in different salt solutions. The addition of a moderate amount of salt (salting in) is required to solubilize this protein. At high concentrations, certain salts compete more favourably for solvent, decreasing the solubility of the protein and thus leading to its precipitation (salting out).**

1. In *gel-exclusion chromatography* a cross-linked dextran without any special attached functional groups is used for the column substrate. Large molecules flow more rapidly through this type of column than small ones. The dextrans have different degrees of cross-linking, making them effective over different size ranges.
2. *Ion-exchange chromatography* makes use of the fact that proteins differ enormously in their affinity for positively or negatively charged columns. However, the cross-linked resins used in amino acid fractionation are rarely used for protein separations because

**Table 8.2. Some Column Materials for Ion-Exchange Chromatography of Proteins**

| *Matrix* | *Functional Groups on Column* |
|---|---|
| Phosphocellulose (PC) | $—PO_3^-$ |
| Carboxymethyl cellulose (CMC) | $—CH_2—COO^-$ |
| Diethylaminoethyl cellulose (DEAE) | $—(CH_2)_2—\overset{+}{N}H(CH_2—CH_3)_2$ |

proteins are too large to penetrate the resin beads. Instead, finely divided celluloses containing either positively or negatively charged groups are most commonly used. The affinity of a protein for the column material is proportional to the salt concentration required to release the protein from the material. Typically, a column is loaded with protein solution at a low ionic strength so that most of the proteins bind to the column. After loading, elution is initiated by gradually increasing the salt concentration of the elution buffer. Proteins are eluted in the order of increasing affinity.

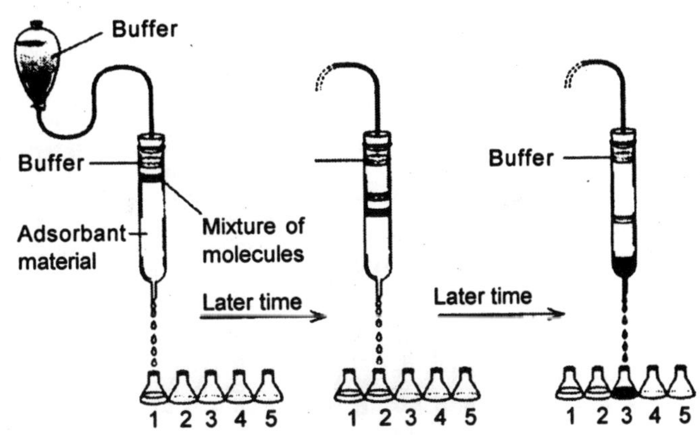


**Fig. 8.3. Collecting fractions during column chromatography. Column material and elution procedure are chosen to effect optimal separation of the desired protein.**

3. Finely divided celluloses may also be used in the column in conjunction with attached hydrophobic groups such as octyl alcohol. In this case, it is proteins with exposed hydrophobic centers that bind to the column with varying affinities. These proteins may be eluted in order of decreasing affinity for the column by increasing the level of free octyl alcohol in the eluting buffer.
4. *Affinity chromatography* makes use of chemical groups that have a special binding affinity for the proteins that are being sought. For example, many enzymes bind preferentially to cofactors, such as adenosine triphosphate or pyridine nucleotides. Often, the separation of such enzymes from other proteins can be achieved on a column that has one of these cofactors attached. As the mixture passes through the column, proteins that have specific binding affinities for the cofactor bind to the column, and other proteins pass through. The cofactor-binding protein can then be eluted with a solution containing the same soluble cofactor.
5. *High-performance liquid chromatography* (HPLC) is not so much a new type of chromatography as a new way of applying old chromatographic techniques, using extremely high pressures. The same principles are involved, but column materials usually consist of more finely divided particles made of physically stronger materials, which can withstand pressures of 5,000-10,000 psi without changing their structure. The column apparatus itself must also be designed to withstand high pressures. The main advantage of HPLC is that superior resolution of eluted substances can be achieved.

## *Gel Electrophoresis*

Gel electrophoresis is the best way to analyze mixtures and assess purity. Gel electrophoresis separates proteins according to their size and their charge. It is almost always performed in aqueous solution supported by a gel system. The gel is a loosely cross-linked network that functions to stabilize the protein boundaries between the protein and the solvent, both during and after electrophoresis, so that they may be stained or otherwise manipulated. The most popular method of electrophoretic separation by gels employs sodium dodecyl sulfate (SDS), This method not only gives an index of protein purity but yields an estimate of the protein subunit molecular weights.

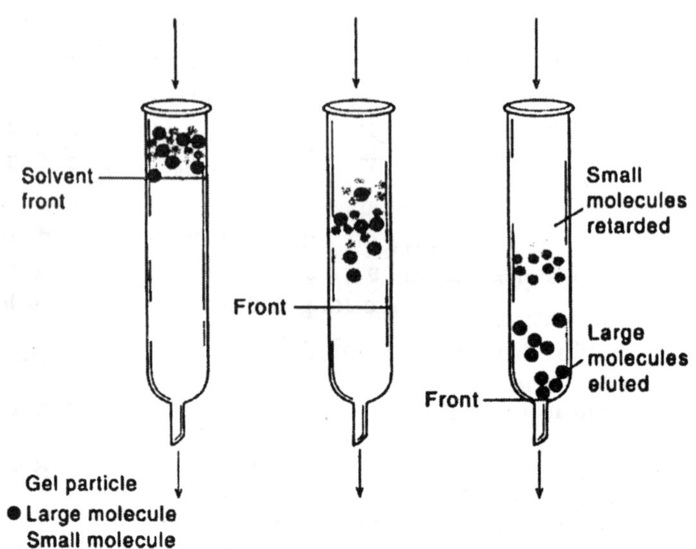


**Fig. 8.4. Polydextran column showing separation of small and large molecules. The column material is immersed in solvent, which penetrates the gel particles. A separation is initiated by layering a small sample containing different-sized proteins on the top of the column. This sample is pushed through the column by opening the stopcock at the bottom and adding further solvent at the top to keep up with the flow. As shown, the small protein molecules can penetrate the gel particles but the big ones cannot. Therefore the big proteins move through the column much more rapidly, and a separation of the two proteins results.**

The mixture of proteins to be characterized is first completely denatured by adding SDS (a detergent) and mercapthemanol and by briefly heating the mixture. Denaturation is caused by the association of the apolar tails of the SDS molecules with protein hydrophobic groups. Any cystine disulfide bonds are cleaved by a disulfide interchange reaction with mercaptoethanol. The resulting unfolded polypeptide chains have relatively large numbers of SDS molecules bound to them. The success of the technique for molecular weight estimation depends on two facts: (1) Each bound SDS molecule contributes one negative charge to the denatured protein complex, so that the charge of the protein in its native state is effectively masked by the more numerous charged groups of the associated detergent molecules. (2) The total number of detergent molecules bound is proportional to the polypeptide chain length or, equivalently, the protein's molecular weght.

As a result, the SDS-denatured protein molecules acquire net negative charges that are approximately proportional to their molecular weights. Another electrophoretic method, called *isoelectric focusing*, is frequently used for characterizing proteins based on differences in their isoelectric points. The apparatus usually consists of a narrow tube containing a gel and a mixture of *ampholytes,* which are small molecules with positive and negative charges.

The ampholytes have a wide range of isoelectric points, and are allowed to distribute in the column under the influence of an electric field. This step creates a pH gradient from one end of the gel to the other, as each particular ampholyte comes to rest at a position coincident with its isoelectric point. At this stage, a solution of proteins is introduced into the gel. The proteins migrate in the electric field until each reaches a point at which the pH resulting from the ampholyte gradient exactly equals its own isoelectrie point.

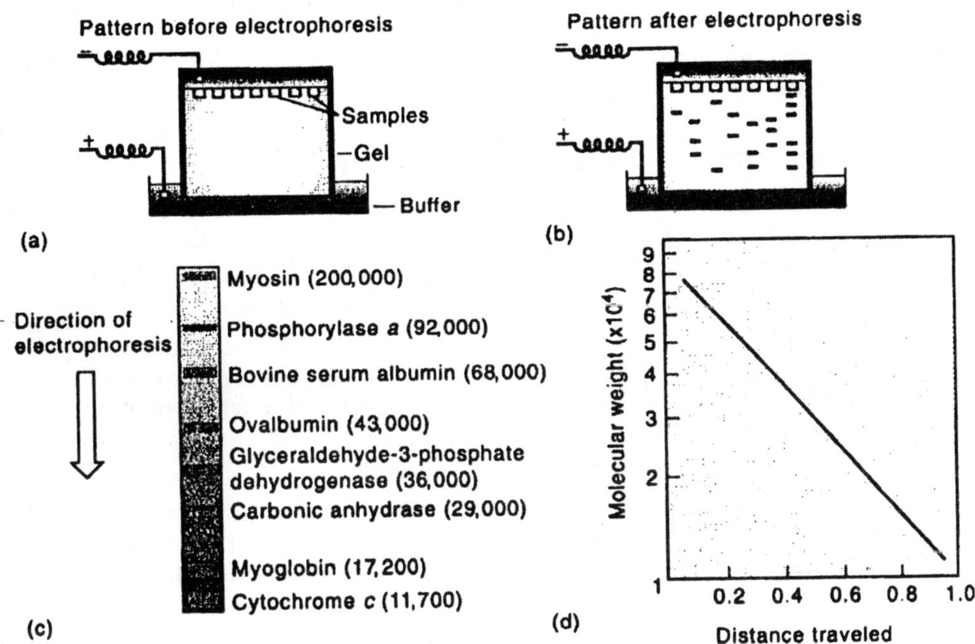


**Fig. 8.5. Gel electrophoresis for analyzing and sizing proteins, *(a)* Apparatus for slab-gel electrophoresis. Samples are layered in the little slots cut in the top of the gel slab. Buffer is carefully layered over the samples, and a voltage is applied to the gel for a period of usually 1-4 h. *(b)* After this time the proteins have moved into the gel at a distance proportional to their electrophoretic mobility. The pattern shown indicates that different samples were layered in each slot. (c) Results obtained when a mixture of proteins was layered at the top of the gel in phosphate buffer, pH 7.2, containing 0.2% SDS. After electrophoresis the gel was removed from the apparatus and stained with Coomassie Blue. The protein and its molecular weight are indicated next to each of the stained bands, *(d)* The logarithm of the molecular weight against the mobility (distance traveled) shows an approximately linear relationship.**

Isoelectric focusing provides a way of both accurately determining a protein's isoelectric point and effecting separations among proteins, the isoelectric points of which may differ by as little as a few hundredths of a pH unit. An elegant extension of the electrophoretic method of separation involves combining isoelectric focusing with SDS gel electrophoresis to produce a two-dimensional electrophoretogram. This technique is most valuable for the analysis of very complex mixtures. First, the sample is run in a one-dimensional pH gradient gel (isoelectric focusing).

The resulting narrow strip of gel, containing the partially separated mixture of proteins, is placed alongside a square slab of SDS gel. An electrical field is imposed so that the sample moves at right angles to its motion in the first gel. Fig. 8.6 shows the separation of total *E. coli* protein into more than 1,000 different components. Although electrophoresis is the method of choice for assessing protein purity, it is not frequently used as a purification step, at least not yet. This is because of the small amounts that can usually be analyzed and because of the denaturing conditions that are often used in electrophoretic analysis.

## Sedimentation and Diffusion

We have seen that sedimentation by centrifugation is used for protein purification. The methods of centrifugation can be used quantitatively for the assessment of protein size and shape. In a centrifugal force field, protein molecules slowly migrate toward the bottom of a centrifuge tube at a rate that is proportional to their molecular weight.

The rate of sedimentation may be recorded by optical methods that do not interfere with the operation of the centrifuge. From this rate, we can obtain the sedimentation constant s. This constant equals the rate at which a molecule sediments, divided by the gravitational field (angular acceleration in a spinning rotor), as defined by the equation

$$s = \frac{dx/dt}{\omega^2 x} \qquad \text{...(8.1)}$$

where $dx/dt$ is the rate at which the particle travels at distance $x$ from the center of rotation, $\omega$ is the angular velocity of the rotor in radians per second (hence $\omega^2 x$ is the angular acceleration), and $t$ is the time of centrifugation in seconds.

The sedimentation constant is usually given in Svedberg units (S); one S = $10^{-13}$s. From the sedimentation constant we can obtain the molecular weight, provided we have certain other information. This additional information includes the frictional coefficient ($f$) of the protein

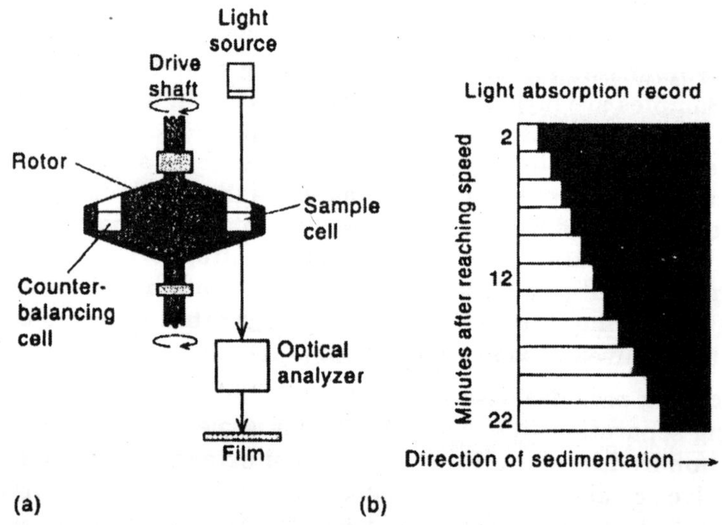


**Fig. 8.6. Apparatus for analytical ultracentrifugation. *(a)* The centrifuge rotor and method of making optical measurements. *(b)* The optical recordings as a function of centrifugation time. As the light-absorbing molecule sediments, the solution becomes transparent.**

and the density of the protein. The coefficient $f$ is bigger for larger proteins and, for proteins of the same molecular weight, it is larger for elongated, rodlike molecules. The density of the protein is important because of the buoyancy factor, $1-\bar{v}_p P_s$, which takes into account the density difference between solvent ($\rho_s$) and the volume of water displaced per gram of protein ($\bar{v}_p$). The equation that relates $s$, $M$, and $f$ is

**Table 8.3. Physical Constants of Some Proteins**

| *Protein* | *Molecular Weight* | *Diffusion Constant ($D \times 10^7$)* | *Sedimentation Constant (s)* | *pI*[a] *(Isoelectric)* |
|---|---|---|---|---|
| Cytochrome *c* (bovine heart) | 13,370 | 11.4 | 1.17 | 10.6 |
| Myoglobin (horse heart) | 16,900 | 11.3 | 2.04 | 7.0 |
| Chymotrypsinogen (bovine pancreas) | 23,240 | 9.5 | 2.54 | 9.5 |
| β-Lactoglobulin (goat milk) | 37,100 | 7.5 | 2.9 | 5.2 |
| Serum albumin (human) | 68,500 | 6.1 | 4.6 | 4.9 |
| Hemoglobin (human) | 64,500 | 6.9 | 4.5 | 6.9 |
| Catalase (horse liver) | 247,500 | 4.1 | 11.3 | 5.6 |
| Urease (jack bean) | 482,700 | 3.46 | 18.6 | 5.1 |
| Fibrinogen (human) | 339,700 | 1.98 | 7.6 | 5.5 |
| Myosin (cod) | 524,800 | 1.10 | 6.4 | — |
| Tobacco mosaic virus | 40,590,000 | 0.46 | 198 | — |

[a]pI = –log I = the isoelectric point, that is, the pH at which the protein carries no net charge (see figure 8.1).

$$s = \frac{M(1-\bar{v}_p\rho_s)}{Nf} \quad ...(8.2)$$

where $N$ is Avogadro's number. Thus, in order to estimate the molecular weight from the sedimentation constant we must have a means of determining $\bar{v}_p$ and $f$.

For most proteins $\bar{v}_p$ is about 0.75 ml/g, so its value does not present much of a problem. The frictional coefficient, however, is a sensitive function of the shape, varying over a wide range, and we must usually know its value if we need a serious estimate of the molecular weight. The value of $f$ is usually found by working with the diffusion constant $D$, which is related to the frictional constant by

$$D = \frac{RT}{Nf} \text{ or } f = \frac{RT}{ND} \quad ...(8.3)$$

where $R$ is the gas constant and $T$ is the absolute temperature. Substituting this expression for $f$ in equation (2) and transposing leads to

$$M_r = \frac{RTs}{D(1-\bar{v}_p\,\rho_s)} \quad ...(8.4)$$

The diffusion constant D is a function of both molecular weight and shape. It can be measured by observing the spread of an initially sharp boundary between the protein solution and a solvent as the protein diffuses into the solvent layer.

Once we know the value of the diffusion constant, we can combine the information with the sedimentation data and calculate the molecular weight of the protein. Representative sedimentation and diffusion data, together with the calculated molecular weights, are presented in table 8.3. The sedimentation constants are usually larger the greater the molecular weight.

Diffusion constants are usually inversely proportional to the molecular weights. Exceptions arise because of proteins with unusual shapes.

For example, the globular protein urease and the rodlike protein myosin have similar molecular weights. Yet their sedimentation and diffusion constants both differ by about a factor of 3. Rapid diffusion can be a highly desirable property for an enzyme that must pass rapidly from one point to another. Such a protein benefits from a globular shape, which is quite common for enzymes. By contrast, a rodlike protein can be advantageous for creating a cytoskeletal boundary within the cytoplasm. Myosin is used for just such purposes in many cell types.

## Protein Purification Procedures

To characterize proteins, we must usually first isolate the protein under study from the complex mixture of proteins found in the organism. Once we decide to purify a particular protein, we must weigh several factors. For example, how much material is needed? What level of purity is required? The starting material should be readily available and should contain the desired protein in relative abundance. If the protein is part of a larger structure, such as the nucleus, the mitochondria, or the ribosome, then it is advisable to isolate the large structure first from a crude cell extraet. Purification must usually be performed in a series of steps, using different techniques at each step.

Some purification techniques are more useful when handling large amount of meterial, whereas othere work best on small amount. A purification procedure is arranged so that ehe techniques that are best for working with large amount are used during early steps in the overall purification. The suitability of each purification step is evaluated in terms of the amount of purification achieved by that step and the percent recovery of the desired protein. Combining techniques introduces new considerations and new problems. If two purification techniques each give 10-fold enrichment for the desired protein when executed independently on a crude extract, it does not mean that they give 100-fold enrichment when combined. In general, they give somewhat less.

As a rule, purification techniques that combine most effectively usually are based on different properties of the protein. For example, a technique based on size fractionation is more effectively combined with a technique based on charge than with another technique based on size fractionation. Throughout the purification process we must have a convenient means of assaying for the desired protein, so we can know the extent to which it is being enriched relative to the other proteins in the starting material.

In addition, a major concern in protein purification is stability. Once the protein is removed from its normal habitat, it becomes susceptible to a variety of denaturation and degradation reactions. Specific inhibitors are sometimes added to minimize attack by proteases on the desired protein. During purification it is usual to carry out all operations at 5°C or below. This temperature control minimizes protease degradation problems and decreases the chances of denaturation. In their natural habitat, proteins are usually surrounded by other proteins and organic factors. When these are removed or diluted as during purification, the protein becomes surrounded by water on all sides. Proteins react differently to a pure aqueous environment; many are destabilized and rapidly denatured.

A common remedial measure is to add 5% – 20% glycerol to the purification buffer. The organic surface of the glycerol is believed to simulate the environment of the protein in the intact cell. Two other ingredients that are most frequently added to purification buffers are mercaptoethanol and ethylenedia-minetetraacetate (EDTA). The mercaptoethanol inhibits the oxidation of protein — SH groups, and the EDTA chelates divalent cations. Divalent cations, even in trace amounts, can lead to aggregation problems or activate degradative enzymes. The following two examples of purification show how various techniques can be effectively combined to produce purified proteins with minimum effort and loss of activity.

## *Purification of an Enzyme with Two Catalytic Activities*

The last two steps in the biosynthesis of the mononucleotide uridine 5′-monophosphate (UMP) are catalyzed by (1) orotate phosphoribosyltransferase (OPRTase) and (2) orotate 5′-monophosphate decarboxylase (OMPDase).

Mary Ellen Jones and her colleagues set out to purify the enzyme or enzymes involved in these two reactions. Their main goal was to determine whether the two reactions are carried out by one protein or more than one. Their findings indicated that the two reactions were both catalyzed by the same enzyme, consisting of a single polypeptide chain. demonstrate this fact, it was necessary to monitor both enzyme activities at each step in the purification and show that both activities copurified. For this purpose, Jones used specific enzyme assays for both enzyme activities.

All fractions were assayed for both enzymatic activities at each stage of the purification. The main data associated with the purification are summarized in table 8.4. This table indicates the total protein obtained in each step, the number of enzyme units[a] for each enzyme, and the ratio of enzyme units to total protein, called the *specific activity*. In the absence of enzyme inactivation, the *specific activity* should be directly proportional to the enrichment. The percent recovery refers to the amount of enzyme activity in the indicated fraction, as compared with the amount present in fraction 1. This number is usually less than 100%.

The apparent losses may reflect actual losses of enzyme during purification, or they may reflect *inactivation* (usually due to unknown causes) of the enzyme during purification. The nine steps involved in the purification of UMP synthase from starting tissue are outlined in figure 8.8. All steps were carried out at 0-5°C. About 200 g of Ehrlich ascites cells, a mammalian tumor rich in the desired enzymes, was suspended in buffer and processed in a tissue homogenizer, which mechanically breaks down the tissue and the cell membranes. Then EDTA and an — SH reagent were added to this total cell lysate. Solid streptomycin sulfate was also added with stirring. Streptomycin sulfate aggregates nucleic acids so that they may be more easily removed by centrifugation. The resulting slurry was subjected to high-speed centrifugation, and the resulting supernatant was carefully decanted for further processing. Preliminary experiments had shown that the desired enzymes were in the 18.5%–28% $(NH_4)_2SO_4$ fraction. This knowledge served as the basis for the next three steps. First 239 g of solid $(NH_4)_2SO_4$ was added to 1040 ml of supernatant.

The resulting precipitate was removed by centrifugation. Then an additional 120g of $(NH_4)_2SO_4$ was added to the supernatant. The resulting slurry was centrifuged. This time the supernatant was discarded after centrifugation, leaving the sediment containing the enzyme activity for further processing. The sediment was resuspended in a dilute buffer for column chromatography. Inspection of table 8.4 indicates only about a twofold increase in specific activity between fractions 1 and 2. The main purification was achieved by two column steps, carried out in series.

**Table 8.4. Outline of Purification of UMP Synthase from Ehrlich's Ascites Carcinoma**

| *Fraction* | *Volume (ml)* | *Protein (mg)* | *OMPDase*[a] | | | *OPRTase*[a] | | | *Ratio OMPDase to OPRTase* |
|---|---|---|---|---|---|---|---|---|---|
| | | | *Unit*[b] | *Sp. Act.* | *Percent Recovery* | *Units*[b] | *Sp. Act.*[c] | *Percent Recovery* | |
| 1. Streptomycin fraction | 1040 | 11,700 | 40.4 | 0.0034 | | 20.5 | 0.0018 | | 2.0 |
| 2. Dialyzed $(NH_4)_2SO_4$ fraction | 144 | 311 | 24.3 | 0.0078 | 60 | 8.7 | 0.0028 | 42 | 2.8 |
| 3. Affinity column eluate (concentrated) | 0.475 | 0.51 | 4.0 | 7.8[d] | 10 | 0.35 | 0.69 | 3.3 | 11.4 |

[a] OMPDase = orotate 5'-monophosphate decarboxylase; OPRTase = orotate phosphoribosyltransferase.
[b] Units refer to (total amount of enzyme activity.)
[c] Specific activity refers to the units of enzyme activity divided by the total protein.
[d] This value represents a 2,300-fold enrichment from fraction 1.

The first column was an affinity column containing an analog of UMP, 6-azauridine 5-monophosphate (azaUMP), covalently attached to an agarose column support system. In dilute buffer, greater than 99% of the protein in fraction 2 is retained on this column. After thorough rinsing of the column with dilute buffer, 5 x $10^{-5}$ M azaUMP was added to the buffer. This addition resulted in the elution of the UMP synthase. Then the column eluant carrying the two enzyme activities associated with UMP synthase was resuspended in pure buffer and passed over a phosphocellulosc column.

Phosphocellulose was chosen because the negatively charged phosphate groups result in the retention of proteins by electrostatic attraction alone; they also resemble the phosphate groups in the naturally occurring enzyme substrate, OMP. Thus, the phosphocellulose column may be thought of as an ion-exchange column and an affinity column combined. In ordinary ion-exchange chromatography, the protein, after column loading, is eluted by increasing the ionic strength with a simple inorganic salt. In this example, Jones used a more specific method to elute the enzyme, which involved adding $10^{-5}$ M azaUMP and 2 x $10^{-5}$ M OMP to the original loading buffer.

The addition does not substantially increase the ionic strength of the buffer. Therefore the only phosphocellulose-bound proteins that are likely to be eluted by this treatment are those with an especially high affinity for either of these nucleotides (azaUMP or OMP). This fact should greatly favour selective elution of those enzymes that carry specific sites for binding these nucleotides. The two column steps together resulted in anenzyme preparation that was approximately 2,300-fold purified over the starting material, as measured by the increase in specific activity. The final product was examined by SDS gel electrophoresis and found to contain a single band with an estimated molecular weight of 51,000.

The denaturing conditions of an SDS gel are expected to dissociate a multisubunit protein. Hence the SDS gel result indicated that the enzyme contains one type of polypeptide chain, but it does not tell us whether the enzyme contains one or more of these chains. Sedimentation analysis on a sucrose density gradient in a nondenaturing buffer indicated a single band with an estimated molecular weight of about 50,000. These two results taken together demonstrate that the enzyme in its native state contains a single polypeptide chain.The purity of the enzyme was also confirmed by isoelectric focusing and two-dimensional electrophoresis, with *isoelectric* focusing in the first direction followed by SDS gel electrophoresis in the second direction. The conclusion drawn from these results was that both of the enzyme activities associated with UMP synthase—OMPDase and OPRTase—are contained in a single protein. The basis for the conclusion was that both enzyme activities are always present in the same fractions throughout the multistep purification. However, inspection of table 8.4 indicates a possible objection to this interpretation. In the columns showing percent recovery, it can be seen that substantial amounts of enzyme activity are lost for both enzymes during purification, but that considerably more activity is lost for the OPRTase. These losses could be due to actual loss of enzymes during purification or to some sort of inactivation of the enzyme sites.

Tissues (Ehrlich's ascites cells)

Step 1 ↓ Homogenized

Crude broken cell lysate

Step 2 ↓ Add streptomycin sulfate (1.5 g/100 ml extract)

**Precipitate**

Step 3 ↓ Centrifuge and discard sediment

FRACTION 1 Supernatant from centrifugation

Step 4 ↓ Add $(NH_4)2SO_4$ to 18.5 weight percent

**Precipitate**

Step 5 ↓ Centrifuge and discard sediment

**Supernatant from centrufugation**

Step 6 ↓ Add more $(NH_4)_2SO_4$ to 28 weight percent

**Precipitate**

Step 7 ↓ Centrifuge and discard supernatant

FRACTION 2 **Sediment (pellet form centrifugation)**

Step 8 ↓ Fractionate resuspended sediment over an azaUMP column

**Eluate containing UMP synthase**

Step 9 ↓ Fractionate UMP synthase containing eluate over phosphocellulose column

FRACTION 3

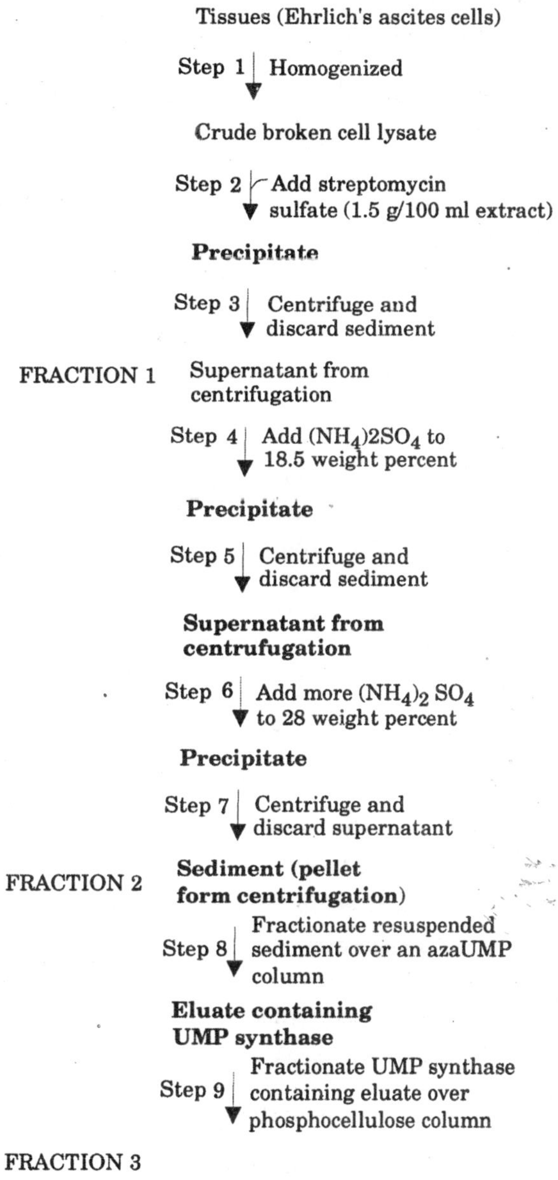

**Fig. 8.7. Outline of purification scheme for UMP synthase from Ehrlich's ascites tumor cells of mice.**

The preferential loss of OPRTase activity is emphasized by the last column in table 8.4, which gives the ratio of the two enzyme activities in the different fractions. Considered alone, these data could indicate that a separate catalytic unit of OPRTase is lost during purification. Jones thinks this is unlikely for two reasons: (1) the activity appears in no fractions other than with OMPDase during purification, and (2) the OPRTase activity is notably unstable. It appears, then, that both enzyme activities exist at distinct sites on a single protein. The greater loss in activity of one enzyme activity over the other is attributed to a greater sensitivity of one reaction site over the other on the enzyme surface.

### *Purification of a Membrane-Bound Protein*

The second purification procedure we examine illustrates an unusual approach to the purification of a membrane-bound protein. The lactose carrier protein of *E. coli* is normally tightly bound to the plasma membrane. This protein is involved in the active transport of the dissaccharide lactose across the cytoplasmic membrane. When lactose carrier protein is present, the intracellular concentration of lactose can achieve levels 1,000-fold higher than those found in the external medium. Ron Kaback devised a simple yet elegant procedure for the purification of this protean. Purification of the membrane-bound lactose carrier protein is a very different problem from the purification of the soluble OMP synthase.

Both the approach to purification and the assays for the protein during purification are quite novel. The assay involves reconstituting a transport system with membranes that are free of lactose carrier protein, then adding the partially purified carrier protein and radioactively labeled lactose. The activity in this assay system is proportional to the transport of radioactive lactose across the membrane in the cell-free reconstituted system. The results of the purification steps are tabulated in table 8.5, and the purification procedure is outlined in Fig. 8.9. In this procedure advantage was taken of the fact that the carrier protein in its native state is firmly bound to the cytoplasmic membrane.

**Table 8.5. Purification of the Lactose Carrier Protein.**

| *Fraction* | *Protein (mg)* | *Percent Recovery (Total Protein)* | *Percent Recovery (Carrier Protein)* | *Purification factor* |
|---|---|---|---|---|
| 1. Membrane fraction | 12.5 | 100 | 100 | 1.0 |
| 2. Urea-extracted membrane | 5.6 | 45 | 76 | 1.7 |
| 3. Urea/cholate-extracted membrane | 2.6 | 21 | 61 | 2.9 |
| 4. Octylglucoside extract | 0.4 | 3.2 | 38 | 12 |
| 5. DEAE[a] column peak | 0.056 | 0.4 | 14 | 35 |

[a]DEAE = diethylaminoethyl cellulose

Thus, the first step consisted of isolating these membranes from the rest of the cell constituents. Starting from the membrane fraction only 35-fold purification was required to achieve pure carrier protein. This rapid result was possible because a special strain of *E. coli,* containing about 100 times the normal carrier protein, was used as starting material for the purification. The high initial content was engineered by putting the carrier protein gene on a multicopy plasmid, which was then inserted into the cell—a procedure. Bacterial cells are much tougher than mammalian cells, requiring a more stringent procedure for cell disruption.

In this case, the cells were placed in a so-called French pressure cell and bled through an orifice from very high pressures (≈ 10,000 psi) to atmospheric pressure. Under these conditions the cells literally explode, fragmenting their membranes and releasing the cytoplasmic contents. The membranes were pelleted by a brief centrifugation, leaving a supernatant containing DNA, ribosomes, and cytoplasmic protein, which was removed by decantation . The pelleted membrane fraction was next resuspended and extracted with 5 M urea and then reextracted with 6% sodium cholate. The pellet obtained by high-speed centrifugation from these two extractions still contained most (61%) of the carrier protein in the more rapidly sedimenting membrane fraction, although 79% of the total membrane-bound protein was released by these treatments. At this point, the carrier protein was released from a suspension of the membranes by addition of the hydrophobic reagent octylglucoside in the presence of *E. coli* phospholipid.

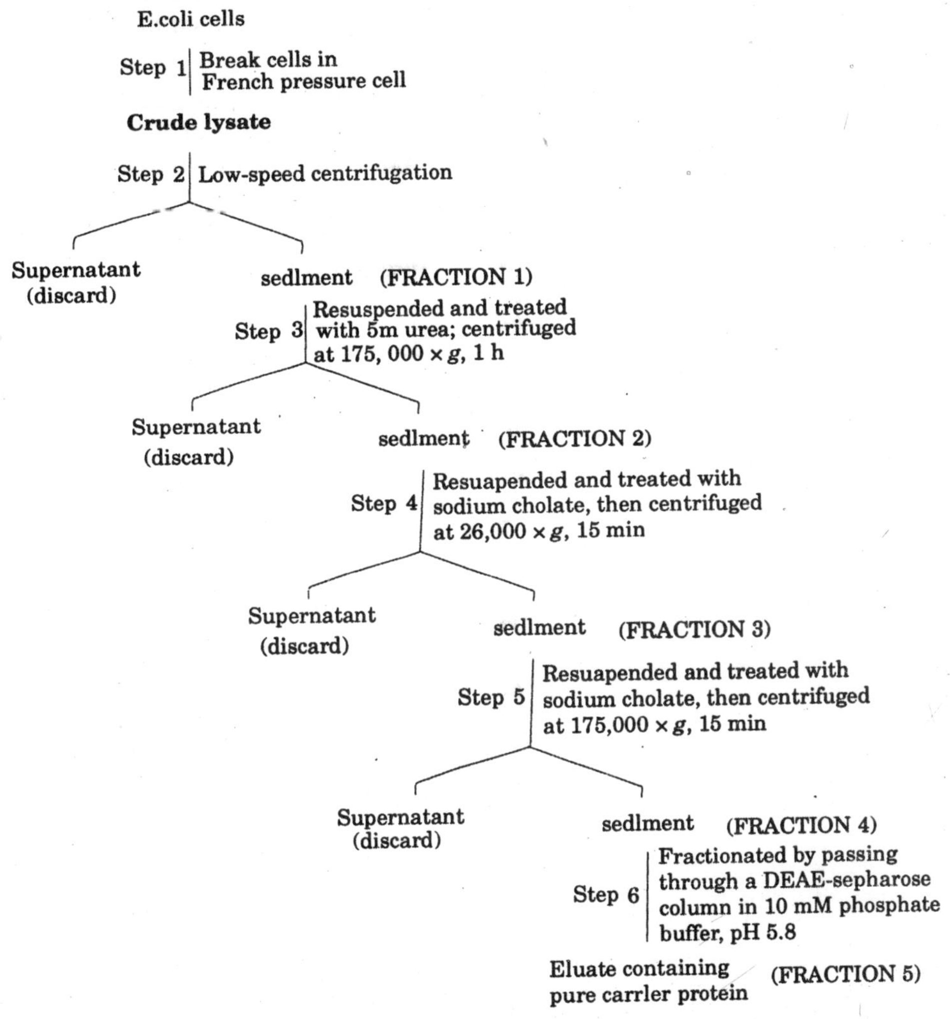


**Fig. 8.8. Outline of purification procedure for lactose carrier protein from *E. coli*.**

It is believed that the octyl part of octylglucoside competes effectively with the membrane for binding to hydrophobic centers on the carrier protein. The *E. coli* phospholipid facilitates dissociation of the carrier protein from the membrane fraction. The solubilized octylglucoside-containing extract was fourfold enriched in carrier protein after a high-speed centrifugation to remove residual membrane and membrane-bound proteins. Finally, the octylglucoside-containing extract was passed over a positively charged diethylaminoethyl (DEAE) sepharose column (sepharose is a form of cross-linked polydextran) in a buffer containing 10 mM potassium phosphate, 20 mM lactose, and 0.25 mg of washed *E.* coli lipid per milliliter.

The carrier protein passed through this column as a symmetrical peak of protein. Most of the remaining protein in the extract adsorbed to the positively charged column. The fractions containing the bulk of the carrier protein activity were judged to be pure by SDS gel electrophoretic analysis. The purified protein contained a single polypeptide chain with an estimated molecular weight of 33,000. The two purifications described here are as different as the two proteins involved.

No two purifications are exactly alike, but the principles of purification, as stated at the outset, are quite similar. Fortunately, an almost endless variety of purification techniques exists. This variety is both helpful and challenging because a great deal of knowledge and creativity are required to exploit it. In addition to professional expertise, the two most important things required to make a purification possible are an unambiguous assay for the protein in question and a means of stabilizing the protein during purification.

# CHAPTER 9

# SYNTHESIS OF VACCINES

Vaccines against infectious diseases are needed to increase global immunization compliance and to decrease the costs of delivery, while expanding participation by vaccine manufacturers. If implemented well, vaccines can provide great economic benefits, with cost-to-benefit ratios of up to 1:10. The technological revolution in vaccine development over the last half of the 20$^{th}$ century has centered on the use of mammalian cell cultures. This technology, now used throughout the developed and developing world, enables the production of more uniform and potent vaccines than is possible with other methods. However, subunit vaccines have penetrated the developing world to a much lesser degree than the developed world. This is for two reasons: IP challenges and technology entry barriers, whose costs increase along with the sophistication of molecular biology techniques. Thus, the development and implementation of new vaccines often involves substantial economic and logistic barriers that are difficult to overcome in many poor countries. Transgenic plant-derived proteins represent a promising strategy for vaccine development that combines innovations in medical science and plant biology for the creation of affordable vaccines.

A growing number of laboratories are investing in the development of plant-derived protein pharmaceuticals, expanding on the seminal works that first put forward this idea. The subject of plant-derived vaccines has been reviewed several times in recent years (e.g. and a comprehensive list of the plant-derived vaccine antigens that have been reported is provided in Table 9.1 and 9.2. While much research is still needed to optimize the production of vaccines in plants and to validate them in large-scale clinical trials, the results to date are very promising and suggest that the technology justifies commercial development. The oral delivery of transgenic plant tissue expressing vaccine antigens, typically mucosally targeted subunits, can promote specific mucosal secretory IgA (slgA) and serum IgG antibody responses via the gut lymphoid system. Although the protective efficacy of a plant-derived vaccine has yet to be determined in humans, some challenge studies in animals have shown promising results.

## VACCINE PRODUCTION IN PLANTS

Three major research strategies have emerged for the production of subunit vaccines in plants. The first, the direct transformation of plants, introduces genes encoding the antigenic proteins of human or animal pathogens stably into the plant genome. In this case, the genes can be inserted into either the nuclear or chloroplast genomes. The resulting plants accumulate

**Table 9.1. Vaccine Antigens Expressed in Plant Virus Epitope Display Systems**

| *Antigen* | *Expression system* | *Immunogenicity* | *Protective challenge* |
|---|---|---|---|
| Canine parvovirus (CPV) VP2 epitope | Epitope display on cowpea mosaic virus in cowpea leaf | Immunogenic in mice when delivered parenter ally or nasally. Dogs developed 3L17-and VP2-specific IgG; sera neutralized CPV *in vitro*. | Protective against lethal challenge in dogs immunized parenterally. |
| Canine parvovirus VP2 epitope | Epitope display on plum pox potyvirus in tobacco leaf | Mice developed CPV-specific antibodies that showed neutralizing activity. Immunogenic in mice and rabbits when delivered parenterally | — |
| Foot and mouth disease virus VP1 epitope | Cowpea mosaic virus in cowpea leaf | No immunogenicity assays performed | — |
| Hepatitis C virus (HCV) HVR[1] epitope of E2 envelope protein | Epitope display on cucumber mosaic virus in tobacco leaf | Cross reactive with a wide range of human anti HVRI antibodies. | — |
| Human immuno-deficinecy virus (HIV) type 1 ELKDWA epitope | Potato virus X in tobacco leaf | Sera from normal and hu-PBL-SCID mice showed anti-HIV-1 neutralizing acitivty. immunogenic in mice when delivered parenter-ally or nassaly. | — |
| Human rhinovirus type 14 VP1 epitope | Epitope display on Cowpea mosaic virus in cowpea leaf | Immunogenic in rabbits when delivered parenterally. | — |
| Mink enteritis virus VP2 epitope | Epitope display on cowpea mosaic virus in cowpea leaf | Immunogenic in mink when delivered parenterally. | Protective against challenge with virulent MEV. |
| *Pseudomonoas aeruginosa* memberane protein F | Epitope display on cowpea mosaic virus in cowpea leaf | Elicited specific antibodies. Immunogenic in mice when delivered parenterally. | Mice protected when challenged with model chronic pulmonary infection with *P. aeruginosa.* |
| *Pseudomonas aeruginosa* membrane protein F | Epitope display on tobacco mosaic virus in tobacco leaf | Elicited specific antibodies against 7 immuno-type strains. Immunogenic in mice when delivered parenterally. | Mice protected when challenged with model chronic pulmonary infection with *P. aeruginosa.* |

**Table 9.1. Continued**

| *Antigen* | *Expression system* | *Immunogenicity* | *Protective challenge* |
|---|---|---|---|
| Rabbit hemorrhagic disease virus VP60 epitope | Epitope display on plum pox potyvirus in tobacco leaf | Rabbits developed specific antibodies that showed neutralizing activity. Immunogenic when delivered to mice parenterally; immunogenic in rabbits when delivered parenterally | Rabbits immunized parenterally survived lethal challenge. |
| Rabies virus glyco-protein (G) and nucleo-protein (N) | Alfala mosaic virus and tobacco mosaic virus in tobacco and spinach leaf | Elicited specific virus-neutralizing antibodies in mice. Immunogenic in mice when delivered orally and parenterally; immunogenic in humans when delivered orally | Moderately protective against lethal challenge infection in mice. |
| Respiratory syncytial virus (RSV) G and F proteins | Alfalfa mosaic virus in tobacco leaf | High levels of serum antibodies specific for RSV-G. Immunogenic in mice when delivered parenterally | Protective against challenge with RSV Long strain. |
| *Staphylocaccus aureus* D2 epitope of fibro-nectin-binding protein (FnBP) | Cowpea mosaic virus in cowpea leaf; Potato virus X in tobacco leaf | Elicited FnBP-specific IgA and IgG. Immunogenic in mice and rats when delivered orally, nasally, or parenterally. | |

the subunit antigens, which can then either be purified from plant tissue for parenteral or mucosal delivery, or the plant tissue can be processed for oral delivery with the antigen as a component of the plant matrix. In the second research strategy, plant viruses have been utilized as pliable genetic platforms for protein expression.

Three formats have been developed, and the one that has undergone the most extensive evaluation is the display of epitopes on the surface of the virus as fusions with the viral coat protein. This epitope-display system usually involves subsequent purification of the viral particle for parenteral delivery. Another viral strategy uses cloned viral genomes to introduce antigenic protein-coding sequences driven by sub-viral promoters, and results in antigen expression as a byproduct of the viral replication cycle. A final variant is the integration of entire viral genomes into plant chromosomes by transformation, with the subsequent transcription of viral RNA as the initiating event in viral replication. A coding sequence for the antigen is included in this viral genome and is thus co-expressed. The potential advantage of viral expression systems compared to stable plant transformation is that viral replication can greatly amplify the template for protein synthesis resulting in high-level protein accumulation.

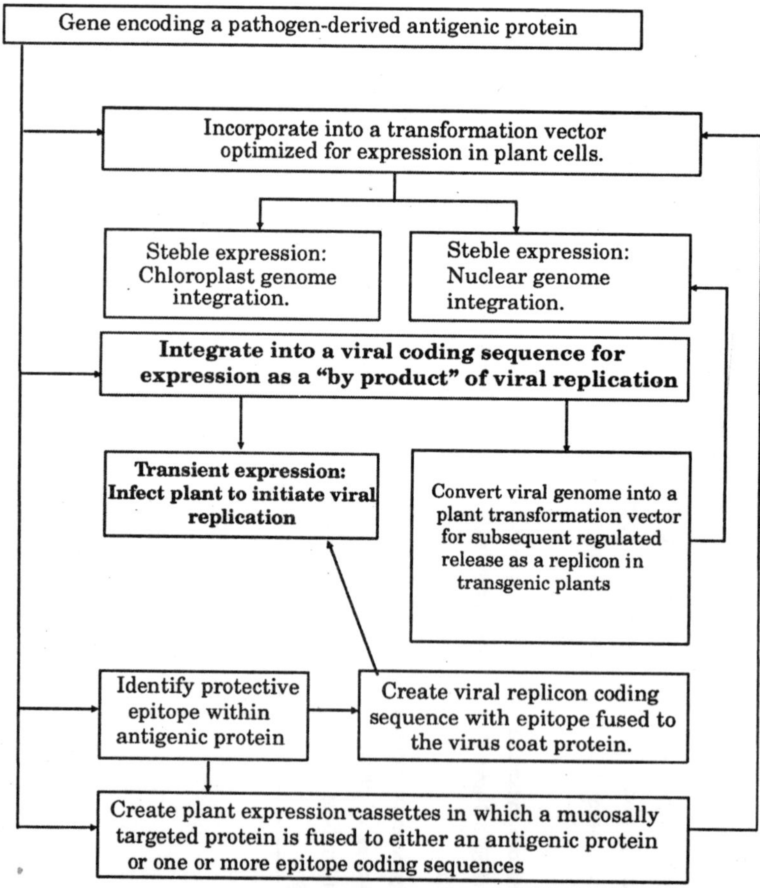


**Fig. 9.1. Plant-derived vaccines: research strategies.**

A third strategy for the production of plant-based vaccines is the design of mucosally targeted fusion proteins. We define this as a separate research strategy to emphasize the goal of creating new mucosal vaccines, even though such fusion proteins can be expressed in transgenic plants or using viral vectors as discussed above. We highlight mucosally targeted proteins because this captures the emerging research emphasis on protein engineering to obtain more effective vaccines, and the potentially unique role of plants for the production and accumulation of novel vaccine products. See Table 9.3 for details of fusion proteins expressed in transgenics.

## SYNTHESIS

Current vaccine technology frequently requires the purification of subunit proteins from mammalian cell cultures or tissues, yeast, fertilized eggs or bacterial fermentation systems to produce the immunogen. Typically, the product requires refrigeration during transport to its final point of use, adding significant cost to the vaccination program. One solution may be to use dried plant extracts containing subunit vaccines, since this provides a product that is stable at ambient temperatures, i.e. equivalent in storage and transport characteristics to dehydrated food products. In addition, the production of antigens in plants may improve product safety by removing animal cell-related contaminants and pathogens.

### Advantages of Plants

Plant-derived pharmaceuticals (PDPs) are proteins or organic compounds produced in plants via recombinant DNA technology, which are used to improve human or animal health. Subunit vaccines represent one category of PDPs that have been validated in a variety of studies, including human clinical trials. Current efforts in product formulation utilize food-processing technology to convert transgenic plant material into dried samples that can be delivered in unit doses with assured product uniformity and quality. Plant-derived vaccines appear to offer several product advantages, including oral delivery, heat stability, lower manufacturing costs for the active ingredient and suitability of the manufacturing technology for use in developing countries.

### Mucosal Immune Responses

Orally delivered, non-replicating subunit vaccines have not yet achieved commercial success using any means of manufacture. Possible hurdles facing the use of orally delivered immunogenic proteins include the likelihood that some proteins will be degraded after ingestion and that some immunogens may not be recognized efficiently at mucosal immune effector sites in the gut. As a result, higher concentrations of immunogen may be required in oral versus parenteral delivery. Although this is a potential limitation, the use of plants as a protein biomanufacturing system offers advantages in that the cost of obtaining the end product is comparatively low.

In addition, empirical evidence suggests that encapsulation of the immunogenic protein within the plant cell matrix during administration provides some protection from degradation by gastric enzymes and acids. The best candidates oral subunit vaccines for production in transgenic plants are the primary antigens of infectious pathogens that aggregate in forms that are recognized at mucosal sites where an immune response is triggered. These include viral surface proteins that co-assemble to form virus like particles (VLPs), and bacterial toxins that naturally aggregate to form mucosally targeted multimeric complexes. In addition, in several laboratories there are also ongoing efforts to produce a variety of fusion proteins that target immunoresponsive mucosal sites.

## Antigens Produced in Plants

Tables 9.1 and 9.2 shows the range of different plant and vector systems that have been used for the expression of antigens. The immunogenicity of PDPs was demonstrated in a number of studies, including those using hepatitis B surface antigen ($HB_SAg$), the S protein of transmissible gastroenteritis coronavirus (TGEV) and human immunodeficiency virus 1 (HIV-1) epitopes. Furthermore, successful challenge trials resulted after immunization with *Pseudemonas aeruginosa* epitopes the FP1 epitope of foot and mouth disease virus (FMDV) and *Escherichia coli* labile enterotoxin B subunit (LTB). Additional antigens expressed in plants include the respiratory syncytial virus (RSV) G and F proteins, the rota-virus VP6 protein and VP7 protein the measles virus hemagglutinin protein and an epitope from the major surface antigen of *Plasmodium falciparum* (PfMSPl). Although the immunogenicity of the plant-derived RSV G and F proteins and the measles virus hemagglutinin protein were demonstrated in animal trials, the immunogenicity of the VP6 antigen and the PfMSPl epitope were not tested.

## Targeting Antigens to Specific Tissues

Tissue-specific targeting of antigens can greatly facilitate the harvesting of proteins. Streatfield *et al.* [28] expressed antigens specifically in maize kernels, Sandhu *et al.* In tomato fruit, and Wright *et al.* In tobacco seeds. Targeting the major glycoprotein (gB) of human cytomegalovirus (hCMV) to transgenic tobacco seeds was previously described and further investigations showed that recombinant gB was almost exclusively deposited in protein storage vesicles in mature tobacco seeds. Sojikul *et al.* and Huang & Mason reported the targeting of a fusion molecule to the endoplasmic reticulum.

## Expression Systems

The achievement of significant expression levels for recombinant proteins produced in plants often depends on the tailoring and optimization of the expression system. Unfortunately, except for a few rules of thumb, insights gained while studying one protein are not always readily transferable to others, and low levels of expression are found in some of these systems. A few studies have attempted to address the low recombinant protein expression levels that may sometimes occur in transgenic plants. The issues of how, where and when to express a transgene in order to achieve maximal accumulation of a functional protein remain priorities in this field. It is especially important in the case of plant-derived vaccines (where the goal is to achieve sufficient dosage with minimal processing of antigenic proteins in edible plant tissue) to ensure that sufficient quantities survive passage through the stomach to elicit a clinically relevant immune response. As discussed above, vaccine antigens can be produced in plants using two different systems: stable genetic transformation and transient expression.

Stable transformation produces a genetic line that can be propagated either by vegetative (stem) cuttings or by seeds, and was the method used in published edible vaccine clinical trials to date. Transient expression involves the use of a recombinant plant virus that carries the vaccine gene and, by systemic infection, causes the plant to express the antigen. The main advantage of transient expression with a plant virus system is that virus replication amplifies the gene copy number, typically resulting in a much higher level of expression than with stable transformation. However, the plant virus systems might in some cases suffer from instability and loss of foreign genes larger than 1 kb.

In addition, the need to inoculate each plant individually makes large-scale production rather labour intensive. Recombinant DNA is integrated into the nuclear or chloroplast genome through stable transformation. *Agrobacterium tumefaciens,* a plant pathogen, is typically used for nuclear transformation, and strains have been engineered with deletions of the virulence genes that cause tumor growth in plants. *A. tumefaciens* efficiently transports DNA into cells and then promotes nuclear chromosomal integration at random sites. Integrated nuclear transgenes are expressed and inherited in typical Mendelian fashion. Some agronomically important plant species (e. g. soybean and most cereal grains) are recalcitrant to *Agrobacterium* transformation, and a biolistic method (microprojectile bombardment) is frequently used for these plants.

DNA coated on micron-sized gold particles is propelled into plant cells using compressed helium gas and becomes incorporated into chromosomal DNA. Foreign DNA that integrates into the nuclear genome yields stable transformants that show Mendelian inheritance of the transgenes. The biolistic method, as traditionally carried out, usually results in higher-copy-number plants compared to those generated by *Agrobacterium,* which can enhance expression. However, excessive copy numbers or very high-level expression of nuclear genes can cause gene silencing, resulting in low protein accumulation. Thus it is important to select transgenic lines that carry only between one and three copies of transgene. The biolistic method is also used to introduce transgenes into the chloroplast genome.

Although such transplastomic plants have been produced for only a few species, the potential for chloroplast transformation in other crop plants is strong. The high chloroplast genome copy number in plant cells enhances recombinant protein expression, as shown for the cholera toxin B (CTB) subunit [52]. DeCosa *et al.* [53] expressed an insecticidal protein in the chloroplasts of mature tobacco leaves at levels up to 45% of total protein. Stable transgene expression was recently described in the chloroplast-derived chromoplasts of tomato fruit. A further advantage is that, as the chloroplast genome is inherited strictly maternally in most plants, there is minimal danger of unintended spread of transgenes by dissemination of pollen. Although chloroplast transformation is not yet a routine procedure, at least 25 different foreign genes have been expressed in transplastomic plants.

Chloroplast transformation is not suitable for some eukaryotic proteins that require post-translational modifications like glcosylation. Gene silencing, however, has not been observed in this system. Seed-specific production of LTB by Streatfield *et al.* resulted in expression levels of up to 1.8% total soluble protein (TSP), and two separate maize breeding programs have increased antigen production by fivefold and tenfold. The investigations of Chikwamba *et al.* regarding the expression of LTB in maize are among the first to include the use of particle bombardment transformation for the production of plant-derived vaccines.

## Mucosally-targeted Fusion Proteins

Most soluble proteins are rapidly hydrolyzed in the intestinal tract. Subunit vaccines are subject to the same proteolytic environment, and as a result some candidates may be ineffective at inducing the mucosal immune system. This is due to degradation prior to reaching immune effector sites in the gut-associated lymphoid tissues (GALT). In order to increase subunit vaccine mucosal immunogenicity, a number of investigators have used protein engineering to create fusion molecules consisting of the antigen (or epitopes) linked to a mucosally active carrier. Two studies have involved the fusion of an anitigen to β-glucuronidase (GUS). Gil *et al.* fused GUS to the 2L21 protective epitope from canine parvovirus, and Dus Santos *et al.* fused GUS to the protective epitope from FMDV. Transformants in both cases were selected on the basis of GUS activity, and both proved to be immunogenic.

Mice immunized orally or parenterally with the GUS-FMDV epitope fusion werecompletely protected against challenge with the native virus. Cholera toxin subunits have also been used to design targeted, plant-derived fusion vaccines. Using a tobacco mosaic virus (TMV) expression system, Nemchinov *et al.* Generated a *Vibrio choleras* CTB fusion to an epitope (HVR1) from the hepatitis C virus (HCV). Tobacco plants inoculated with the recombinant TMV produced the HVR1 epitope fused to a functionally active, pentameric CTB, which reacted with HVRl-specific monoclonal antibodies and sera from individuals infected with virus from four of the major genotypes of HCV. Nasal immunization of mice with a crude plant extract containing the recombinant CTB-HVR1 elicited both anti-CTB serum antibodies and anti-HVRl serum, antibodies that specifically bound to HCV-like particles. Sojikul *et al.* reported the expression of a fusion protein designed to improve the expression of the hepatitis B surface antigen (HBsAg). This was achieved by mimicking the process of HBsAg targeting to the endoplasmic reticulum of human liver cells during HBV infection.

The gene encoded a recombinant HBsAg modified to contain an N-terminal signal peptide from soybean vegetative storage protein A (VSPaS). This signal peptide directed the HBsAg-VSPaS fusion protein to the endoplasmic reticulum and resulted in greater fusion protein stability, enhanced protein accumulation and formation of more VLPs. Moreover, the HBsAg-VSPaS fusion stimulated higher levels of serum IgG when administered parenterally in mice compared to native HBsAg.

## Multivalent and Multicomponent Vaccines

Among vaccine products, those that stimulate several facets of the immune system such as the induction of strong humoral and mucosal responses as well as effective cellular immune responses – are highly desirable. Combination vaccines targeting multiple pathogens in one formulation are likewise preferred. Therefore, developing both multivalent and multicomponent plant-based vaccines provides both efficacious and cost-effective immunization strategies. This can be accomplished through the development of recombinant plants carrying transgenes that encode the antigens of several pathogens, either by direct simultaneous transformation, sexual crosses of individually transformed lines, or the blending of separately transformed plant tissues. One approach for the production of multicomponent vaccines was taken by Yu and Langridge and Arakawa *et al.* They fused peptides containing important protective epitopes derived from two enteric pathogens - enterotoxic *E. coli* (ETEC), which causes bacterial traveler's diarrhea, and rotavirus, which causes acute viral gastroenteritis – to the A2 and B subunits of cholera toxin, respectively.

**Table 9.2. Vaccine Antigens Expressed in Plant Systems: Subunit Antigen Expression in Transgenics.**

| *Antigen* | *Expression system* | *Immunogenicity* | *Protective challenge* |
|---|---|---|---|
| Bacillus anthracis protective antigen | Tobacco leaf | No immunogenicity assays | — |
| Bovine herpes virus type 1 (BHV1) glycopro tein D (gDc) | Tobacco leaf | Sustained and specific humoral and cellular im mune response in mice and cows when delivered once parenterally. | Cows challenged with infectious dose of BHV1 exhibited milder symptoms over a shorter duration. |
| Enteropathogenic *E. coli* pilus subunit A (BfpA) | Tobacco leaf | Immunogenic when delivered orally in mice. Anti-BfpA in fecal matter. | — |
| Enterotoxigenic E. coli B subunits of the heat labile toxin (LTB) | Maize seed | Elicited neutralizing antibodies. Immunogenic when administered orally. Serum and secretory immune responses in humans. | Partially protective in mouse gut fluid assay. |
| Enterotoxigenic *E. coli* B subunits of the heat labile toxin (LTB) | Potato tuber | Receptor-binding activity. Immunogenic in mice and humans when administered orally. | Partially protective in mouse gut fluid assay. |
| Enterotoxigenic *E. coli* B subunits of the heat labile toxin (LTB) | Tomato fruit and leaf | Offspring were passively immunized through oral immunization of parents. | — |
| *E. coli* 0157:H7 Intimin epitope | Tobacco callus | Mucosally immunogenic in mice when primed parenterally and boosted orally | Mive immunized parenterally and orally boosted showed reduced duration of colonization upon challenge. |
| Foot and mouth disease virus (FMDV)VP1 epitope | Arabidopsis leaf | Serum antibodies reacted strongly with intact FMDV particles. Immunogenic in mice when administered parenterally. | Protective aginst challenge with virulent FVDM. |
| Foot and mouth disease virus VP1 epitope | Potato | FMDV-VP1-specific antibody response. Immunogenic. | Protective against challenge. |
| Foot and mouth disease virus VP1 epitope | Alfalfa leaf | Mice developed specific antibody response to synthetic peptide, VP1 epitope. and intact FMDV particle. Immunogenic in mice when adminis tered parenterally or orally. | Mice protected against challenge with FMDV virus |

| *Antigen* | *Expression system* | *Immunogenicity* | *Protective challenge* |
|---|---|---|---|
| Hepatitis B surface antigen (HBsAg) | Cherry tomatillo leaf and fruit | Immunogenic in mice as an oral booster after parenteral priming. | — |
| Hepatitis B surface antigen (HBsAg) | Lettuce leaf | Immunogenic in humans when administered orally; 2 of 3 human subjects developed very modest serum antibody response titers in primary immunization. | — |
| Hepatitis B surface antigen (HBsAg) | Lupin callus | Immunogenic in mice when administered orally | — |
| Hepatitis B surface antigen (HBsAg) and middle protein | Potato tuber | Immunogenic in mice when administered orally. | — |
| Hepatitis B surface antigen (HBsAg) | Soybean cell culture | No immunogenicity assays performed. | — |
| Hepatitis B surface antigen (HBsAg) | Tobacco leaf | Mice vaccinated parenterally produced all IgG subclasses and IgM. | — |
| Hepatitis E virus (HEV) ORF2 epitope | Tomato leaf and fruit | No immunogenicity assays performed. | — |
| Human cytomegalo virus glycoprotein B | Tobacco seed | gB-specific antigenic activity. | — |
| Human immuno-Deficiency virus (HIV) type 1 p24 protein | Tomato bushy stunt virus in tobacco leaf | No immunogenicity assays performed. | — |
| Human immuno-deficiency virus (HIV) type 1 gp 120 (v3 loop) protein | Alfalfa mosaic virus in tobacco leaf | Elicited specific virus-neutralizing antibodies in mice when delivered parenterally. | — |
| Human immuno-deficiency virus (HIV) type 1 gp 120 (v3 loop) protein | Tomato bushy stunt virus in tobacco leaf | HIV epitope detected by V3-specific mono-clonal antibodies and human sera from HIV positive patients. Immunogenic in mice when delivered parenterally. | |
| Human papilloma virus type 11 major capsid protein L1 | Tobacco leaf, potato tuber | Reacted well with assembled capsids and isolated L1 capsomers Immunogenic in mice when administered parenterally with sub-sequent oral boosting. | — |

**Table 9.2. continued...**

| *Antigen* | *Expression system* | *Immunogenicity* | *Protective challenge* |
|---|---|---|---|
| Human papilloma virus (HPV) type 16 onco-protein E7 | Potato X in tobacco leaf (complete reading frame) | Immunogenic in mice when administered with surrogate (C3 cells). | Mice protected moderately when challenged parenterally. |
| Human papilloma virus (HPV) type 16 major capsid protein L1 | Tabacco leaf | Nutralizing monoclonal antibodies bounds plant-produced particles. | |
| Human papilloma virus (HPV) type 16 major capsid protein L1 | Tobacco and potato | Weak, transient anti L1 antibody response in 3 of 24 mice when administered orally. Immunogenic when administered orally after parenteral boost. | — |
| Infectious bronchitis virus (IBV) S1 glycoprotein | Potato tuber | Antibodies neutralized IBV from mice and chickens. Immunogenic in chickens when administered orally. nasally, nasally, and parenterally. | Protective after 3 immunizations in chickens when challenged with virulent IBV. |
| Measles virus hemagglutinin (H) protein | Tobacco leaf | Mice produced serum anti-H antibodies that Neutralized measles virus in vitro. Immunogenic in mice when delivered orally or parenterally. | — |
| Measles virus hemagglutinin (H) protein | Carrot leaf and root | IgG1 and IgG2 cross reacted strongly with measles virus and neutralized virus *in vivo*. Immunogenic in mice when delivered parenterally. | — |
| Measles virus hemagglutinin (H) protein | Carrot leaf and root | H-specific antibody isotypes and subclasses in mouse sera; mouse sera neutralized wt virus. Immunogenic in mice when administered parenterally. | — |
| Murine hepatitis virus glycoprotein S 5B19 epitope | Tobacco mosaic virus in tobacco leaf | Mice developed serum lgG and IgA specific for 5B19. Immunogenic in mice when delivered nasally or parenterally. | Protective in mice immunized parenterally or orally when challenged with MHV strain JHM. |
| Norwalk virus (NV) capsid protein | Potato tuber; tobacco leaf | Mice and humans developed serum lgG and secretory lgA specific for recombinant NV. Immunogenic in mice and humans when administered orally. | |
| *Plasmodium falciparum* (PfMSP) | Tobacco leaf | Antigenic characteristics identical to *E. coli* expressed protein. Immunogenic characteristics identical to *E. coli* expressed protein. | |
| Porcine epidemic diarrhea virus (PEDV) spike protein | Tobacco leaf | Systemic and mucosal immune reponse in mice immunized orally. | — |

| *Antigen* | *Expression system* | *Immunogenicity* | *Protective challenge* |
|---|---|---|---|
| Rabbit hemorrhagic disease virus VP60 epitope | Potato leaf | VP60 specific antibodies elicited in rabbits. Immunogenic in rabbits when administered parenterally. | Rabbits protected when challenged with virulent RDHV. |
| Rabies virus glyco-protein (G) | Tomoto leaf and fruit | Intact protein forms. | |
| Respiratory syncytial virus fusion protein | Tomato fruit | Mice developed serum and mucosal RSV-F specific antibodies. Immunogenic in mice when delivered orally. | — |
| Rinderpest virus hemagglutinin(H) protein | Pigeon pea leaf | — | —<br>— |
| Rinderpest virus hemagglutinin (H) protein | Tobacco leaf | High titers of neutralizing antibodies detected in immunized mice | |
| Rotavirus VP6 protein | Potato tuber | Anti-GAR antibodies detected. Immunogenic in mice when delivered parenterally. | — |
| Rotavirus VP6 protein | Tomato callus | No immunogenicity assays. | |
| Rotavirus VP7 protein | Potato tuber | Serum IgG and IgA specific for VP7; IgA had VP7-neutralizing activity. Immunoginic in mice when delivered orally. | — |
| Tetanus toxin fragment C | Tobacco leaves | Systemic IgG and local IgA responses. Immuno-genic in mice when administered nasally. | — |
| Transmissible gastro-enteritis coronavirus N-terminal domain of the spike glyco-protein (S) | Arabidopsis leaf | gS elicited virus-neutralizing antibodies. Immunogenic when administered parenterally. | - |
| Trunsmissble gastro-enteritis coronavirus N-terminal domain of the spike glyco-protein(S) | Maize seed | Orally immunized piglets had a memory im-mune response leading to rapid accumulation of virus-neutralizing antibodies on challenge. | Reduced symptoms in swine immunized orally and challenged with virulent TGEV. |

**Table 9.2. continued...**

| *Antigen* | *Expression system* | *Immunogenicity* | *Protective challenge* |
|---|---|---|---|
| Transmissible gastro-enteritis coronavirus N-terminal domain of the spike glyco-protein (S) | Potato tuber | Mice developed serum antibodies specific for gS. Immunogenic in mice when delivered parenter-ally or orally. | - |
| Transmissible gastro-enteritis coronavirus N-terminal domain of the spike glyco-protein (S) | Tobacco leaf | TGEV-specific antibodies (in pigs) showed low levels of virus-neutralizing activity. Immuno-genic in pigs when administered parenterally. | — |
| *V. cholerae* B subunits of the cholera toxin (CTB) | Potato | Receptor-binding activity. Immunogenic in | Mice challenged with CT showed up to 60% reduction in diarrheal fluid accumulation in small intestines |
| *V. cholerae* B subunits of the cholera toxin (CTB) | Tovacco leaf chloroplast | No immunogenicity assays performed. | — |
| B subunits of the cholera toxin (CTB) | Tomato loaf and fruit | No immunogenicity assays performed. | — |
| Canine parvovirus (CPV) VP2 epitope fused to GUS | Arabidopsis leaf | Specific antibodies detected and immunigenic in mice when delivered parenterally. | — |
| Enterotoxigenic *E. coli* epitope and totavirus epitope fused to CTB | Potato tuber | Mice developed detectable levels of serum and intestinal antibodies. Immunogenic in mice against ETEC, rotavirus, andV. cholerac when delivered orally. | Symptoms reduced in passively immunized mouse neonates following rotavirus challenge. |
| *E. coli* B subunit of the heat labile enterotoxin (LTB) fused toan immu-nicontraceptive epitope | Tomato fruit and leaves | No immunigenicity assays performed. | — |

**Table 9.2. continued...**

| *Antigen* | *Expression system* | *Immunogenicity* | *Protective challenge* |
|---|---|---|---|
| Hepatitis B surface antigen (HBsAg)fused to soybean vegetative protein (VSPaS) | Tobacco leaf | VSPaS-HBsAg generatee higher levels of serum lgG than native HBsAg in mice. | — |
| Hepatitis C virus (HCV) HVR1 epitope fused to *v. choleare* CTB | Tobacco leaf | Reacted with HRV1-specific monoclonal antibodies and sera from individuals infected with HCV. HVR1 epitope formed fused to CTB was immunogenic in mice when administered nasally. | — |
| Mannuheimia haemolytica A1 leukotoxin 50 fused to green fluourscent protein | White clover leaf | Rabbits produced antibodies that recoginized and neutralized leukotoxin. Immunogenic in rabbits after drying when delivered parenterally. | — |
| Ricin B fusee to green fluorescent protein (GFP) | Tobacco leaf and hairy root culture | GfP-specific lgG present in serum; lg A in serum and fecal matter. | — |
| Rotavirus enterotoxin protein NSP4 fused to CTB | Potato | Fusion proteins assembled into cholera holotoxin-like structures that retained enterocyte-binding affinity. Orally immunized mice generated levels of serum and mucosal antibodies specific for the native antigen. Induced TH1 immune response. | — |

The two recombinant cholera toxin subunit fusions were expressed from a single bidirectional promoter, ensuring coordinated expression for the two gene fusions and potentially facilitating the assembly of the chimeric holotoxin. In this approach, cholera toxin provides a scaffold for presentation of the protective epitopes, acts as a mucosal targeting molecule without toxic effects due to use of the nontoxic A2 and B subunits, and is itself a vaccine candidate. The recombinant protein represents a trivalent vaccine that can elicit significant mucosal and humoral responses against *Vibrio cholerae,* ETEC, and rotavirus. Mice, orally immunized with potatoes expressing these recombinant antigens, developed immune memory B cells as well as helper T cell type 1 (Thl) responses, which are indicators of successful immunization. Further, pups of immunized dams were protected from challenge with rotavirus, with a significantly lower morbidity rate as compared to controls. These results provide robust evidence for a vaccine strategy employing chimeric microbiological proteins expressed in plants and for multicomponent vaccines.

## Stability and Processing

Lee *et al.* carried out preliminary stability studies on dehydrated plant tissues. Clover plants expressing the *Mannheimia haemolytica* A1 leukotoxin 50 fusion protein were harvested and allowed to dry at room temperature and ambient humidity for 1– 4 days. After 3 days, the clover tissue retained approximately 20% of its initial fresh weight, but no significant degradation of the fusion protein was observed. Hence, the fusion protein did not require refrigeration for stability. The clover-derived fusion protein induced, in injected rabbits, an immune response that recognized and neutralized the native antigen in modified neutral red cytotoxicity assays.

A comprehensive stability study was conducted by Smith *et al.* on HBsAg from soybean cells. Under optimum conditions, antigen stability was maintained for at least 1 month after isolation of the surface antigen, although excess detergent rendered the antigen susceptible to proteolytic degradation by contaminating plant enzymes. This proteolysis was counteracted by the addition of skimmed milk or its protein component, which stabilized the antigen for up to 2 months. Also, by altering the sodium ascorbate concentration or buffer pH, the proportion of HBsAg displaying monoclonal-reactive epitopes increased between 8- and 20-fold.

Although antigen stability will have to be investigated on a case-by-case basis, it is obvious that simple *in vitro* manipulations may prove valuable in increasing the immunogenicity and stability of plant-derived antigens. Castanon *et al.* investigated the minimal processing of a potato-derived rabbit hemorrhagic disease virus (RHDV) vaccine consisting of the VP60 gene product. Harvested potatoes were peeled, cut into pieces, lyophilized, powdered, stored and used in trials within 3 months of collection. Extracts made from the potato powder were used in rabbits as a subcutaneous primer and intramuscular booster. The rabbits immunized with the transgenic potato elicited specific antibody responses and were protected against challenge with virulent RDHV. Freeze-dried tomato fruit expressing LTB was orally immunogenic in mice, indicating the utility of the tomato system.

# CLINICAL TRIALS

The results of five clinical trials involving orally delivered plant-derived vaccines have been published. With the exception of LTB, each study used different recombinant antigens, and each reported the stimulation of serum antibodies, and in some cases, mucosal antibodies.

## Enterotoxic *E. coli* and *Vibrio cholerae*

Two of the most widely spread and well-studied enterotoxigenic forms of bacterial diarrhea are ETEC and *Vibrio cholerae.* The toxins they produce, labile toxin (LT) and cholera toxin (CT) respectively, are very similar in primary sequence, structure, and mechanism of action. They are homologous multi-subunit proteins in which the non-toxic B subunit mediates $GM_1$ ganglioside binding, and thus are candidates for vaccines that can neutralize toxin activity. The B-subunits of these toxins, LTB and CTB, are among the most potent oral immunogens known.

Oral delivery efficiently causes the accumulation of specific serum (lgG,lgA) and mucosal (slgA) antibodies [73]. Both LT and CT also function as mucosal adjuvants, stimulating antibody production against co-delivered antigens. The ganglioside-binding activity of the LTB pentamer is required both for its mucosal immunogenicity and for the adjuvanticity of the holotoxin. Tacket *et al.* performed the first human clinical trial with a transgenic plant-derived vaccine. Fourteen volunteers ingested either 100 g of transgenic potato, 50 g of transgenic potato, or 50 g of non-transgenic potato. The LTB content of the tubers varied between 3.7 and 15.7 $\mu g\ g^{-1}$ of tuber weight and the doses were given on days 0, 7 and 21. Volunteers reported only a few instances of minor side effects (nausea, cramps or diarrhea), and the raw potato was well tolerated overall. Ten of 11 volunteers who ate the potatoes expressing LTB developed at least 4-fold increases in levels of toxin- neutralizing serum IgG against LTB ; none of the volunteers who ate the placebo potatoes showed any increase in anti-LTB antibodies. Five of 10 volunteers showed at least 4-fold increases in anti-LTB IgA, detected in stool samples. These data compared favourably with an earlier study in which volunteers were challenged with $10^9$ ETEC cells (Tacket *et al,* unpublished). This study was significant because it was the first ever to examine an edible plant vaccine in humans, and showed great potential for this new strategy. A more recent clinical study using LTB expressed in processed corn seed produced similar results to the potato study.

## Norwalk Virus

Tacket *et al.* reported a clinical trial performed using transgenic potato tubers carrying the gene for Norwalk virus capsid protein (NVCP), which assembled into Norwalk virus like particles (VLPs). Twenty adult volunteers ingested either two or three doses each of 150 g of raw transgenic potato tuber containing 215-750 μg NVCP (expression was variable). Only about 50% of the NVCP subunits assembled into VLPs in the potato cells, possibly as a result of low recombinant protein concentration. Thus, the effective amount of administered potato vaccine was 125-375 μg per dose. Unassembled subunits are likely to be much less stable in the gastrointestinal tract and thus less immunogenic. However, 19 of 20 subjects in the experimental group showed significant increases in the numbers of IgA-antibody forming cells (AFCs), ranging from 6-280 per $10^6$ peripheral blood mononuclear cells (PMBC), and 6 of 20 subjects in this group developed increases in IgG AFCs. Four volunteers shincreases in serum IgG anti-NVCP liters, four showed increased serum IgM and six showed increased IgA in their stool samples (a 17-fold mean increase). Although the antibody responses were less impressive than those obtained with LTB, the study showed that a plant-derived non-replicating antigens other than LTB and CTB could stimulate human immune responses after oral delivery.

Insect cellderived 250-μg doses of purified NVLP showed more effective seroconversion. Thus it is likely that part of the potato-delivered NVCP was unavailable for uptake in the gastrointestinal tract. More recent studies in transgenic tomato fruits with a plant-optimized NVCP gene resulted in higher expression levels and more potentimmune responses in mice fed freeze-dried tomatoes (X. Zhang and H.S. Mason,unpublished results). A clinical trial is planned with dried tomato powder formulated in gelatin capsules (D. Kirk, H.S. Mason, and C. Tacket, trial investigators) in order to evaluate safety and immunogenicity.

### Hepatitis B Virus

There is one published report concerning the immunogenicity in humans of orally delivered HBsAg expressed in plants. Two of three volunteers who ate two 150-g doses of transgenic lettuce (containing -1-2 μg HBsAg per dose) developed a modest protective (>10 IU $I^{-1}$) serum antibody titer after the second dose. The serum antibody titers declined rapidly after 4 weeks, probably due to the very low antigen dosage. However, the study showed that presumably naive subjects could be seroconverted by the oral delivery of plant-expressed HBsAg. In the US, a clinical trial was carried out at the Roswell Park Cancer Institute, Buffalo, NY, to test the oral immunogenicity of recombinant potatoes expressing HBsAg.

The study was limited to volunteers who had previously been vaccinated and seroconverted with the standard injectable yeast-derived HBsAg. The trial involved 33 volunteers who ate either two (days 0 and 28) or three (days 0, 14 and 28) 100-g doses of HBsAg potato tubers containing approximately 1 mg HBsAg per dose, while a group of 10 volunteers ate nontransgenic potatoes only. The potato HBsAg vaccine boosted serum IgG antibody titers in more than half of the volunteers (Thanavala, Mason and Arntzen, unpublished results), which suggests that the oral delivery of plant-produced HBsAg could be a viable delivery system for an HBV boosting vaccine.

### Rabies Virus

Yusibov *et al.* delivered to 14 human volunteers spinach expressing epitopes from the rabies virus glycoprotein and nucleoprotein fused to the coat protein of alfalfa mosaic virus (AlMV). Five of the fourteen subjects had previously received a conventional rabies vaccine. Three of those five and all nine of the initially naive subjects displayed significant antibody responses to the rabies virus. None of the control individuals showed a significant elevation in rabies-specific antibodies. The conventional rabies vaccine was administered to the nine initially naive subjects seven days after completing the oral vaccination. Three of these volunteers produced neutralizing antibodies against rabies virus, although none of the five control subjects did. This study showed a clear indication that the orally delivered rabies vaccine has potential as an oral booster for the conventional rabies vaccine.

## ISSUES AND CHALLENGES

### Development and Licensing of Plant-derived Vaccines

Conventional pharmaceutical development has been estimated to costs roughly between $100 million and $800 million per product, and takes over 12 years. The development of conventional vaccines may be somewhat less costly, although new recombinant DNA-derived subunit vaccines produced in fermentation-based systems are likely to be similar to protein pharmaceuticals in development costs. At present, none of the major pharmaceutical companies is directing funding towards the development of plant-derived vaccines for infectious diseases. This may reflect :

- doubts about the potential for significant return on investment;
- uncertainties in the regulatory processes;
- limited human clinical trial data that establish required dosages, timing of delivery, and evaluation of possible adverse immunological effects;
- a lack of personnel with sufficient expertise in plant biology.

As a result, the opportunity to produce vaccines in plants represents a classic example in which the reliance on market forces for the development of health products is failing. Participation of both the public sector and the non-profit sector will be essential to provide leadership and investment support to unlock the potential of plant-derived vaccines. A principal justification for the public sector promotion of plant-based vaccines is the significant favourable characteristics of this technology for the manufacture of vaccines against rare and neglected diseases. Developing new pharmaceuticals for these diseases is not a high priority for the major drug companies due to the low profit margins.

## Confronting GM Food Issues

The plant-based production of vaccines is a potentially transformative technology, but the use of a similar technology for agricultural biotechnology has stimulated significant public debate, especially focused on genetically modified foods (GM foods). Knowledgeable public debate is valuable, but the debates about GM foods have not always been based upon scientific considerations, and consequently the debates have become polarized.

Because plant-derived pharmaceuticals are not intended for use as food products, the crops that produce them must have special stewardship to ensure containment. In addition, the proteins they produce will have to be separated and purified in processing facilities dedicated to that purpose. The use of crop species that are currently part of the food supply for the production of oral vaccines will necessitate genetic separation from the food supply as an essential parameter for production of these materials. It is likely that global health organizations such as the Pan-American Health Organization (PAHO) and WHO will plan an essential role in the process of transferring the technology on a global scale as new regulatory frameworks are implemented.

## Edible Vaccines

In 2003, the UN Human Poverty Index showed that 1.2 billion of the developing world's 4.8 billion people were living on less than $1 per day, while a further 2.8 billion were living on less than $2 per day. The global poverty level, and the incidence of disease, is increasing dramatically every year. The World Health Organization estimates that approximately 170 million people worldwide are infected with hepatitis C virus (HCV), with 3-4 million new cases each year, and that more than one third of the world's population is infected with hepatitis B virus (HBV). In Asia, the prevalence of chronic hepatitis B and C is very high (about 150 million and 110 million people infected with HBV and HCV, respectively). A large majority of HCV-infected patients have severe liver cirrhosis and currently there is no vaccine available for thisdisease.

In addition, the rising cost of treatment for severe illnesses calls for the more economical production of therapeutic proteins. For example, the annual demand for insulin-like growth factor 1 (IGF-1) per cirrhotic patient is 600 mg (1.5-2 mg per day) and the cost of IGF-1 per mg is $30,000. Furthermore, the current annual cost of interferon therapy for viral hepatitis is $26,000 (Cowley & Geoffrey, Newsweek, April 22, 2002). It is evident that agricultural scale production of therapeutic proteins and vaccines is necessary to meet this large demand at an affordable cost. Plants have been used in medicine for many centuries but it is now possible to exploit plants as bioreactors for the production of human therapeutic proteins. The use of plants to produce biopharmaceuticals has several advantages, including the ability of plants to supply large quantities of therapeutic proteins to patients in a cost efficient manner.

In addition, plant-derived products are less likely than those derived from animal cells to be contaminated with human pathogenic microorganisms, since plants do not act as hosts for human infectious agents. Plants offer a number of advantages over other transgenic systems. For instance, the use of a cell fermenter or bioreactor is unnecessary and the need for post-translational modification and purification can be eliminated, thus lowering the production costs. The need for fermentation facilities has seriously limited the introduction of new thera peutic proteins in the past.

However, it has been estimated that one tobacco plant is able to produce more recombinant protein than a 300-liter fermenter of *E. coli* (Crop-Tech, VA) reflecting tobacco's biomass production of 40 tons of leaf fresh weight per acre (based on multiple mowings per season) and the production of up to one million seeds per plant. Consequently, the quantity of recombinant protein that can be harvested is only limited by the number of hectares that can be planted with transgenic crops. Furthermore, human proteins can be expressed without codon optimization, and fully assembled proteins can be produced since post-translational modifications such as glycosylation and the formation of disulfide bonds are possible in plant cells, especially within particular cellular compartments.

## HUMAN PROTEINS IN PLANTS

The first human protein produced in plants – growth hormone – was expressed as a fusion with the *Agrobacterium tumefaciens* nopaline synthase gene product. It was not until 1989 that the expression of antibodies in tobacco demonstrated the ability of plants to produce therapeutic proteins for human use. Transgenic tobacco, maize, soybean and alfalfa plants are expected to yield over 10 kg of therapeutic protein per acre, and this should reduce production costs by approximately 90% compared to other systems. However, with the exception of enzymes like phytase in tobacco (14% total soluble protein (TSP), and eubacterial glucanase in *Arabidopsis thaliana* (26% TSP, expression levels are generally less than 1% TSP in nuclear transgenic plants, especially for human therapeutic proteins. Such examples include the B subunit of *Escherichia coli* heat-labile enterotoxin (LTB) (0.01% TSP), hepatitis B virus envelope surface protein (0.01% TSP), human cytomegalo-virus glycoprotein B (0.02% TSP, transmissible gastroenteritis coronavirus glycoprotein S (0.06% TSP), human serum albumin (0.02% TSP), human protein C (0.001% TSP), human epidermal growth factor (0.0001% TSP), erythropoietin

(0.026% TSP), and human interferon-β (0.000017% of fresh weight). Therefore, strategies are required to increase expression levels, allowing commercial exploitation of the therapeutic and human proteins mentioned above. Such strategies have been the focus of transgenic plant research over the past several years.

## THE TRANSGENIC CHLOROPLAST SYSTEM

The world of plant genetic engineering research was revolutionized when it was reported that *Bacillus thuringiensis* (Bt) Cry2Aa2 protein could accumulate to a level of 46.1% TSP in transgenic tobacco chloroplasts. This was not only the highest level of recombinant protein ever achieved in plants, but a complete bacterial operon was expressed successfully for the first time resulting in the formation of stable Cry2Aa2 crystals. These results were possible because native bacterial genes were expressed at levels several hundred-fold higher in the chloroplast compartment than in nuclear transgenic plants. Chloroplast transformation not only results in higher protein expression levels compared to nuclear genetic engineering, but also provides several other advantages. Even though transgenic chloroplasts may be present in pollen, foreign genes do not escape to other crops because chloroplast DNA in not passed into the egg cell.

In most crop plants, plastid genes are inherited uniparentally in a strictly maternal fashion. Although pollen from such plants contains metabolically active plastids, the plastid DNA itself is lost during the process of pollen maturation and hence is not transmitted to the next generation. Also, the chloroplast can be a good place to store proteins, or their biosynthetic products, that might be harmful to the host plant and adversely affect its physiology if allowed to accumulate in the cytoplasm. This has been demonstrated in the case of the cholera toxin B subunit (CTB), a candidate oral subunit vaccine for cholera, because it accumulated in large quantities within transgenic plastids but was non-toxic to the plant; in contrast, even very small quantities of LTB were toxic when expressed in the cytosol. Similarly, trehalose, which is used in the pharamaceutical industry as a preservative, was very toxic when it accumulated in the cytosol but was non-toxic when it was compartmentalized within plastids. Transgenes are integrated into the spacer regions of the chloroplast genome by homologous recombination using chloroplast DNA flanking sequences. This allows site-specific integration and thus eliminates the position effects that are frequently observed in nuclear transgenic plants. As a result, it is not necessary to screen numerous putative transgenic lines to select those with high-level transgene expression.

All chloroplast transgenic lines express the same level of foreign protein, within a range of physiological variations. Yet another advantage is the lack of transgene silencing in chloroplast transgenic plants, which is a serious concern in nuclear transformation. It has been shown that there is no transcriptional gene silencing in chloroplast transgenic lines despite the accumulation of transcripts at a level 169-fold higher than in nuclear transgenic plants. Similarly, there is no transgene silencing at the translational level, despite the accumulation of foreign protein at levels up to 47%of the total plant protein in chloroplast transgenic lines. Purification costs can account for much of the expense involved in biopharmaceutical production. In the case of insulin, for example, chromatography alone accounts for 30% of the production costs and 70% of the set-up costs.

It is estimated that the oral delivery of properly folded and fully functional biopharmaceuticals in plant tissues could potentially reduce production costs by 90%. For such oral delivery to be successful, antibiotic selection should be avoided and the biopharmaceutical protein must be expressed in edible parts of the plant that require no cooking or processing. Antibiotic-free selection using a gene of plant origin has been developed recently". Alternatively, antibiotic resistance genes can be eliminated using direct repeats or theCre-*liox*P system. Another requirement for oral delivery is the ability to express foreign proteins in plastids that are present in non-green tissues. One such example is the expression of an antibiotic resistance gene *(aadA)* in tomato chromoplasts. More recently, stable and highly efficient plastid transformation has been achieved in the non-green tissues of carrot plants. Using carrots as the source of an edible vaccine or therapeutic protein provides many advantages, such as the uniform source of cell culture (which is one of the essential requirements for producing therapeutic proteins, i.e. a homogeneous single source of origin), rapid division, the large biomass using bioreactors, and direct delivery of a precise dose of vaccine antigens or biopharmaceuticals.

Furthermore, when carrots are used as the delivery vehicle, cooking is unnecessary and this preserves the structural integrity of therapeutic proteins during consumption. Most importantly, carrot plants do not flower in the first year. Therefore, there are no flowers or reproductive structures during the harvest and this provides complete gene containment, preventing outcross via pollen and gene flow via seeds. One of the most important features of the transgenic chloroplast system is that chloroplasts are able to carry out the processing of eukaryotic proteins, including the correct folding of subunits and the formation of disulfide bridges. Functional assays showed that chloroplast-derived cholera toxin B subunit (CTB) binds to the intestinal membrane $G_{M1}$-ganglioside receptor, confirming correct folding and disulfide bond formation in the plant-derived CTB pentamers. Formation of disulfide bonds within a single polypeptide, resulting in a fully functional human therapeutic protein, has also been demonstrated for human somatotropin (growth hormone), interferon-α and interferon-γ.

Human therapeutic proteins as small as 20 amino acids or as large as the anthrax protective antigen (83 kDa) have been expressed in transgenic chloroplasts. The ability to express operons or multi-gene cassettes allows the engineering of foreign pathways and the expression of multi-subunit therapeutic proteins. The successful assembly of a monoclonal antibody in transgenic chloroplasts has demonstrated the ability of this system to produce complex multi-subunit proteins and has also indicated the presence of chaperones required for the assembly of such proteins within plastids. These observations usher in a new era for the production of therapeutic proteins via chloroplast genetic engineering. The rest of this chapter provides examples of therapeutic proteins expressed in transgenic chloroplasts.

## Chloroplast-derived Human Antibodies

The most widely studied therapeutic proteins produced in plants include monoclonal antibodies for passive immunotherapy and antigens for use as oral vaccines. Antibodies against dental caries, rheumatoid arthritis, cholera, *E. coli* diarrhea, malaria, certain cancers, Norwalk virus, HIV, rhinovirus, influenza, hepatitis B virus and herpes simplex virus have been produced in transgenic plants. However, the anti- *Streptococcus mutans* secretory antibody for the prevention of dental caries is the only plant-derived antibody currently in Phase II clinical trials. Until recently, most antibodies were expressed in tobacco, potato, alfalfa, soybean, rice and wheat. It has been estimated that for every 170 tons of harvested tobacco, 100 tons represents harvested leaves. A single hectare could thus yield 50 kg of secretory IgA.

**Table 9.3. The production of therapeutic proteins for human or animal health in transgenic chloroplasts**

| *Biopharmaceutical proteins / Vaccine* | *Gene* | *Site of Integration* | *Promoter* | *5′/3′ regulatory elements* | *% tsp expression* | *Laboratory* |
|---|---|---|---|---|---|---|
| Elastin derived polymer | EG121 | *tml/tmA* | Prrn | T7gene10/ TpsbA TpsbA | ND | Daniell |
| Human somatotropin | HST | trnV/rpsl2/7 | Prrn[a], PpsbA[b] | T7gene10[a] or psbA[b]/Trpsl6 | 7.0%[a] and 1.0% [b] | Monsanto |
| Cholera toxin | *CtxB* | *trnl/tmA* | Prrn | ggagg/TpsbA | 4% | Daniell |
| Antimicrobial peptide | MSI-99 | *tml/tmA* | Prrn | ggagg/TpsbA | 21.5-47% | Daniell |
| Insulin like growth factor | *JGF*-1 | *trnl/tmA* | Prrn | PpsbA/TpsbA | 33% | Daniell |
| Interferon alpha 5 | *INfα5* | *tml/tmA* | Prrn | *PpsbA/TpsbA* | ND | Daniell |
| Interferon alpha 2b | *INFα2B* | *trnl/tmA* | Prrn | PpsbA/TpsbA | 19% | Daniell |
| Human Serum Albumin | *Has* | *tml/tmA* | Prrn[a], PpsbA[b] | ggagg[a], psbA[b]/ TpsbA | 0.02 %[a], 11.1%[b] | Daniell |
| Interferon gamma | *IFN-g* | *rbcL/accD* | *PpsbA* | PpsbA/TpsbA | 6% | Reddy |
| Monoclonal antibodies | *Guy's 13* | *tmI/trnA* | Prrn | ggagg/TpsbA | ND | Daniell |
| Anthrax protective antigen | Pag | *tml/tmA* | Prrn | *PpsbA/TpsbA* | 4–5% | Daniell |
| Plague F1~V fusion antigen | CaFl~VLcrV | *tmF1/trnA* | Prrn | *PpsbA/TpsbA* | 14.8% | Daniell |
| CPV VP2 protein | *CTB-2L21[a], GFP-2L21[b]* | *tml/trnA* | Prrn | *PpsbA/TpsbA* | 31.1%[a], 22.6%[b] | Daniell/ Veramendi |

Furthermore, it has been estimated that the cost of antibody production in plants half that in transgenic animals and 20 times lower than in mammalian cell cultures Based on the chloroplast's ability to produce bioactive and fully assembled proteins, a codon-optimized and humanized gene encoding a chimeric monoclonal antibody (IgA/G, Guy's 13) under the control of a specific 5'-untranslated region, was used successfully to synthesize and assemble monoclonal antibodies in transgenic tobacco chloroplasts, including the formation of disulfide bridges. Guy's 13 was developed to prevent dental caries, which is caused by *Streptococcus mutans* Daniell and Wycoff confirmed integration of the chimeric antibody gene into the chloroplast genome using PCR and Southern blot hybridization analysis. Protein expression was also confirmed by western blot analysis. Expression of the individual heavy and light chains as well as the fully assembled antibody was shown indicating the presence of chaperones for proper protein folding and enzymes for formation of disulfide bonds within transgenic chloroplasts.

## Chloroplast-derived Biopharmaceuticals

### *Human Serum Albumin*

Human serum albumin (HSA) is currently obtained by the fractionation of blood serum. It accounts for 60 % of the total protein in blood serum and it is the most widely used intravenous protein. Because of the low expression of HSA in nuclear transgenic plants (approximately 0.02% TSP it would be desirable to produce this biopharmaceutical protein in transgenic chloroplasts. To achieve higher expression levels, the HSA gene was placed under the translational control of three different regulatory elements: a Shine-Dalgarno sequence (SD), the 5′ *psbA* region, and the cry2Aa2 UTR. In seedlings transformed with the SD-HSA construct, HSA accumulated to a level of only 0.8% TSP, while 1.6% TSP was achieved using the 5' *psbA* control region, and 5.9% TSP was achieved using the cry2Aa2 UTR. On the other hand, HSA reached a maximum of 0.02%, 0.8% and 7.2% TSP in transgenic potted plants when the transgene was regulated by the SD sequence, *cry2Aa2* UTR and 5′ *psbA* sequence, respectively. This demonstrated that excessive proteolytic degradation could reduce overall yields unless compensated by enhanced translation Furthermore, purification of HSA was possible by centrifugation, due to the formation of inclusion bodies within transgenic chloroplasts.

Inclusion bodies were precipitated by centrifugation and easily separated from the majority of cellular proteins present in the soluble fraction by a single centrifugation step, which may eliminate the need for expensive affinity columns or chromatographic techniques. Because the 5′ *psbA* region is light dependent, the accumulation of HSA protein should be increased after continuous illumination. Fernandez-San Millan *et al.* monitored HSA accumulation by ELISA. A maximum of 50 h of continuous light produced an HSA yield of 11.1% TSP in mature leaves. This is the highest expression of H S A demonstrated so far and is 500-fold higher than previous reports of HSA expression in the leaves of nuclear transgenic plants.

Due to the high local concentration of the protein, inclusion bodies were formed and this protected it from proteolytic cleavage. Another positive result from this study was the knowledge gained about the regulatory elements (SD, 5′ *psbA* region and the *cry2Aa2* UTR). These serve as a model system for enhancing the expression of foreign proteins that are highly susceptible to proteolytic degradation and provide advantages during purification through the use of current purification techniques for inclusion bodies.

### *Human Insulin-like Growth Factor-1*

Insulin-like growth factor-I (IGF-I) is a naturally occurring, single chain polypeptide with three disulfide bonds. It is produced in the liver and has a molecular mass of 7649 Da. IGF-I is not only critical for the growth of muscle and other tissues, but its therapeutic potential currently is being evaluated in diabetes, IGF-I induced neuroprotection, and in the promotion of bone healing. This therapeutic protein is commercially valuable because cirrhotic patients require 600 mg of IGF-1 per year at a cost of 530,000 per mg. In the past, IGF-1 has been expressed in *E. coli* but the mature protein cannot be produced in this system because disulfide bonds do not form in the *E. coli* cytoplasm. Since IGF-I has a high eukaryotic codon content, chloroplast codon optimization was performed by Ruiz and a synthetic IGF-1 gene was created to further increase expression levels.

Integration of the IGF-1 gene into the tobacco chloroplast genome was achieved through homologous recombination and was confirmed using PCR and Southern blot analysis. Western blotting and chemiluminescence techniques were used to verify the high levels of IGF-I expression in transgenic chloroplasts. ELISA tests were used to quantify IGF-1 in transgenic chloroplasts containing the native and synthetic *IGF*-1 genes, revealing levels as high as 32%

TSP. Most importantly, these observations showed that the chloroplast translation machinery is quite flexible, unlike the bacterial translation machinery, which could only translate the synthetic chloroplast-codon optimized *IGF-1* gene.

## Human Interferon (IFNα2b)

Malignant carcinoid tumors, a symptom of carcinoid syndrome, presented a therapeutic challenge for many years until the introduction of interferon treatment in 1982 by Oberg and Eriksson. They reported that 47 out of 111 patients (42%) treated with interferon-α (median dose of 6 mega-units (MU) of interferon-α, five times weekly) demonstrated a significant biochemical response and 15% demonstrated more than 50% reduction in tumor size. Survival analysis demonstrated a median survival period of only 8 months in the group of patients treated with chemotherapy, compared to 80+ months ($p < 0.001$) in the groups treated with interferon-α. However, alpha interferons have other therapeutic uses, such as the inhibition of viral replication and cell proliferation, enhancement of the immune response, and most recently, the treatment of patients suffering from West Nile virus.

The Food and Drug Administration first approved a specific subtype of interferon-α (IFNα2b) in 1986 for the treatment of hairy cell leukemia in the United States and the recombinant IFNα2b now on the market is produced using an *E. coli* expression system. In plants, nuclear transformation has resulted in very low expression levels (0.000017%) TSP in transgenic tobacco and rice. Due to necessary *in vitro* processing and purification, the average cost of treatment is $26,000 per year. The injection of IFNα2b causes side effects such as fatigue, weight loss and anemia due to diarrhea, as well as flushing and bronchoconstriction. Treatment with alpha interferons became less popular when it was found that up to 20% of patients produce anti-IFNα antibodies, which reduce the effectiveness of the treatment because IFNα2b aggregates with human serum albumin in the blood. However, oral administration could be beneficial in the treatment of various infectious diseases. In the Daniell laboratory, rerambinant IFNα2b containing a polyhistidine purification tag and a thrombin cleavage site was expressed in transgenic tobacco chloroplasts, for use as an oral therapeutic.

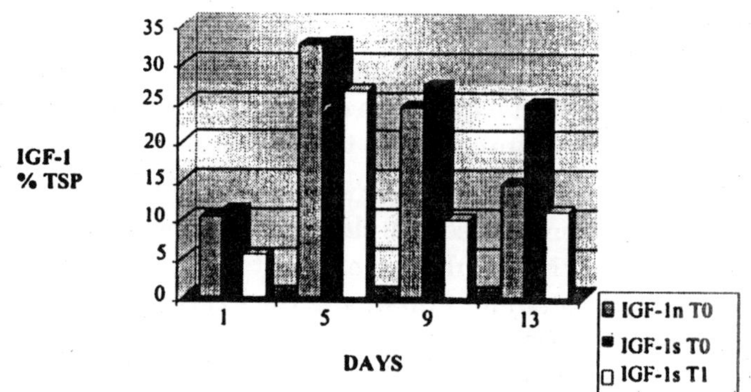


**Fig. 9.2. Expression of IGF-1 in transgenic chloroplasts after continuous light exposure for 13 days. IGF-1 expression is shown as a percentage of the total soluble protein. IGF-1 n is the native gene and IGF-1s is the chloroplast codon-optimized gene.**

For comparison, the gene cassette was integrated into the chloroplast genome of cv. Petit Havana and into a low-nicotine variety of tobacco, LAMD-609. Western blots performed using a monoclonal antibody to detected monomers and multimers of IFNα2b in both tobacco varieties, and disulfide bond formation was confirmed. Integration of the transgene into both chloroplast genomes was confirmed by Southern blot analysis.

In the Petit Havana transgenic lines, chloroplast genome homoplasmy occurred in the first generation, and this corresponded to the highest level of IFNα2b expression. ELISAs revealed IFNα2b levels of up to 18.8% TSP in Petit Havana and up to 12.5% TSP in LAMD-609. IFNα2b activity was confirmed by the ability of the recombinant protein to protect HeLa cells against the cytopathic effects of encephalomyocarditis virus (EMC) and through the identification of interferon-induced transcripts. Chloroplast derived IFNα2b was found to have the same activity as commercially produced IntronA. The mRNA levels of two genes induced by IFNα2b (2'-5' oligoadenylate synthase and *STAT-2)* were tested by RT-PCR using primers specific for each gene.

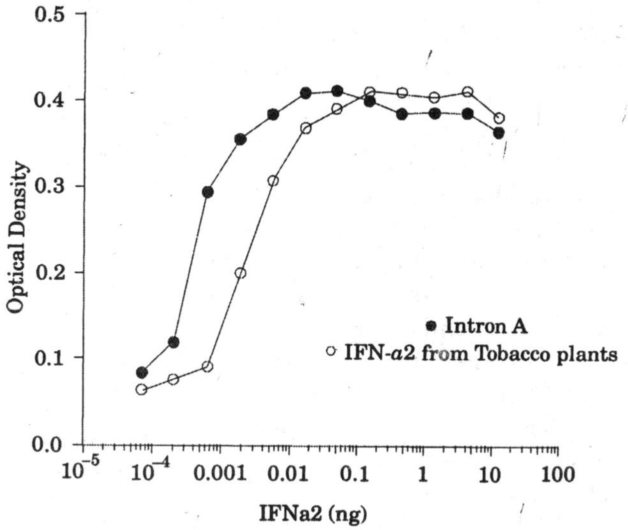

Chloroplast-derived IFNa2b induced the expression of both genes in a manner similar to commercial IFNα2b. These levels of expression and functionality are ideal for IFNα2b purification and for the further use of oral IFNα2b delivery in pre-clinical studies.

**Fig. 9.3. Demonstration of IFNα2b functionality by the ability of IFNα2 b to protect HeLa cells against the cytopathic effect of encephalomyocarditis virus (EMC). Chloroplast derived IFNα2b was as active as commercially produced Intron A.**

## *Anti-Microbial Peptides (AMPs): MSI-99*

Due to the increasing number of drug-resistant bacteria arising from the misuse of antibiotics and anti-microbial agents, new strategies to combat or prevent different human pathogens need to be developed. Research on defense peptides secreted from the skin of the African clawed frog *(Xenopus laevis)* has resulted in the discovery of magainin, an amphipathic α-helix-containing peptide, which has been investigated as a broad-spectrum topical agent, a systemic antibiotic, a wound healing agent, and an anticancer drug. Magainin has an affinity for negatively charged phospholipids in the outer leaflet of the prokaryotic membrane. A sequence encoding the magainin analog MSI-99, a synthetic lytic peptide, was integrated into the chloroplast genome of tobacco variety Petit Havana and expressed at high leyels (21.5-47%TSP)The minimum inhibitory concentration of MSI-99 was investigated.

Based on total inhibition of bacterial and fungal cells, MSI-99 was most effective against P. *syringae,* requiring only 1 ug per 1000 bacteria. The amount of antimicrobial peptide required to kill bacteria was used to estimate the level of expression in transgenic plants, since the lytic activity of antimicrobial peptides is concentration dependent. The effectiveness of the chloroplast-derived lytic peptide was tested *in vitro,* using a multi-drug resistant Gram-negative bacterium, *Pseudomonas aeruginosa,* which is an opportunistic pathogen of plants, animals and humans. Cell extracts prepared from $T_r$generation plants resulted in 96% inhibition in growth of this pathogen. These results may provide an alternative method for combating drug-resistant human pathogenic bacteria. In addition, the lytic peptide may also be useful as a treatment for patients suffering from cystic fibrosis, who are extermely susceptible to *P. aeruginosa infections*. As stated above, AMPs such as magaining have been studied as possible anticancer agents because of their preference for negatively charged phospholipids.

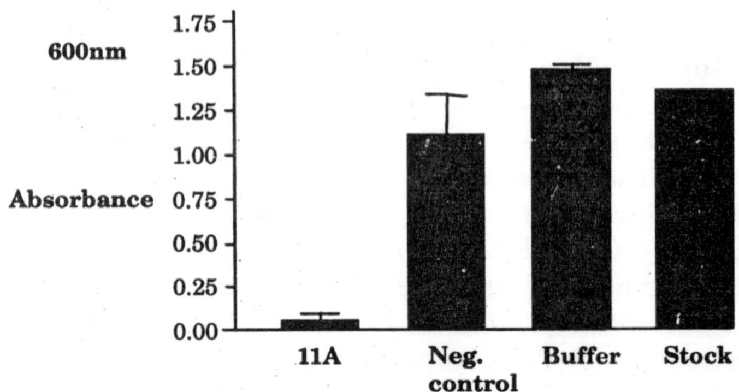


**Fig. 9.4. In vitro bioassay of $T_1$ generation chloroplast transgenic line against P. *aeruginosa*. Bacterial cells from an overnight culture were diluted to $A_{600}$ 0.1-0.3 and incubated for 2 hours at 25°C with 100 μg of total protein extract. One ml of LB was added to each sample and incubated overnight at 26 °C. Absorbance was recorded at 600 nm. Data was analyzed using a GraphPad Prism.**

It has been reported that the outer leaflet of melanoma and colon carcinoma cells contain 3-7 fold higher levels of phosphatidylserine than normal cells. Previous studies have shown that analogs of magainin were effective against hematopoietic cancer, melanoma, sarcoma and ovarian teratoma lines. The transgenic chloroplast system could thus provide sufficient expression levels for the commercially feasible production of AMPs as anticancer drugs and as antimicrobial agents.

## Chloroplast-derived Vaccine Antigens

Edible vaccines are currently being developed for a number of human and animal diseases, including measles, cholera, foot and mouth disease, hepatitis B and C. Many of these diseases ar likely to require booster vaccinations or multiple antigens to induce and maintain protective immunity. This problem could be solved, given the ability of plant to express more than one transgene, allowing the delivery of multiple antigens for repeated inoculations. Other advantages of plant-based vaccines include the reduced neeed for medical personnel and sterile injection conditions, heat stability, as explained by Webster et al. it is unlikely that an edible vaccine would lead to oral tolerance because it is achieved by dose-specific oral antigen delivery. To explain the concept of oral therapeutic protein delivery, hepatitis B surface antigen (HBsAg) was used for oral therapeutic protein delivery, hepatitis B surface antigen (HBsAg) was used for oral immunization.

No primary immune response was detected after two 300-μg doses of yeast-derived HBsAg. Howerver, a primary response began after two servings oftrangenic potatoes containgin 85-300μg of HBsAg. The main advantage of expressing biopharamaceuticals in plants is bioencapsulation. since the proteins are presented inside plant material, degradation occurs only slowly and a larger quantity of the protein can survive proteolysis. For this reason, less of the recombinant protein is needed to achieve the same theropeutic effect as a non-orally delivered protein.

## *Cholera Toxin B Subunit (CTB)*

The first step in developing a highly expressed edible vaccine containing an adjuvant is the successful expression of CTB in transgenic plants. Since CTB has previously been expressed in nuclear transgenic plants at levels of 0.01% TSP (leaves) and 0.03% TSP (tubers), expression of this adjuvant should increase if the correspinding gene is integrated into the chloroplast genome. Integration of an unmodified CTB-coding sequence into the chloroplast genome, confirmed by PCR and Southern blot analysis, resulted in the accumulation of CTB as funcitional oligomers, at levels of up to 4.1% TSP in transgenic chloroplast. $C_{MI}$-Ganglioside binding assays confirmed that chloroplast-derived CTB binds to the intestinal membrane receptor of cholera toxin establishing unequivocally that chloroplasts are capable of forming disulfide bridges during the assembly of foreign proteins. These results and the stable transformation of plastids in edible plant parts confirm the feasibility of producing in large quantities edible vaccines, of adjuvant-antigen fusion proteins, capable of eliciting an immune response in humans. In has been often shown that an adjuvant increases the immune response and in turn raises the antibody titer.

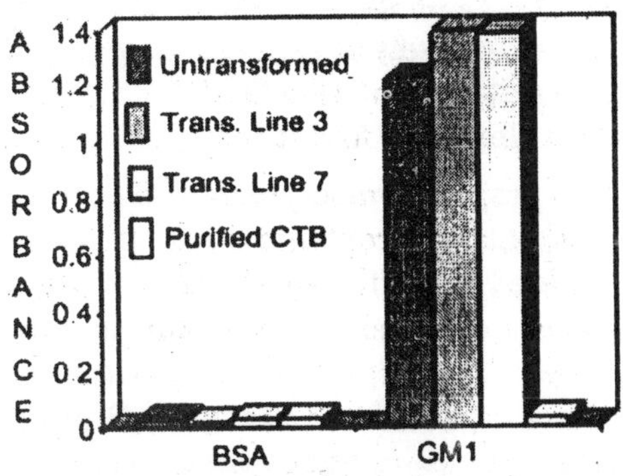


**Fig. 9.5. CTB-$G_{M1}$-ganglioside binding ELISA assay. Plates, coated first with $G_{M1}$-ganglioside and bovine serum albumin (BSA), respectively, were irrigated with total soluble plant protein from chloroplast transgenic lines (3 and 7) and 300 mg of purified bacterial CTB. The absorbance of the $G_{M1}$-ganglioside-CTB-antibody complex in each case was measured at 405 nm. Total soluble protein from untransformed plants was used as the negative control.**

## *Bacillus anthracis Protective Antigen*

The anthrax bioterrorist attacks that followed the events of September 11$^{th}$ 2001 resulted in a renewed interest *Bacillus anthracis,* the causative agent of this disease. Research has focused on the development of better vaccines than the one currently available. It has been estimated that the aerosolized release of 100 kg of anthrax spores upwind of Washington DC would cause mortalities of 1,30,000-30,00,000. Nonetheless, wild-type *Bacillus anthracis* is susceptible to conventional antibiotics, including penicillin, oxyfloxacin and ciprofloxacin. The problem lies not with the bacterial infection itself, but with three proteins released by the bacteria-protective antigen (PA, 83 kDa), lethal factor (LF, 90 kDa) and edema factor (EF, 89 kDa) -known as anthrax toxins. The Centers for Disease Control list *Bacillus anthracis* as a category A biological agent and estimate the cost of an anthrax attack to exceed $26 billion per 100,000 exposed individuals. Concerns regarding vaccine purity, the need for multiple injections, and the limited supply of PA, underscore the urgent need for an improved vaccine.

Therefore, in the Daniell laboratory, the PA gene *(pag)* was inserted into a chloroplast vector along with the *psbA* regulatory signals for enhanced translation. As in other investigations, integration of transgenes into the chloroplast genome was confirmed by PCR and Southern blot analysis. Crude plant extracts contained up to 2.5 mg of full length PA per gram of fresh leaf tissue and showed exceptional stability (several months in stored leaves or crude extracts). Maximum levels of expression were observed in mature leaves under continuous illumination. Co-expression of the ORF2 chaperonin from *Bacillus thuringiensis* did not increase PA accumulation or fold it into cuboidal crystals in transgenic chloroplasts.

Trypsin, chymotrypsin and furin proteolytic cleavage sites present in PA were protected in transgenic chloroplasts because only PA 83 was detected. Both CHAPS and SDS detergents extracted PA with equal efficiency and PA was detected in the soluble fraction. Chloroplast-derived PA bound to anthrax toxin receptor, heptamerized, and bound to lethal factor, resulting in macrophage lysis. Up to 25 μg of functional PA per ml of crude extract was observed. With an average yield of 172 mg of PA per plant, 400 million doses of vaccine (free of contaminants) could be produced per acre of transgenic tobacco, using a low yielding experimental cultivar in a greenhouse, which could be further enhanced 20-fold in commercial cultivars cultivated in the field (with multiple harvests).

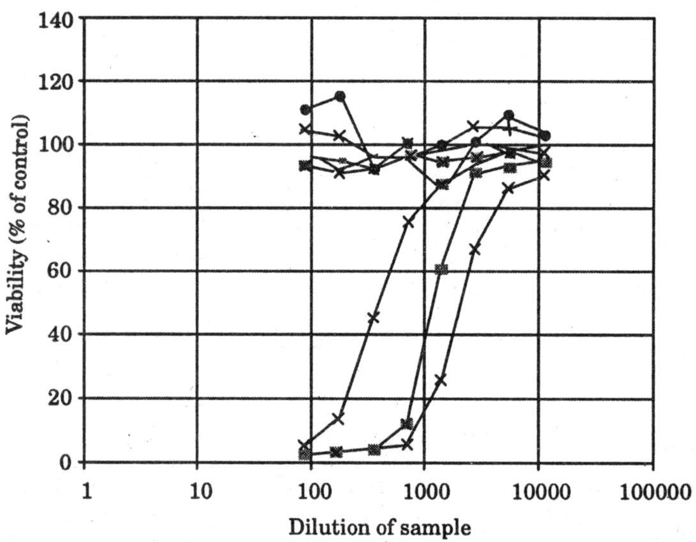


**Fig. 9.6. Macrophage cytotoxic assays of plant extracts. Supernatant samples were tested from T1-generation of chloroplast transgenic line pLD-JWl (proteins were extracted in buffer containing no detergent and MTT was added after 5 hours). pLD-JWl (extract stored for 2 days); pLD-JWl (extract stored for 7 days); PA 5 μg ml$^{-1}$; —x— Control wild type (extract stored for 2 days); Control wild type (extract stored for 7 days; Control wild type no lethal factor (LF) (extract stored for 2 days); —•— Control wild type no LF (extract stored for 7 days); Control pLD-JWl no LF (extract stored for 2 days); Control pLD-JWl no LF (extract stored for 7 days).**

## *Yersinia pestis* F1~V Fusion Antigen

*Yersinia pestis,* the Gram-negative bacterium responsible for plague, is also listed as a category A biological agent by the Centers for Disease Control. *Y. pestis* is naturally transmitted from one animal host to another either directly or via a flea vector. This pathogen has been used for centuries as a biological warfare agent and it raises concerns today as one of the microorganisms with potential for use against civilian or military populations. If used as a warfare agent, the pneumonic form of plague would be the most probable outcome of infection and is devastating due to the rapid onset of the disease, the high mortality rate, and the rapid spread of the disease. If a person with pneumonic plague is not treated with antibiotics within 18 hours of the onset of symptoms, he or she will die. Therefore, immunization against aerosolized plague has been an area of interest in current research. The current, killed whole unit vaccine is only slightly effective against the bubonic form of plague and not immunoprotective against pneumonic and septicemic'

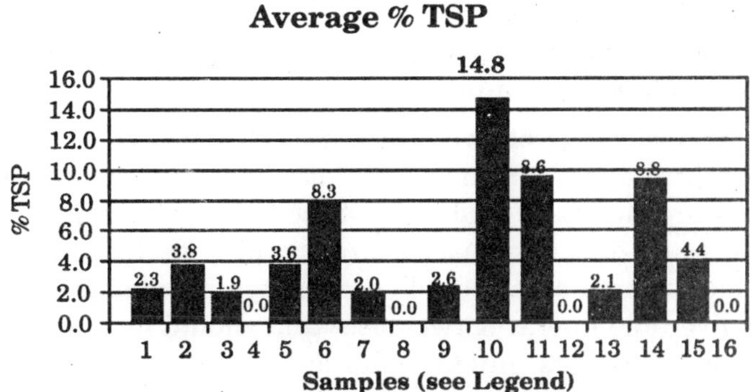


**Fig. 9.7. Enzyme linked immunoassays, with protein yields expressed as % TSP and μg of F1V per gram of fresh leaf material. For the continuous illumination experiment, leaf material was sampled on days 0, 1, 3 and 5. Young, mature, and old samples were taken for each experiment. The figure shows average levels of F1V antigen for: Samples 1-4 (DayO: (1) young, (2) mature, (3) old, and (4) wild type leaf samples); Samples 5-8 (Day 1: (5) young, (6) mature, (7) old and (8) wild type leaf samples)); Samples 9-12 (Day 3: (9) young, (10) mature, (11) old and (12) wild type leaf samples); Samples 13-16 (Day 5: (13) young, (14) mature, (15) old &. (16) wild type leaf samples).**

Several subunit vaccines have been evaluated for immunogenicity and protective efficacy against *Y. pestis*. CaFl and LcrV are the two most favourable choices. F1 is a capsular protein located on the surface of the bacterium with anti-phagocytic properties. The V antigen is a component of the *Y. pestis* Type III secretion system and it may form part of an injectosome. Expression of these antigens as fusion proteins in *Salmonella typhimurium* has been shown to be immunogenic in mice, but has not proven to be safe. However, when these antigens are produced in *E. coli,* they have been shown to be safe and to provide protection against subcutaneous and aerosol challenges of *Y. pestis* in mice. Expressing subunit vaccines in chloroplasts is very promising for three reasons: (*i*) subunit vaccines, which do not express active toxins, are safe and do not multiply; (*ii*) bacterial and many viral genes have a high AT content allowing for high-level expression in the chloroplast; and (iii) oral delivery of vaccines yields high mucosal IgA liters along with high systemic IgG titers, enabling the immune system to fight off infectious agents at important entry points (e. g. the lungs and vagina). Investigations in the Danielllaboratory demonstrated expression of the F1 and V antigens in transgenic chloroplasts as a recombinant F1~V fusion protein, consisting of the F1 protein fused at its carboxyl terminus to the amino terminus of the entire V antigen.

The gene encoding the fusion proteins was obtained from the U.S. Army Medical Research Institute of Infectious Diseases (USAMRIID) and integrated into the chloroplast genome of Petit Havana and LAMD-609 varieties using a universal transformation vector. Expression levels of F1~V in the two varieties were compared by western blotting and ELISA, after continuous illumination for up to five days. For the continuous illumination analysis, leaf material was sampled on days 0, 1, 3 and 5, and samples were taken from young, mature and old leaves. As stated above, the *psbA* 5' UTR, which was used to control the fusion gene, is light regulated, and continuous illumination therefore enhanced translation. This led to a buildup of F1~V protein until the plant became stressed by the unnatural light period.

Mature leaves showed the highest yields, with the largest amount of F1~V accumulating on the third day of continuous illumination, producing an average yield of 14.8% TSP. Functional assays are currently underway. Since plague remains endemic in some regions of the world, and because of the increased threat that *Y. pestis* could be used as a biological warfare agent, the development of improved vaccines against plague is a high priority. The ideal vaccine should be deliverable in one or two doses and should have the ability to produce high-titer and long-lasting antibodies quickly. Moreover, such a vaccine should protect against aerosolized transmission of *Y. pestis*.

### *Canine Parvovirus (CPV) VP2 Protein*

Canine parvovirus (CPV) infects dogs and other Canidae such as wolves, coyotes, South American dogs and Asiatic raccoon dogs, producing haemorrhagic gastroenteritis and myocarditis. CPV-1 was the first strain of canine parvovirus to be discovered, and was initially described in 1967. It did not pose much of a threat except to newborn puppies. However, the CPV-2 strain, which appeared in the US in 1978, appeared to be a mutated form of the feline parvovirus (more commonly known as feline distemper virus). Infected animals shed CPV-2, which is hardy in the environment, in large numbers, and this led to rapid worldwide dissemination. Attempts to shield puppies from exposure were completely futile. A second mutant, CPV-2 a, was recognized in 1979 and was found to be more aggressive.

The vaccine was in short supply and many veterinarians had to use feline distemper vaccine as a substitute, since it was the closest one available. The most common form of the virus today is CPV-2 b and young animals are commonly vaccinated with an attenuated whole virus vaccine. Advances in molecular biology have lead to the identification of CPV antigens that are capable of eliciting a protective immune response. A linear antigenic peptide (2L21) from the VP2 capsid protein (amino acids 1-23) of CPV was selected and its DNA sequence was successfully introduced into tobacco chloroplasts. The 2L21 synthetic peptide, chemically coupled to a KLH carrier protein, has been extensively studied and has been shown to protect dogs and minks against parvovirus infection very effectively. This peptide was expressed in nuclear transgenic plants as an N-terminal fusion protein with β-glucuronidase (GUS) but expression levels were inadequate. Therefore, the 2L21 peptide, fused either to the cholera toxin B subunit (CTB) or the green fluorescent protein (GFP) was expressed in chloroplast transgenic plants, and accumulated to levels up to ten-fold higher than those previously reported in nuclear transformation.

The expression levels were dependent on plant age. Both young and senescent plants accumulated lower amounts of proteins than mature plants, showing that the time of harvest is important when scaling up the process of protein production. The maximum level of CTB-2L21 was 7.49 mg $g^{-1}$ fresh weight (equivalent to 31.1% TSP) and that of GFP-2L21 was 5.96mg $g^{-1}$ fresh weight (equivalent to 22.6% TSP). The inserted epitope was detected with a CPV-neutralizing monoclonal antibody, indicating that the epitope is correctly positioned at the C-terminus of the fusion proteins. The resulting chimeric CTB-2L21 protein retained an ability to form pentamers, possessed the $G_{M1}$-ganglioside binding characteristics of the native CTB, and induced antibodies that were able to recognize the VP2 protein of CPV following the intraperitoneal immunization of mice. These results show that plant derived CTB-2L21 recombinant protein is immunogenic via intraperitoneal administration, as shown by its ability to induce a humoral response that cross-reacts with the native VP2 protein.

Additional experiments are underway to check the ability of this fusion protein to induce specific immune responses after mucosal delivery. This is the first report of an animal vaccine epitope expressed in transgenic chloroplasts.

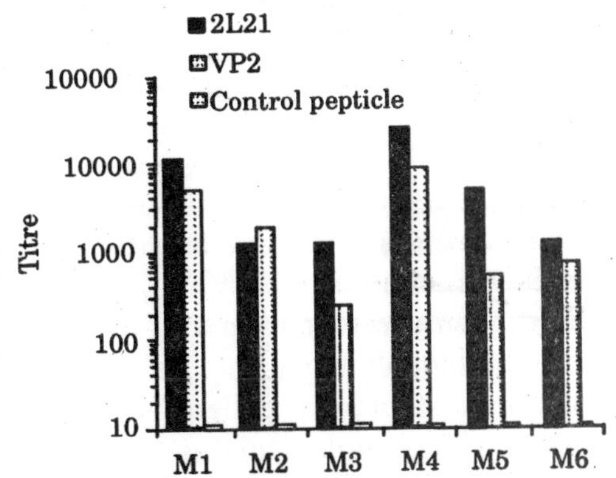


**Fig. 9.8. Titers of antibodies at day 50 induced by plant-derived CTB-2L21 recombinant protein. Balb/c mice were intraperitoneally immunized with leaf extract from CTB-2L21 transgenic plants. Animals were boosted at days 21 and 35. Each mouse received 20 µg of CTB-2L21 recombinant protein. Individual samples of mouse serum were titrated against 2L21 synthetic peptide,VP2 protein and a control peptide (amino acids 122-135 of hepatitis B virus surface antigen). Titers were expressed as the highest serum dilution lo yield twice Ihe absorbance mean of preimmune sera. M1-M6: mice 1 lo 6; 2L21: epitope from the VP2 protein of the canine parvovirus; CTB: cholera toxin B; VP2: protein of the canine parvovirus that includes the 2L21 epitope.**

## PURIFICATION STRATEGIES

The purification of therapeutic proteins expressed in transgenic systems is a major challenge. For example, insulin produced in *E. coli* is purified by chromatography, which accounts for 30% of operating expenses and 70% of set up costs. Therefore, new strategies for biopharmaceutical purification are currently being investigated. To overcome the high cost of purification, protein-based polymers have been produced in *E. coli* through genetic engineering. For example, a synthetic polymer gene encoding the polypeptide $(GVGVP)_{121}$ was overexpressed in *E. coli* to the extent that polymer inclusion bodies occupied 80-90% of the cell volume. The inverse temperature transition property of this polymer makes it an ideal fusion protein for the purification of therapeutic proteins. The protein is soluble in water at temperatures below 25 °C, but the polymer aggregates into a more-or-dered, viscoelastic state called a coacevate at 37 °C.

Indeed, it is known that polypeptides, proteins, protein-based polymers and any polymers in which the hydrophobic and polar residues are in the correct balance, will fold and assemble (i.e. become more ordered) as the temperature increases and will disassemble and unfold (i.e. become less ordered) as the temperature decreases. This inverse temperature transition property makes it easier and cheaper to harvest polymers in aqueous solutions simply by increasing the temperature and this avoids cumbersome purification procedures and the use of enzymes and organic solvents which may alter the quality of the polymer or any biopharmaceutical protein. The idea of producing biopolymers in plants instead of bacteria seems very appealing since a further decrease in production costs could be achieved.

Genetic engineering of plants should allow the production of polymers at much lower costs and in higher volumes than are possible with microbial fermentation. Because of these advantages, a large protein based polymer $(GVGVP)_{121}$ was expressed in transgenic tobacco chloroplasts. Research in the Daniell laboratory confirms the possibility of producing recombinant insulin using this inexpensive purification approach and offers the prospect of low-cost treatment for diabetic patients. The proinsulin gene was fused to a gene encoding the synthetic biopolymer $(GVGVP)_{40}$, expressed in *E. coli* and the fusion protein was purified by raising the temperature from 4 °C to 42 °C, utilizing the inverse temperature transition property. At 4°C, the biopolymer exists as an extended molecule, but when incubated at 42 °C it folds into dynamic structures called β-spirals that further aggregate by hydrophobic association to form twisted filaments. Raising the temperature to 42 °C does not affect the tertiary structure of the proinsulin polypeptide. This has been demonstrated previously by the production of human insulin in bacteria using a temperature-responsive promoter. Production of the recombinant protein was induced by temperature shift from 30 °C to 42 °C with, out any adverse effects on protein stability. The fusion protein was purified twice using the inverse temperature transition property of the biopolymer, and the protein was analyzed by SDS-PAGE, western blotting and immunodetection using mouse anti-human proinsulin primary antibody.

Negative staining of the SDS-polyacrylamide gels with 0.3 M $CuCl_2$ confirmed the presence of the proinsulin-polymer fusion protein. Copper-stained gels with a black background appear to have dark bands against a light, semiopaque background when illuminated obliquely from above. The same gel was then stained with Commassie R-250, which would stain only the proinsulin but not the polymer. Commassie R-250 does not stain polymer proteins without a fusion, since the sulphonic acid groups on the dye only form ion pairs with lysine and arginine, which are not present in $(GVGVP)_{40}$. Since Commassie R-250 staining was successful, this confirmed the presence of the fusion protein. Thus, the polymer-proinsulin fusion protein has been confirmed by western blot, and two different staining methods.

CHAPTER

10

# MINERAL NUTRIENTS

Higher plants are autotrophic organisms that can synthesize all of their molecular components from inorganic nutrients obtained from the local environment. For many mineral nutrients, this involves uptake by the plant and incorporation into the carbon skeletons that are essential for growth and development. This conversion of the mineral nutrients into carbohydrates, lipids, and amino acids is termed *nutrient assimilation*. For some nutrients, assimilation requires complex biochemical reactions. For example, nitrate ($NO_3^-$) is first reduced to ammonia ($NH_3$) and then introduced into intermediary metabolism through a series of enzyme-mediated steps. Other nutrients, especially the macronutrient and micronutrient cations are assimilated by forming complexes with organic compounds, such as $Ca^{2+}$-pectate associated with the cell wall.

In leguminous plants that can assimilate nitrogen from the atmosphere, this process involves an intricate interaction between the host plant and the nitrogen fixing bacteria with which it lives in a symbiotic relationship.ln this chapter, the basic reactions by which nutrients (sulfur, nitrogen, phosphate, oxygen, and cations) are assimilated will be discussed with an emphasis on the assimilation of nitrogen.

## Nitrogen Assimilation

Nitrogen is a key element in many of the compounds present in plant cells. It is found in the nucleoside phosphates and amino acids that form the building blocks of nucleic acids and proteins, respectively. Availability of nitrogen for crop plants is an important limiting factor in agricultural production, and the importance of nitrogen is demonstrated by the fact that only oxygen, carbon, and hydrogen are more abundant in higher plant cells.

## Nitrogen Cycle

Nitrogen is present in many forms in the biosphere. The atmosphere contains vast quantities of molecular nitrogen ($N_2$), about 78% of the atmosphere by volume. For the most part, this large reservoir of nitrogen is not directly available to plants. Acquisition of nitrogen from the atmosphere requires the breaking of an exceptionally stable triple covalent bond between two nitrogen atoms ($N \equiv N$), and higher plants do not have the capacity to carry out this reaction directly. On the other hand, nitrogen gen present in the form of nitrate or ammonia is readily assimilated by higher plants. The consumption of plants by animals allows nitrogen to move further up the food chain, and the nitrogen is returned to the soil through the death and decomposition of plants and animals. These processes are a part of the nitrogen cycle. The conversion of molecular nitrogen into other forms such as ammonia or nitrate is known as *nitrogen fixation,* and it can occur as a result of both natural and industrial processes.

Under the conditions of elevated temperature (about 200°C) and high pressure (about 200 atmospheres), molecular nitrogen will combine with hydrogen to form ammonia. The extreme conditions are required to overcome the high activation energy of the reaction. This reaction, called the Haber process, is a starting point in the production of a variety of industrial and agricultural products. Worldwide industrial nitrogen fixation amounts to about 50 million tons annually, equivalent to about 12% of the global nitrogen removed from the atmospheric pool each year. Natural processes account for a larger amount of nitrogen fixation. Of the total amount fixed, about 10% occurs as a result of lightning.

Lightning causes the formation of hydroxyl free radicals, free hydrogen atoms, and free oxygen atoms from water vapour and oxygen in the atmosphere, and these reactive species attack molecular nitrogen to form nitric acid ($HNO_3$), which is carried down to the soil by rain. A much smaller amount of nitrogen fixation occurs in the stratosphere as a result of photochemical reactions involving gaseous nitric oxide and ozone. The nitric acid produced in this manner eventually finds its way into the lower atmosphere, from which it is removed by rain. The remaining 90% of the nitrogen made available by natural processes is fixed through the action of micro-organisms in a process termed *biological nitrogen fixation.*

Biological Nitrogen Fixation is carried out by free-living bacteria, including cyanobacteria, and by bacteria that are in symbiotic association with plants. These organisms contain enzyme systems that can catalyze chemical reactions involving molecular nitrogen. From an agricultural standpoint, biological nitrogen fixation is an important source of soil nitrogen; on a worldwide basis, the demands for nitrogen in crop production cannot be met through the use of chemical fertilizers. The reactions of the nitrogen cycle that follow nitrogen fixation involve interconversions of fixed forms nitrogen and the return of nitrogen to the atmosphere. Organic forms of nitrogen such as amino acids that are returned to the soil in wastes and by the decay of dead plants and animals are converted to ammonia in a process called *ammonification.*

Bacteria and fungi in the soil transform the amino nitrogen into free ammonia, which can react with other components in the soil to form ammonium salts. This released ammonia can also be oxidized first to nitrite ($NO_2^-$) and then to nitrate ($NO_3^-$) by the process of *nitrification.* Oxidation of ammonia to nitrite is carried out by soil bacteria of the *Nitroso* group *(Nitrosomonas),* while the further oxidation from nitrite to nitrate is carried out by bacteria of the *Nitro* group *(Nitrobacter).* Nitrogen present as nitrate can be readily used by plants or it can be returned to the atmosphere through *denitrification,* a process in which nitrate is reduced to nitrogen.

Many bacteria use denitrification as a source of respiratory energy. when oxygen is absent and either $NO_3^-$, $NO_2^-$, or $N_2O$ is present. Denitrification is an important biological process because, in its absence, nitrate would rapidly accumulate in the oceans. Nevertheless, when the amount of nitrogen fixed annually (125 million tons) is compared with amount of nitrogen lost to the atmosphere by denitrification (93 million tons), a net gain is observed. This gain presumably represents fixed nitrogen that is slowly accumulating in soil, lakes, and the ocean.

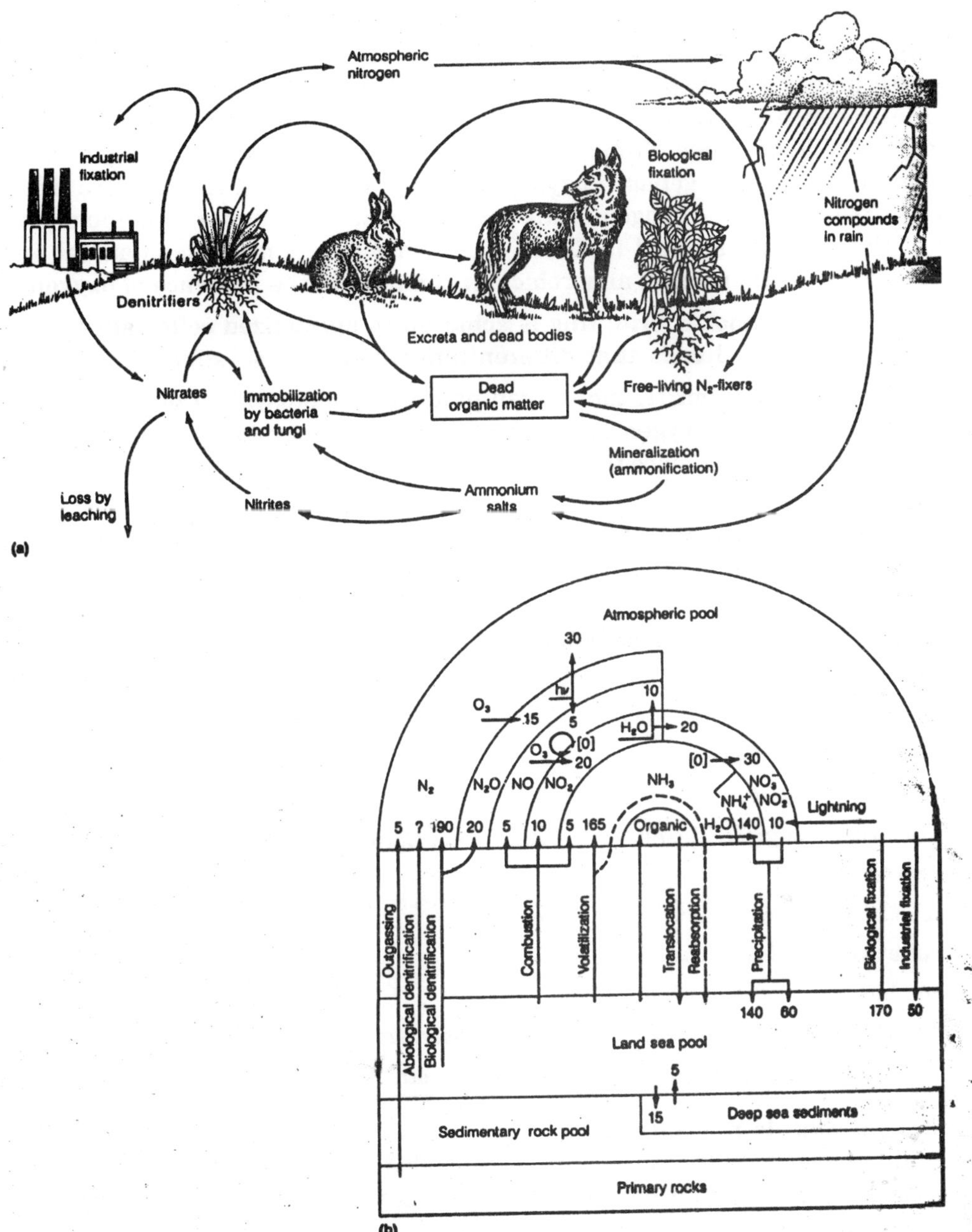


**Fig. 10.1. Nitrogen cycles through the atmosphere as it changes from a gaseous form to reduced ions before being incorporated into organic compounds in living organism. (a) Some of the steps involved in the nitrogen cycle. (b) Estimated amounts of nitrogen transfereed from one global pool to another expressed in millions of metric tons per year. *hv* represents photochemical reactions.**

## Biological Nitrogen Fixation

Nitrogen-fixing bacteria occur either as free-living microorganisms or in symbiotic association with higher plants, primarily those of the family Leguminosae. In the symbiotic relationship, nitrogen fixed by the bacterium is acquired by the host plant in exchange for other nutrients. The wide range of microorganisms known to carry out nitrogen fixation are listed in Table 10.1. The nitrogenase enzymes involved in nitrogen fixation are irreversibly inactivated by oxygen, so the nitrogen-fixing process must occur under anaerobic conditions. Thus, each of the organisms listed in Table 10.1 either functions under natural anaerobic conditions or can create an internal anaerobic environment in the presence of oxygen.

In the filamentous cyanobacteria, this is achieved in specialized cells called *heterocysts.* Heterocysts are thick-walled cells that differentiate when heterocystous cyanobacteria are deprived of $NH_4^+$. Photosystem II, the oxygen-producing photosystem of photosynthesis, is absent from heterocysts, so' oxygen is not generated. Heterocysts appear to represent an adaptation for nitrogen fixation, since they are widespread among cyanobacteria that are capable of fixing nitrogen aerobically.

**Table 10.1. Some Examples of Organisms that can Carry out Nitrogen Fixation**

*Symbiotic nitrogen fixation*

1. Leguminous plants, *Parasponia: Rhizobium, Bradyrhizobium*
2. Alder (tree), *Ceonothus* (shrub), Casuarina (tree): *Frankia* (actinomycete)
3. Tropical grasses: *Azospirillum* (actinomycete)
4. *Azolla* water fern: *Anabaena*

*Nonsymbiotic nitrogen fixation*

1. Cyanobacteria (blue-green algae): Nostoc, *Anabaena, Calothrix,* etc.
2 Other bacteria
   (*a*) Aerobic: *Azotobacter, Beijerinckia, Derxia,* etc.
   (*b*) Facultative: *Bacillus, Klebsiella,* etc.
   (*c*) Anaerobic:
      (*i*) Nonphotosynthetic: *Clostridium, Methanococcus* (archaebacterium), etc.
      (*ii*) Photosynthetic: *Rhodospirillum, Chromatium,* etc.

Cyanobacteria that do not have heterocysts can carry out nitrogen fixation only under anaerobic conditions such as those that occur in flooded fields. In Asian countries, nitrogen-fixing cyano-bacteria of both the heterocyst and nonheterocyst types are a major means of maintaining an adequate nitrogen supply to the soil in rice fields. The microorganisms fix nitrogen when the fields are flooded and decompose and release it when the fields dry out. An important contributor to the available nitrogen of flooded - rice fields is the tiny water fern *Azolla,* which maintains a symbiotic association with the nitrogen-fixing cyano-bacterium *Anabaena.*

The *Azolla- Anabaena* complex can fix as much as 1.2 kg of atmospheric nitrogen per acre per day Free-living bacteria that are capable of fixing nitrogen can either be aerobic, facultative, or anaerobic. Aerobic nitrogen-fixing bacteria such as *Azotobacter* are thought to maintain reduced oxygen conditions (*micro-aerobic* conditions) through their high levels of respiration. Others, like *Gloeotheca,* evolve $O_2$ photosynthetically during the day and fix nitrogen during the night. Facultative organisms, which are able to grow under both aerobic and anaerobic conditions, generally fix nitrogen only under anaerobic conditions.

For anaerobic nitrogen-fixing bacteria, oxygen does not pose a problem; these anaerobic organisms can be either photosynthetic, as in the case of *Rhodospirillum,* or non-photosynthetic, as in the case of *Clostridiuni.* In symbiotic nitrogen fixation, such as the association of nitrogen-fixing bacteria with the roots of legumes, the amount of free oxygen is reduced by the high respiration rates of the symbionts and by the presence of an oxygen-binding heme protein, *leghemoglobin.*This protein is present in the cytosol of infected nodule cells at rather high concentrations (700 $\mu M$ in soybean nodules) and gives the nodules their pink colour.

The globin portion of leghemoglobin is produced by the host plant in response to infection by the bacteria and the heme portion is produced by the bacterial symbiont. The rate of nitrogen fixation is closely correlated with the concentration of leghemoglobin. The protein has a very high affinity for oxygen; half-saturation takes place at 10-20 nM ($10^{-9}$ M) oxygen, compared with 126 nM for the $\beta$ chain of human hemoglobin. It is thought that leghemoglobin buffers the oxygen concentration within the root nodule, allowing respiration without inhibition of the nitrogenase.

In addition, leghemoglobin acts as an oxygen carrier, facilitating oxygen diffusion across a thin, unstirred layer surrounding the respiring symbiotic bacterial cells. Leghemoglobin thus functions at the cellular level in a manner analogous to hemoglobin in animal circulatory systems.

## Symbiotic Nitrogen Fixation

The symbiotic association of nitrogen-fixing bacteria with plant roots generally occurs in multicellular structures called nodules. The best-characterized symbiotic association involving nodules is the one occurring on the roots of leguminous plants. These nodules are produced by the host plant root upon infection by nitrogen-fixing gram-negative bacteria of the genus *Rhizobium.* Members of the family Leguminosae (Fabaceae) are the only plants capable of forming nitrogen-fixing nodules with *Rhizobium,* with the single exception of the nonlegume genus *Parasponia,* a tropical tree.

The ability to fix nitrogen in association with *Rhizobium* may account for the fact that the legumes are the third-largest angiosperm family, with a wide distribution from the tropics to the arctic zones. Infectious *Rhizobium* species exhibit some specificity for different leguminous host plants. Such host selection implies that the *Rhizobium* species can recognize its plant host and vice versa. *Rhizobium* typically occurs free-living in the soil. The first stage in the formation of the symbiotic relationship is the migration of the bacteria toward the *rhizosphere,* the area immediately surrounding the roots. This is a chemotactic response and is probably mediated by chemical attractants, especially flavonoids, produced by the roots Rhizobia have evolved to use flavonoid ompounds released by legume roots as positive regulaors of the genes involved in nodulation *(nod* genes), In soybean, the rhizobia are thought to secrete factors that stimulate subepidermal cell divisions in the root. The dividing cells in the root evidently potentiate the root hairs above them to become target sites for infection. Once in the rhizosphere, the bacteria attach to the root surface, particularly to the root hairs.

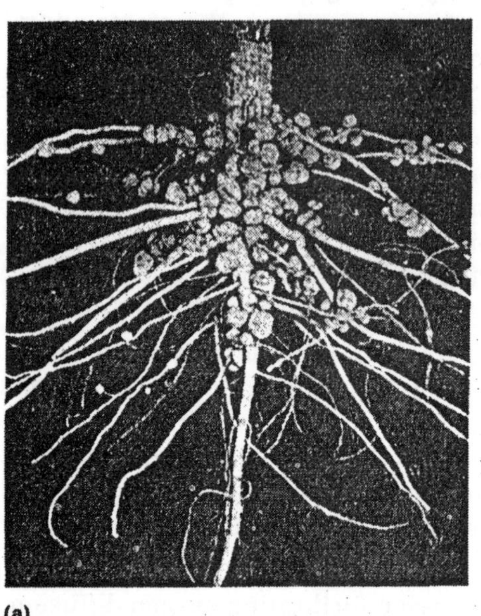

(a)

(b)

**Fig. 10.2. Root nodules on soybean (a) and alder tree (b). The nodules in soybean are a result of infection by *Rhizobium japonicum;* those in alder (Alnus) are caused by *Frankia,* a bacterium that belongs to the actinomycetes. The associations between actinomycetes and their hoste are called actinorhizal association. All known actinorhizal associations occur in shrubs or trees.**

**Table 10.2. Host Preference Among Rhizobium Species**

| *Rhizobium species* | *Preferred host genus* |
|---|---|
| *R. leguminosarum* | *Pisum (pea), Vicia (broad bean), Lens (lentil), Cicer* (chick pea) |
| *R. Trifolii* | *Trifolium* (clover) |
| *R. phaseoli* | *Phaseolus* (kidney bean) |
| *R. meliloti* | *Medicago* (alfalfa), *Melitotus* (sweet clover) |
| *R. japonicum* | *Glycine* (soybean) |

The mechanism of attachment is still poorly understood. According to an early idea, legume roots secrete sugar-binding proteins called lectins, which interact with the surface polysaccharides of *Rhizobium* cells and facilitate binding to the root cell walls. Lectins may explain the association between clover and *Rhizobium trifolii,* but in the case of the *pea-Rhizobium leguminosarum* association a different, calcium-dependent protein has been implicated. It is now thought that multiple mechanisms of attachment are probably involved, some general and some host-specific.

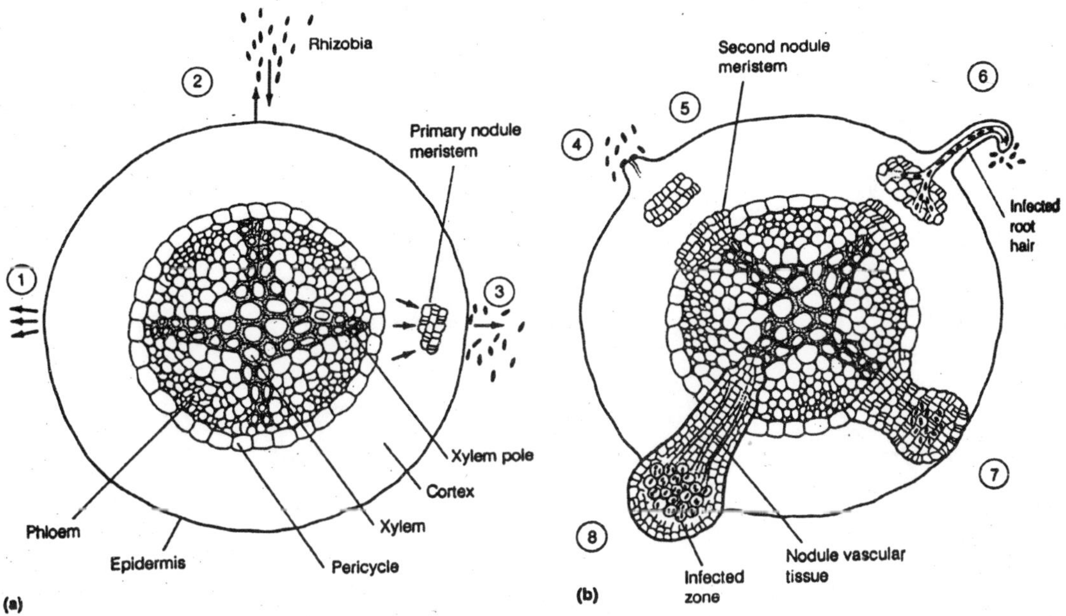


**Fig. 10.3. Stages in the initiation and development of a soybean root nodule, (a) Events involved in the imitation of the nodule: (1) the root excretes substances; (2) these substances attract rhizobia and stimulate them to produce cell-division factors; and (3) cells in the root cortex divide to form the primary nodule meristem. (b) Stages of infection and nodule formation: (4) bacteria attach to the root hair; (5) cells in the pericycle near the xylem poles are stimulated to divide; (6) the infection thread forms and extends inward as the primary nodule meristem and the pericycle continue to divide; (7) the two masses of dividing cells fuse into a single clump while the infection thread continues to grow; and (8) the nodule elongates and differentiates, including the vascular connection to the root stele. Bacteroids are released intp the cells in the center.**

## Formation of an Infection Thread

The rhizobia that bind to emerging root hairs induce a pronounced curling of these cells, trapping colonies of rhizobia within the coils. Curling of root hairs is thought to be stimulated by bacterial production of the plant hormone auxin. The root hair cell wall becomes degraded in these regions, allowing the bacterial cells direct access to the outer surface of the plant plasma membrane The next step involves the formation of a remarkable structure called the infection thread.

The *infection thread begins as an internal tubular extension of the plasma membrane caused by the fusion of* Golgiderived membrane vesicles at the site of infection. It grows at the tip, much as a stalagmite grows upward from the floor of a cave as the result of the dripping of mineralrich water; that is, secretory vesicles “drip” onto the end of the tube. As the membrane tube extends inward, a thin layer of cellulosic cell wall is deposited.

The wall material lines the inner surface of the infection thread and is physically connected with the host cell wall. This makes sense when we consider that the inner surface of the infection thread membrane is topologically equivalent to the outer surface of the host cell plasma membrane.

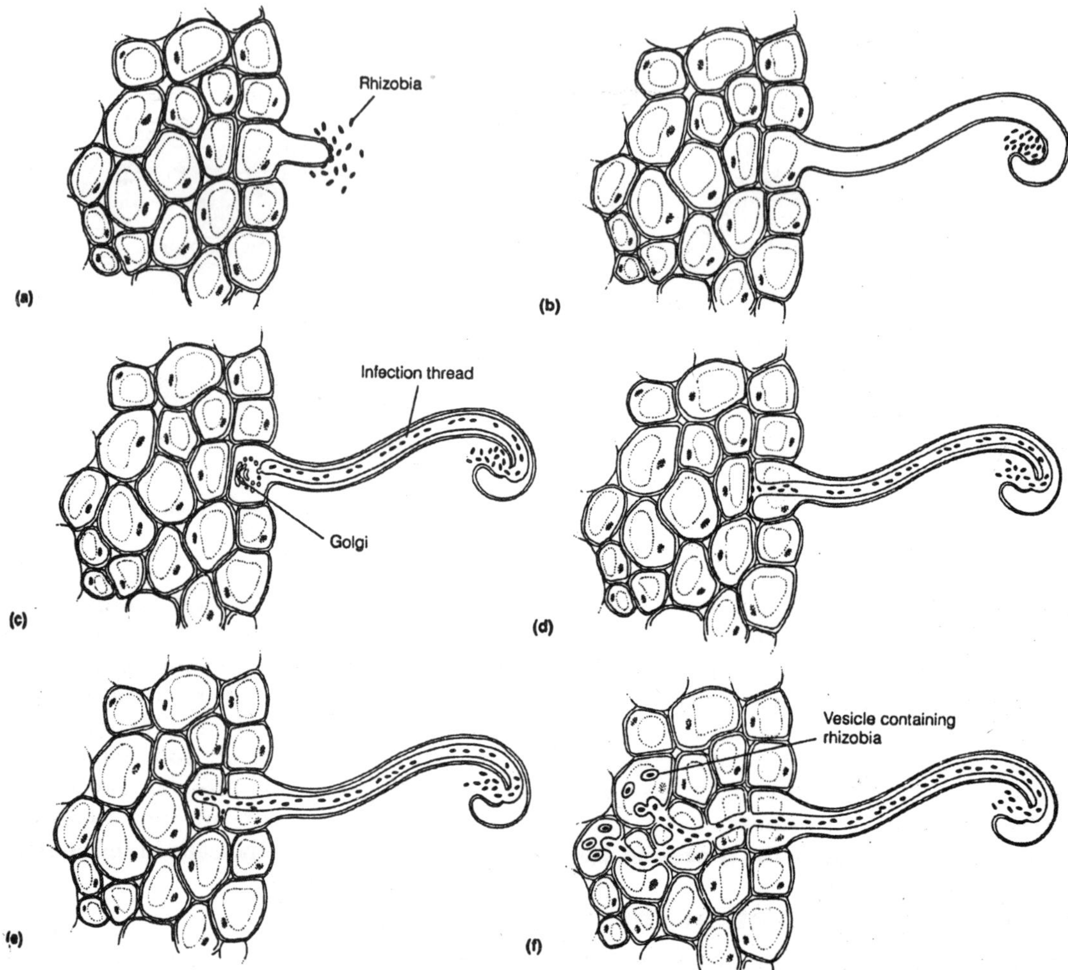


**Fig. 10.4. The infection process, (a) Rhizobia bind to emerging root hair, (b) In response to factors produced by the bacteria, the root hair exhibits abnormal curling growth, and rhizobia cells proliferate within the coils, (c) Localized root hair wall degradation leads to infection and formation of the infection thread from Golgi secretory vesicles, (d) The infection thread reaches the end of the cell, and its membrane fuses with the plasma membrane of the root hair cell, (e) Rhizobia are released into the apoplast and penetrate the compound middle lamella to the subepidermal cell plasma membrane, leading to the initiation of a new infection thread, which forms an open channel with the first. The two infection threads are actually discontinuous because the infection thread membrane is absent from the apoplast. (f) The infection thread extends and branches until it reaches target cells, where vesicles containing bacterial cells are released into the cytosol.**

The growth of the infection thread is closely correlated with the proliferation of enclosed rhizobia, which still technically occupy an extracellular compartment. Branching of the infection thread enables the bacteria to spread to many cells. At the same time, cortical cells near the epidermis undergo rapid divisions and give rise to the body of the nodule.

When the infection thread reaches the base of the root hair cell it fuses with the plasma membrane, releasing free bacteria into solubilized pockets of the apoplast The infection thread membrane fuses with the host cell plasma membrane. The bacteria in the apoplast apparently degrade the cell wall to gain access to the plasma membrane of the subepidermal cell, and the infection process repeats itself In this way, infection threads spread from cell to cell, allowing the bacteria to penetrate deep within the developing nodule without ever entering the cytoplasm of the host. In the final step in the process, the *Rhizobium* cells are internalized into the cytoplasm of the host cells by endocytosis. This occurs when the infection thread reaches specialized cells within the nodule.

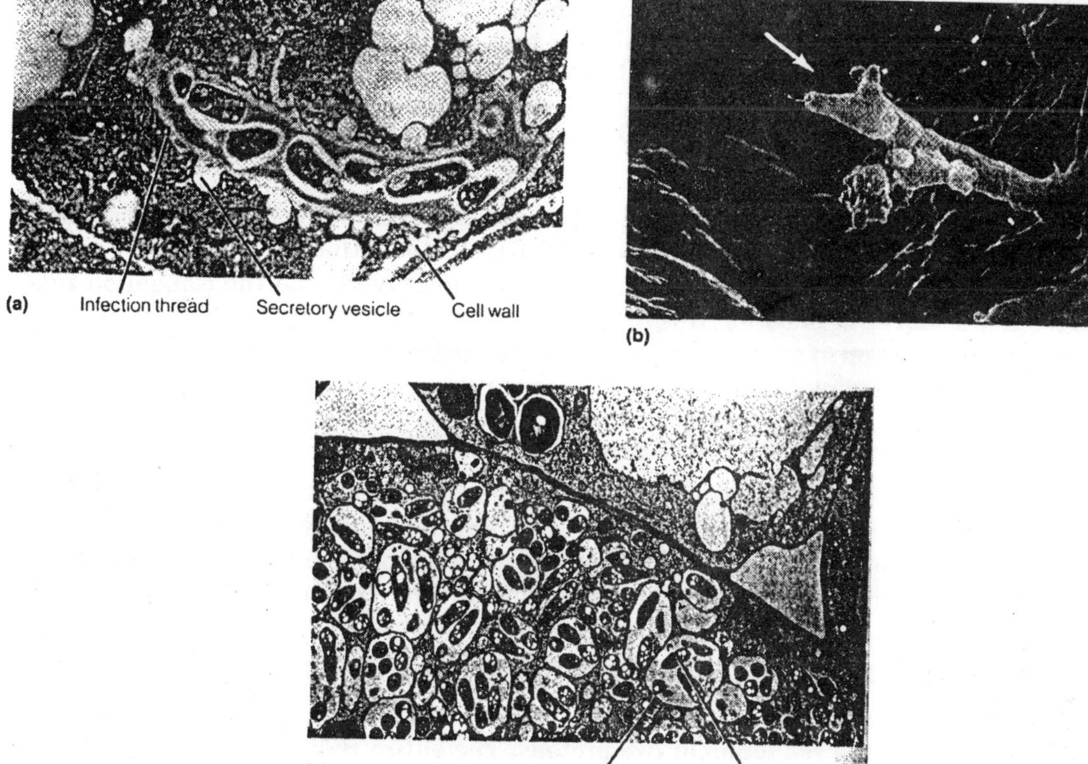


**Fig. 10.5. Electron micrographs of infection threads, (a) Growth of the infection thread containing bacteria. Note the fusion of vesicles (v) with the infection thread (it). The plasma membrane and cell wall of the root cell are continuous with the infection thread. Each bacterium is surrounded by a halo of capsular polysaccharides. The bacteria continue to divide as the thread grows (1cm = 1μm). (b) Scanning electron micrograph of an infection thread which is in the early stages of branching near the tip, indicated by the arrow (3750 x ). (a and b courtesy of Dr. Ann Hirsch, UCLA.) (c) Electron micrograph of bacteroids surrounded by a plant-derived peribacteroid membrane in an infected soybean root nodule cell (2.5cm = 1μm)**

The tip of the infection thread vesiculates, releasing *Rhizobium* cells packaged in a membrane derived from the host cell plasma imembrane. At first the bacteria continus to divide, and the surrounding membrane increases in Surface area to accommodate this growth by fusion with smaller vesicles. Very soon, however, the bacteria stop dividing; they enlarge and differentiate into nitrogen-fixing endosymbiotic organelles called bacteroids. This differentiation is thought to be triggered by a signal from the plant. The membrane surrounding the bacteroids is called the peribacteroid membrane response to signals from the bacteria, the host cells undergo marked morphological and biochemical changes, including the production of leghemoglobin.

The nodule as a whole develops such features as a vascular system, which facilitates the exchange of fixed nitrogen produced by the bacteroids for nutrients contributed by the plant, and a layer of cells to exclude $O_2$ from the root nodule interior. In some temperate legumes (e.g., peas), the nodules are elongated and cylindrical due to the presence of a nodule meristem. The nodules of tropical legumes, such as soybeans and peanuts, lack a persistent meristem and are spherical in shape. Plants other than legumes can also develop symbiotic relationships with nitrogen-fixing organisms. With some plants, such as the deciduous tree alder (*Alnus*), this involves the formation of root nodules by the actinomycete, *Frankia*. The nodules in *Alnus* occur in clusters, which are formed by repeated branching.

Grasses can also develop symbiotic relationships with nitrogen-fixing organisms, but these associations do not involve the production of root nodules. Instead, the nitrogen-fixing bacteria (actinomycetes) seem to be anchored in the root surfaces, mainly around the elongation zone and the root hairs. Bacteria such as *Azospirillum* are being intensively studied for possible application in the cultivation of corn and other grasses The potential nitrogen-fixation properties of these bacteria could be exploited to reduce the use of nitrogen fertilizer.

## The Biochemical Process of Nitrogen Fixation

When organisms carry out nitrogen fixation, the end product, as in industrial nitrogen fixation, is ammonia. The ammonia is subsequently incorporated into organic compounds such as glutamine or glutamate and is utilized by the microorganism, or by the host plant in the case of a symbiotic association. The overall reaction for the fixation of molecular nitrogen into ammonia is

$$N_2 + 8e^- + 8H^+ + 16ATP \rightarrow 2NH_3 + H_2 + 16ADP + 16P_i \quad ...(10.1)$$

Note that the reduction of $N_2$ to $2NH_3$, a six-electron transfer, appears to be coupled to the evolution of $H_2$ form the reduction of two protons. The entire reaction is catalyzed by the enzyme nitrogenase. Nitrogenase is an enzyme complex that can be separated into two components, the MoFe protein and the Fe protein, neither of which has catalytic activity by itself. The Fe protein is the smaller of the two components and has two identical subunits of 30,000 to 72,000 Da, depending on the organism. It contains one 4Fe-4S iron sulfur cluster that participates in the redox reactions involved in the conversion of $N_2$ to $NH_3$. The Fe protein is extremely sensitive to oxygen and is irreversibly inactivated by $O_2$ with a half-life of 30-45 s.

The MoFe protein has four subunits with a total molecular mass of 180,000 to 235,000 Da, depehding on the species. It has two molybdenum atoms per molecule in two Mo-Fe-S clusters and a variable number of Fe-S clusters. The MoFe protein is also inactivated by oxygen, with a half-life of 10 min. In the overall reaction catalyzed by nitrogenase the nonheme iron protein ferredoxin serves as an electron donor to the Fe proetein, which in turn reduces the MoFe protein. ATP hydrolysis is involved in this step.

The MoFe protein reduces the $N_2$ substrate. Nitrogenase can reduce a number of substrates (Table 10.3), although under natural conditions it reacts only with $N_2$ and $H^+$. One of the reactions catalyzed by he nitrogenase, the reduction of acetylene, is used in estimating nitrogenase activity by gas chrpmotography.

Direct measurements of $N_2$ fixation pcquircs cxpensive instrumentation that is not readily available, but the reduction of acetylene to ethylene can easily be measured. Acetylene reduction involves two electrons, as opposed to the eight electrons required for 'the reduction of $N_2$ and $2H^+$. Therefore, four ethylene molecules correspond to the reduction of one $N_2$ molecule.

**Table 10.3. Reactions Catalyzed by Nitrogenase**

| | |
|---|---|
| $N_2 \rightarrow NH_3$ | Denitrogen fixation |
| $N_2O \rightarrow H_2 + H_2O$ | Nitrous oxide reduction |
| $N_3^- \rightarrow N_2 + NH_3$ | Azide reduction |
| $C_2H_2 \rightarrow C_2H_4$ | Acetylene reduction |
| $2H^+ \rightarrow H_2$ | $H_2$ production |
| $ATP \rightarrow ADP + P_i$ | ATP hydrolytic activity |

The acetylene reaction has some limitations however, such as ethylene production by some microorganisms. For these reasons, the acetylene method is valuable for comparative studies, whereas direct measurements of $N_2$ reduction by mass spectroscopy are required for precise quantitation of nitrogen energetics of nitrogen fixation is complex. Production of $NH_3$ from $N_2$ and $H_2$ is an exergonic reaction with a $\Delta G^{o\prime}$ of –27 KJ $Mol^{-1}$.

However, industrial production of $NH_3$ from $N_2$ and $H_2$ is a very energy -intensicve process because of the activation energy required to break the triple bond in $N_2$. It is also clear that the enzymatic reduction of $N_2$ by nitroganase requires alarge investment of ATP although the specific feree energy changes of the reactions involved remain to be esta lished. The energy cost of nitrogen fixation cna also be evealuated on the basis of carbohydrate utilization by an intact plant. Calculations show that 12g of organic carbon are consumed per gram of $N_2$ fixed. Based on Equation 10.1, the $\Delta G^0$, for the overall reaction for biological nitorgen fixation has been estimated as –203 KJ $mol^{-1}$.

Since the overall reaction is highly.

Since the overall reaction is highly exergonic, ammonium production is limited by the rather slow turnover rate of the nitroenase complex. Under natural conditions, the reduction of $H^+$ to $H_2$ gas can be substantial and can compete with $N_2$ for electrons from nitrogenase.

In *Rhizobium,* 30 to 60% of the energy supplied to nitrogenase may be lost as $H_2$. Thus, this reaction would limit the efficiency of nitrogen fixation by these microorganisms. However, some *Rhizobium* strains contain an enzyme, hydrogenase, that can split $H_2$ and recycle electrons for $N_2$ reduction, thus increasing the efficiency of nitrogen fixation.

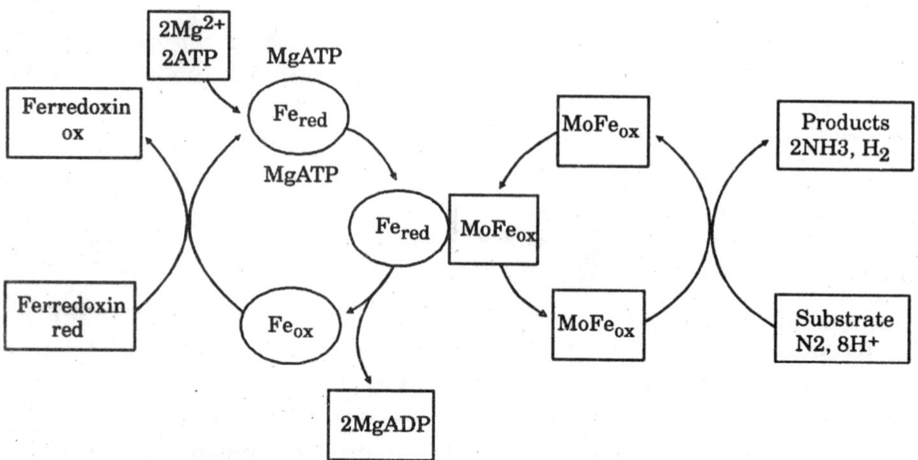


**Fig. 10.6. The reaction catalyzed by nitrogenase. Ferredoxin reduces the Fe protein, while binding of ATP takes place. The Fe protein reduces the MoFe protein, and the MoFe protein reduces the nitrogen. The flow of electrons is shown by the darker arrows.**

## The Nitrate Assimilation Pathway

Nitrate is the principal source of nitrogen that is available to higher plants under normal field conditions; hence, the nitrate assimilation pathway is the major point of entry of inorganic nitrogen into organic compounds. In most temperate soils, ammonia is rapidly converted to nitrate through nitrification by bacteria. Only in poorly *aerated* soils with inadequate drainage, where the bacteria that normally carry out nitrification cannot grow well, does ammonia accumulate to a significant extent. Although nitrate is generally the major form of nitrogen available to plants, some plants directly utilize ammonia under certain conditions, such as when growing in acid soils in which nitrification by microorganism is inhibited.

## Inducible Transport System

Before entering the assimilation pathway, nitrate must be taken up by the cell. In plants not previously exposed to nitrate, nitrate stimulates its own uptake after a lag period. This fact and the inhibition of nitrate uptake by protein and RNA synthesis inhibitors suggest that nitrate induces its own specific transport system. Nitrate uptake is inhibited by cyanide, anaerobic, conditions, and dinitrophenol, an uncoupler of oxidative phosphorylation, suggesting that metabolic energy is required.

Efforts to purify a plasma membrane nitrate carrier are under way. Once inside the celf, excess nitrate may be taken up by the vacuole. Several types of channels on the tonoplast potentially could facilitate vacuolar nitrate accumulation. Accumulation of high concentrations of nitrate in vacuoles of fodder or agricultural crops, such as spinach, is economically important because of its toxicity to livestock and humans. In the liver, nitrate is reduced to nitrite, which is very reactive and combines strongly with hemoglobin, rendering it unable to bind oxygen.

## Nitrite by Nitrate Reductase

Following its uptake from the soil by plant roots, nitrate is reduced to ammonia prior to assimilation into organic compounds. The first step in this process is the reduction of nitrate to nitrite. This reaction, which occurs in the cytosol, can be written as

$$AH_2 + NO_3^- \rightarrow A + NO_2^- + H_2O$$

for the redox couple

$$NO_3^- + 2H^+ + 2e^- \rightarrow NO_2^- + H_2O$$

and is carried out by the enzyme nitrate reductase. The electron donor of nitrate reductases in higher plants may be either NADH or NADPH, but the most common form of the enzyme utilizes NADH according to the equation:

$$NO_3^- + NAD(P)H + H^+ \xrightarrow{2e^-} NO_2^- + NAD(P)^+ + H_2O \quad ...(10.2)$$

The NADH nitrate reductases of higher plants appear to be homodimers, that is, composed of two identical subunits. The molecular mass of the holoenzyme ranges from 200,000 to 270,000 Da, depending on the species, while the subunit molecular mass ranges from 100,000 to 120,000 Da. All eukaryotic nitrate reductases contain three prosthetic groups: FAD, heme, and a molybdenum complex. It is thought that one FAD, one heme, and one Mo are present per subunit. Molybdenum is bound to the enzyme via a complex with a small organic molecule called a *pterin,* which acts as a chelator of the metal. Nitrate reductase is the principal molybdenum-containing protein in vegetative tissues. Under conditions of molybdenum deficiency, the activity of nitrate reductase in cells is much reduced. The sequences of nitrate reductase mRNA from barley, *Arabidopsis,* and tobacco have now been determined from clones isolated from cDNA libraries. By comparing the deduced amino acid sequences with those

A sequence found at the C-terminal end of the protein was similar to the FAD binding domain of cytochrome-$b_5$ reductase, and another sequence near the middle of the protein was similar to the heme-binding domain of cytochrome $b_5$. Finally, two short sequences near the N terminus were similar to the 'molybdenum-binding site of the enzyme, sulfite oxidase.

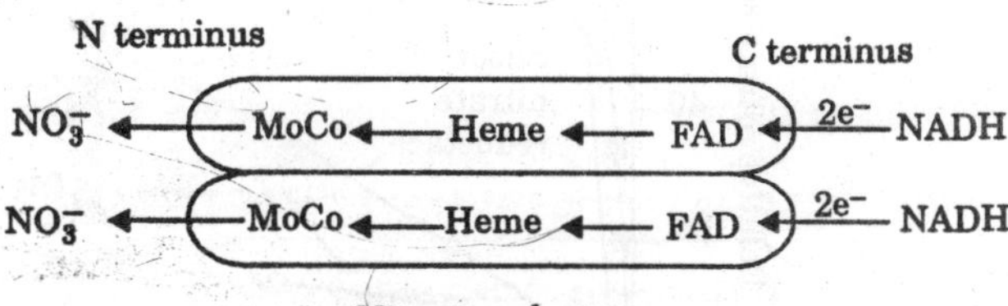


**Fig. 10.7. A model of the nitrate reductase dimer of higher plants, illustrating the three binding domains: molybdenum complex (MoCo), heme, and FAD. NADH binds at the FAD-binding region of each subunit and initiates a two-electron transfer from the carboxyl (C) terminus, through each of the electron transfer components, to the amino (N) terminus. Nitrate is reduced at molybdenum complex near the amino terminus. of other proteins that bind FAD, heme, or molybdenum, Crawford et al. (1988) were able to assign three functional domains to the subunit.**

## Nitrate Reductase

The synthesis of nitrate reductase is regulated by its substrate; that is, nitrate induces the de novo synthesis of the enzyme. The steady-state level of nitrate reductase is determined by the rate of degradation (turnover) as well by the rate of synthesis. A newly synthesized nitrate, reductase protein has a half-life of only a few hours in the cell. Thus, when nitrate is withdrawn, the level of nitrate reductase rapidly decreases. In all plants studied so far, the induction of nitrate reductase by nitrate is preceded by an increase in nitrate reductase mRNA.

In barley seedlings, nitrate reductase mRNA was detected approximately 40 min after addition of nitrate, and maximum levels were attained by about 2 h. Slightly different patterns were observed in shoots and roots. The pattern of mRNA accumulation contrasts with the gradual linear increase in nitrate reductase activity. Although nitrate is the primary inducer of nitrate. reductase, light is also required for the maximum effect. Somers et al. (1983) showed that barley seedlings treated with nitrate in the dark accumulated very low levels of nitrate reductase.

Exposure of the seedlings to white light resulted in a strong promotion of nitrate reductase activity. The effect of white light can be seen at the level of mRNA. However, light alone is insufficient to induce nitrate reductase in most cases; the role of light is to enhance the effect of nitrate.

## Nitrite Is Reduced to Ammonium

Nitrate is reduced in the cytosol .to nitrite by nitrate reductase. Rather than accumulating to toxic levels in the cytosol, the nitrite is rapidly transported into plastids—chloroplasts in green leaves and proplastids in nongreen tissues. Each type of plastid has its own assimilatory *nitrite reductase*. Chloroplasts contain a ferredoxin-nitrite reductase, while proplastids contain an NAD(P)H-nitrite reductase activity. Both enzymes catalyze the six-electron reduction of nitrite to ammonium with the overall reaction:

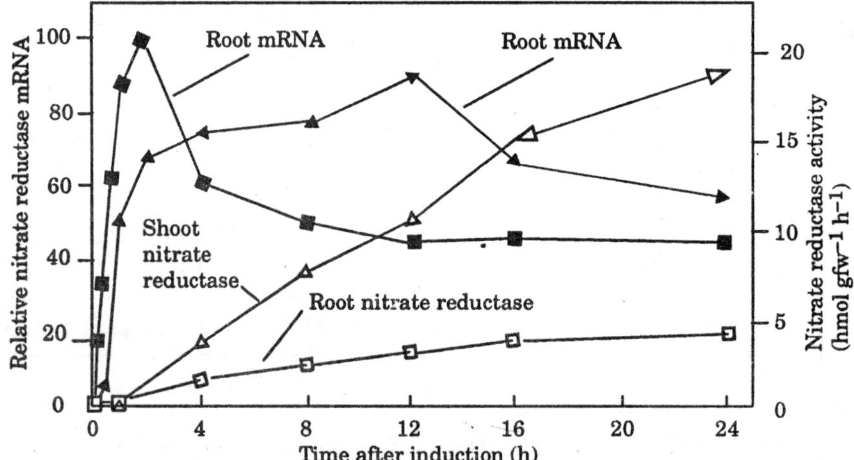


**Fig. 10.8. The pattern of nitrate reductase and nitrate reductase mRNA induction in shoots and roots of barley. The induction of mRNA precedes the appearance of the enzyme in both cases, although the precise timing is slightly different in the two tissues.**

$$NO_2^- + 6e^- + 8H^+ \rightarrow NH_4^+ + 2H_2O \quad ...(10.3)$$

Ferredoxin (Fd) is the electron donor in photosynthetic tissues, as shown in the following reaction:

$$NO_2^- + 6Fd_{red} + 8H^+ \xrightarrow{6e^-} NH_4^+ + 6FD_{ox} + 2H_2O \quad ...(10.4)$$

where the subscripts red and ox stand for reduced and oxidized, respectively. In contrast, nongreen tissues utilize NADH or NADPH supplied by respiration as the electron donor for nitrite reduction :

$$NO_2^- + 3NAD(P)H + 5H^+ \xrightarrow{6e^-} NH_4^+ + 3NAD(P)^+ 2H_2O \quad ...(10.5)$$

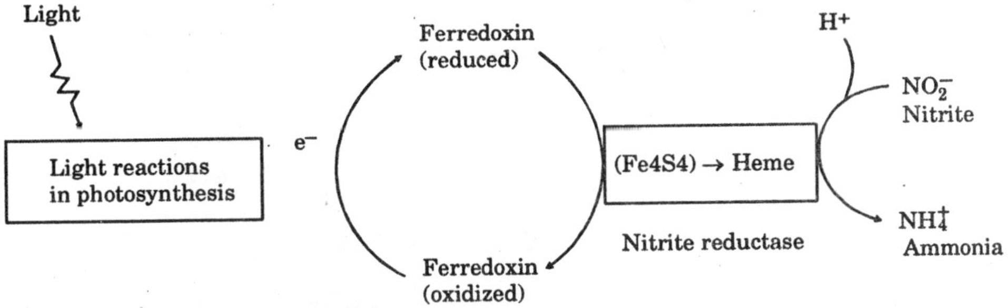

Light

Ferredoxin (reduced)

$H^+$

$NO_2^-$ Nitrite

$e^-$

Light reactions in photosynthesis

(Fe4S4) → Heme

$NH_4^+$ Ammonia

Nitrite reductase

Ferredoxin (oxidized)

**Fig. 10.9. Model for coupling photosynthetic electron flow, via ferredoxin, to the reduction of nitrite by nitrite reductase. The enzyme contains two prosthetic groups, $Fe_4S_4$ and heme, which participate in the reduction of nitrite to ammonium.**

A "ferredoxin-like" protein may also be involved in transferring electrons to the enzyme in proplastids. The best-characterized nitrite reductase is the photosynthetic enzyme of spinach. It consists of a single polypeptide of 60,000 Da and contains two prosthetic groups: an iron-sulfur cluster ($Fe_4S_4$) and a specialized heme. Kinetic experiments suggest that nitrite is bound to the coupled $Fe_4S_4$-heme pair on the enzyme and is reduced directly to ammonium, without accumulation of nitrogen compounds of intermediate redox states. Although localized in plastids, nitrite reductase, like nitrate reductase, is nuclear-encodec.

The nitrite reductase of wheat leaves is synthesized on cytosolic ribosomes as a 64,000-Da precursor, which is processed to 60,500-Da during transit across the chloroplast membranes.

## Nitrate Metabolism

In most plant species both the roots and the shoot have the capacity to carry out nitrate metabolism. However, the degree to which nitrate reduction takes place in the roots or the leaves depends on a number of factors, including the level of nitrate supplied to the roots, plant species, and plant age. In many plants, when the amount of nitrate supplied to the roots is low, nitrate reduction takes place primarily in the roots. When the supply of nitrate is increased, a greater proportion of nitrate metabolism will occur in the leaves because root nitrate reduction becomes limiting under these conditions.

However, species differ in the amount of nitrate metabolized in the roots, even under similar conditions of nitrate supply. When the xylem exudate from the stems of various plants grown with their roots exposed solely to nitrate was examined, the proportion of nitrate to other nitrogen compounds varied from species to species.

In plants such as the cockleburr (*Xanthium sp.*), nitrate metabolism appears to be restricted to the shoots, whereas the roots of the white lupine (*Lupinus* sp.) have the capacity to metabolize nitrate into organic forms. In all cases in which nitrate is metabolized by the roots, nitrogen is translocated in an organic form, primarily as amino acids (see next section), and not as free ammonia. This is undoubtedly due to the toxicity of the latter compound.

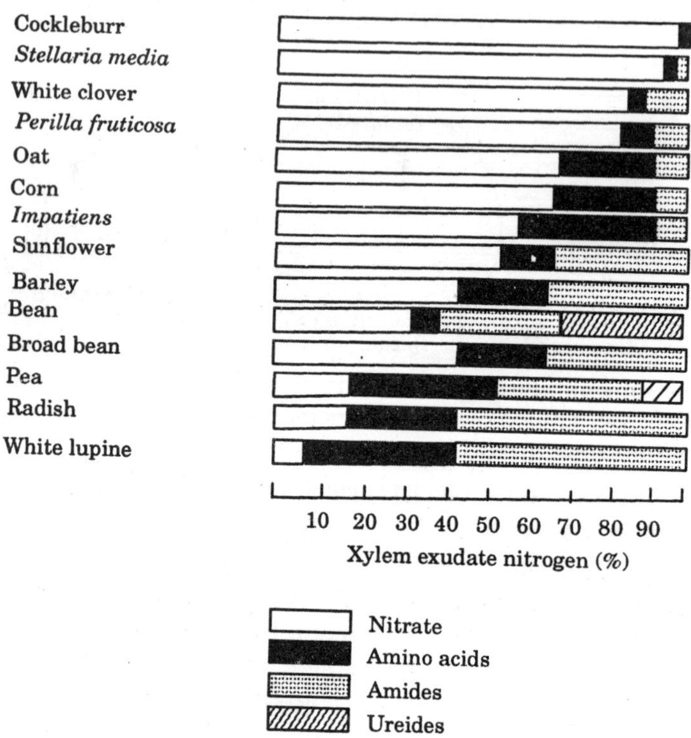


**Fig. 10.10. Relative amounts of nitrate and other nitrogen compounds in the xylem exudate of various plant species. The plants were grown with their roots exposed to nitrate solutions, and xylem sap was collected by severing the stem. Note the presence of ureides, specialized nitrogen compounds, in bean and pea.**

## Organic Compounds

Ammonia and ammonium ions do not accumulate in plant cells but are instead rapidly incorporated into organic compounds. Structural diagrams of a number of the compounds involved in ammonia metabolism and their pathways are shown in Fig. 10.12 Both ammonia and ammonium ions (generally grouped together as ammonia-nitrogen) are toxic to plant cells if present in high enough concentrations. Only certain plants such as *Begonia* are able to accumulate ammonia-nitrogen, and this occurs within the large central vacuole. The assimilation of ammonia-nitrogen into carbon compounds can proceed through two possible pathways.

The first pathway involves the reductive animation of α-ketoglutarate to produce glutamate:

$$\alpha\text{-Ketoglutarate} + NADH + NH_3 \rightarrow \text{glutamate} + NAD^+ + H_2O \qquad ...(10.6)$$

This reaction is catalyzed by glutamate dehydrogenase (GDH), which is found in chloroplasts as well as in mitochondria. In the mitochondria, the enzyme would have available an abundant supply of NADH and α-ketoglutarate continually produced by the tricarboxylic acid (TCA) cycle. The second pathway for assimilation of ammonia involves a reaction with glutamate to form its amide, glutamine.

$$\text{Glutamate} + NH_3 + \text{ATP} \rightarrow \text{glutamine} + \text{ADP} + P_i \quad ...(10.7)$$

This reaction is catalyzed by glutamine synthetase (GS) and requires input of energy in the form of ATP. Glutamine can then be converted to glutamate by the following reaction:

$$\text{Glutamine} + \alpha\text{-ketoglutarate} \rightarrow 2 \text{ glutamate} \quad ...(10.8)$$

which is catalyzed by glutamate synthase (GOGAT[1]). Glutamine synthetase is localized in the chloroplast and cytosol in the leaves and in the cytosol in cells of the roots. Glutamate synthase is localized in the chloroplasts in the leaves and in plastids in roots. Three different forms (isoenzymes) of glutamate synthase may exist: a photosynthetic form using ferredoxin as electron donor and two nonphotosynthetic forms using either NADH or NADPH. Both of these possible pathways result in the same final product, glutamate, but which is the more important pathway in terms of the assimilation of ammonia-nitrogen? Studies of the individual enzymes have revealed that glutamate dehydrogenase, the enzyme of the first pathway, has a relatively high $K_m$ value for ammonia.

The $K_m$ value corresponds to the concentration of ammonia at which the enzyme operates at one-half the maximal rate. This $K_m$ value is within the concentration levels considered to be toxic for plant cells. Therefore, this enzyme would not contribute to the metabolism of $NH_3$ at the much lower concentrations of this compound expected in the cell. Glutamine synthetase has a much lower $K_m$ value for ammonia. This difference in the range of ammonia concentrations at which these two enzymes operate has led most investigators to consider the second pathway the more logical one for the incorporation of ammonia-nitrogen. Even stronger evidence for the GS/GOGAT pathway of ammonia incorporation has come from studies using $^{15}N$-labeled precursors and specific inhibitors of these enzymes.

In kinetic experiments in which $^{15}NO_3^-$ or $^{15}NH_4^+$ was used as the precursor, labeled nitrogen appeared first in the amide group of glutamine, followed by other amino acids such as glutamate, alanine, serine, glycine, aspartate, and asparagine. When specific inhibitors of glutamine synthetase (methionine sulfoximine) or glutamate synthase (azaserine) were used with $^{15}N$-labeled precursors, the label was not incorporated into amino acids.

## Transamination Reactions

Once assimilated into glutamate, the nitrogen can be further incorporated into other amino acids through transamination reactions. These reactions are carried out by enzymes known as aminotransferases. An example of a transamination reaction is

$$\text{Glutamate} + \text{oxaloacetate} \rightarrow \text{aspartate} + \alpha\text{-ketoglutarate} \quad ...(10.9)$$

in which the amino group of glutamate is transferred to the carboxyl atom of the α-keto acid. All transamination reactions require pyridoxal phosphate (vitamin $B_6$) as a cofactor. Aminotransferases are found in the cytosol, chloroplasts, and microbodies. The enzymes localized in the chloroplasts may have a significant role in amino acid biosynthesis, since rapid incorporation of radioactive carbon-14 into glutamate, aspartate, alanine, serine, and glycine occurs when plant leaves are exposed to labeled carbon dioxide.

(a) Major Ammonia Assimilation Pathways

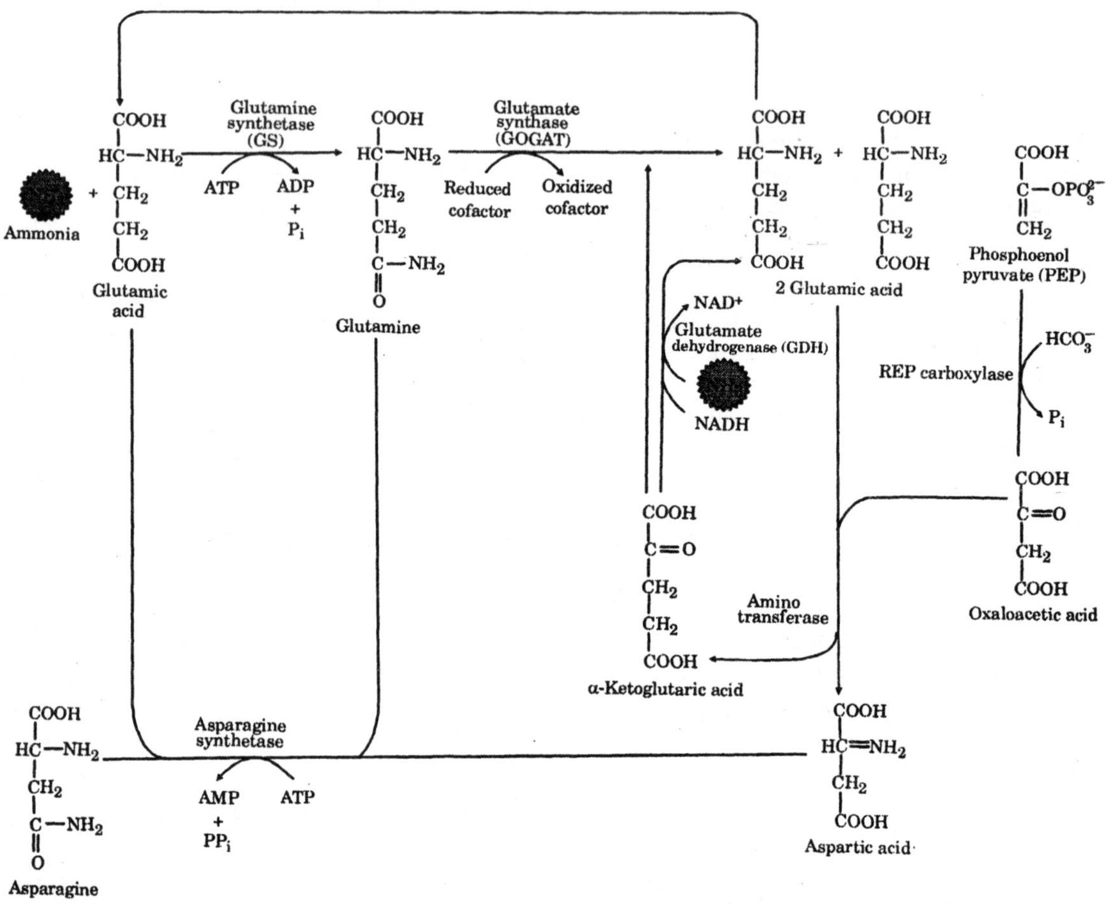


(b) Ureides

Allantoic acid

Allantoin

Citrulline

**Fig. 10.11. Structure and pathways of compounds involved in amonia metabolism. (a) Ammonia can be assimilated either by combination with glutamate to form the amide glutamine (glutamine synthetase) or by reductive amination of α-ketoglutarate to produce glutamate (glatamate dehydrogenase). Two glutamates are produced from glutamine and α-ketaglutarate (glutamate synthase). A reduced cofactor is required for the reaction: ferredoxin in green leaves and NADPH in root tissues. since the formation of glutamine in an uphill reaction requiring ATP, glutamate production helps to drove the reaction forward. Further reactions yield apsartate and the amide asparagine. (b) Structures of the major ureide compounds found in nitrogen-fixing plants.**

## Nitrogen-Fixing Plants

Fixed nitrogen is converted to organic form in the root nodules before being transported to the shoot via the xylem. Based on the composition of the xylem sap, nitrogen-fixing legumes can be divided into amide exporters or ureide exporters. Temperate-region legumes, such as pea (*Pisum*), clover (*Trifolium*), broad bean (*Vicia*), and lentil (*Lens*), tend to be exporters of amides, principally asparagine and glutamine. Legumes of tropical origin, such as soybean (*Glycine*), kidney bean (*Phaseolus*), peanut (*Arachis*), and southern pea (*Vigna*), export nitrogen in the form of *ureides*. The three major ureides are allantoin, allantoic acid, and citrulline.

Allantoin is synthesized in peroxisomes from uric acid, and allantoic acid is synthesized from allantoin in the endoplasmic reticulum. The site of citrulline synthesis from the amino acid ornithine has not yet been determined. All three compounds are ultimately released into the xylem and transported to the shoot, where they are rapidly metabolized. During catabolism, ammonium is released and enters the assimilation pathway.

## Sulfur Assimalation

The sulfur present in the carbon constituents of higher plant cells comes mostly from sulfate acquired from the soil by plant roots. The bulk of the soil sulfate comes from the weathering of parent rock material; however, with industrialization, an additional source of sulfate is atmospheric pollution. The burning of fossil fuels releases several gaseous forms of sulfur, including sulfur dioxide ($SO_2$) and hydrogen sulfide ($H_2S$), which find their way to the soil in rain. When dissolved in water, sulfur dioxide is hydrolyzed to become $H_2SO_4$, a strong acid, which is the major source of acid rain. Plants can also metabolize sulfur dioxide taken up in the gaseous form through the stomata.

However, prolonged exposure (>8 h) to high atmospheric concentrations (>0.3 ppm) causes extensive tissue damage because of the formation of sulfuric acid. The major sulfur-containing organic compounds in higher plant cells are the amino acids cysteine and methioning: Uptake of sulfate into roots is an active process thought to be mediated by specific carrier proteins. Once inside the cell, sulfate is either converted to organic compounds via the assimilation pathway or transported as free sulfate into the vacuole through anion channels in the tonoplast. Vacuolar sulfate has been shown to make up 99% of the total sulfur in the duckweed, *Lemna minor.* The vacuole is thus an important site of accumulation of excess sulfate in certain species.

## Assimilation into Carbon Compounds

Like nitrate, the sulfate acquired by plant cells must be reduced before it can be incorporated into organic compounds. This reduction process involves the formation of several intermediate sulfur compounds prior to incorporation of sulfur into amino acids. In higher plants, the enzymes responsible for the assimilatory reduction of sulfate are located in the chloroplasts of leaf cells and in the pro-plastids of root cortical cells. In plants that utilize the C4 pathway of photosynthesis, the bundle sheath chloroplasts are the main sites of sulfate reduction.

## The Activation of Sulfate

Among the intermediate sulfur compounds are two compounds, adenosine 5'-phosphosulfate (APS) and 3'-phosphoadenosine 5'-phosphosulfate (PAPS), that represent the activated forms of sulfur. Only activated forms of sulfate can be reduced and converted into other organic compounds. Sulfate activation involves adenosine 5′-triphosphate (ATP) and sulfate in the following successive reactions:

$$\text{ATP} + \text{sulfate} \rightarrow \text{APS} + \text{PP}_i \qquad \Delta G^{\circ\prime} = 45 \text{ kJ mol}^{-1} \qquad ..(10.10)$$

$$\text{APS} + \text{ATP} \rightarrow \text{PAPS} + \text{ADP} \qquad \Delta G^{\circ\prime} = -25 \text{ kJ mol}^{-1} \qquad ...(10.11)$$

which are catalyzed by the enzymes ATP sulfurylase and APS kinase, respectively. Although the first reaction, as written, is not energetically favourable for product formation, its coupling to the favourable reaction involving PAPS formation drives sulfate activation. An additional driving force for sulfate activation is the subsequent hydrolysis of pyrophosphate ($PP_i$) to 2 moles of inorganic phosphate.

$$PP_i + H_2O \rightarrow 2P_i \qquad \Delta G^{\circ\prime} = -33.5 \text{ kJ mol}^{-1} \qquad ...(10.12)$$

This energetically favourable reaction is catalyzed by an enzyme known as an inorganic pyrophosphatese. From PAPS, sulfate may be directly assimilated into organic compounds such as sulfolipids, which contain sulfur bound as a sulfate ester. However, assimilation of sulfur into amino acids and proteins requires reduction to the level of sulfide.

## Reduction of Sulfate to Sulfide

The next step in the assimilation of sulfur is the actual reduction of the activated form to the level of sulfide. Eight electrons are required, and the oxidation number of sulfur changes from + 6 to – 2. However, the exact mechanism of APS reduction is poorly understood. Because of the energetic constraints discussed in the previous section, APS should not accumulate significantly in plant cells. Therefore, for some time it was assumed that PAPS was the activated form of sulfate that was reduced to sulfide. However, other studies of sulfur reduction have suggested that PAPS is first converted back to APS, which then serves as the substrate for reduction. This reaction involves the hydrolysis of the 3′-phosphate groups of PAPS by a magnesium-dependent 3′-phosphonucleotidase. Therefore, in reductive sulfur metabolism the formation of PAPS may be primarily a means of carrying out the energetically unfavourable reaction of APS formation at the expense of ATP hydrolysis.

In the reduction of APS to sulfide, reduced ferredoxin acts as an electron donor for the reaction:

$$\text{APS} + 8\text{Fd}_{red} + 5\text{H}^+ \rightarrow \text{sulfide} + \text{AMP} + 8\text{Fd}_{ox} + 3\text{H}_2\text{O} \qquad ...(10.13)$$

There is evidence that this overall reaction may involve the production of sulfite, $SO_3$, as a protein-bound intermediate. Although the reduction of APS to sulfide can take place in the roots, the rate is generally several-fold higher in the leaves and is light-stimulated. This stimulation by light could be due to the enhanced production of reduced ferredoxin by the activity of the photosystems. In nonphotosynthetic cells, NAD(P)H is the likely electron donor.

## Incorporation of Sulfide into Cysteine and Methionine

The sulfide produced by the reduction of APS does not accumulate in plant cells but is instead rapidly incorporated into the sulfur-containing amino acids (cysteine and methionine). The formation of cysteine involves the addition of sulfide to the three-carbon acceptor] O-acetylserine in a reaction catalyzed by O-acetylserine sulfhydrylase:

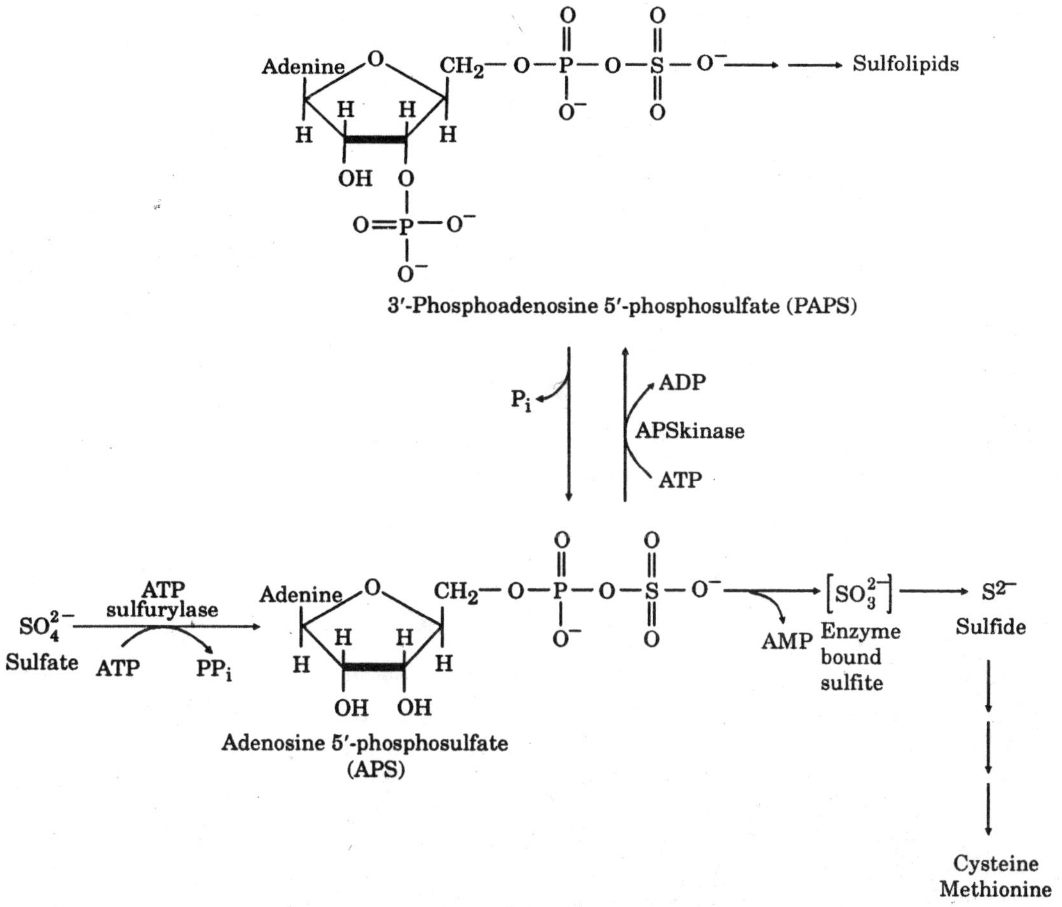

3′-Phosphoadenosine 5′-phosphosulfate (PAPS)

Adenosine 5′-phosphosulfate (APS)

**Fig. 10.12. The structure and synthesis of the two activated sulfur compounds, APS and PAPS. The enzyme ATP sulfurylase cleaves pyrophosphate from ATP and replaces it with sulfate. APS is used in the synthesis of sulfur-containing amino acids. PAPS is formed from APS by the enzyme APS kinase.**

$$O\text{-Acetylserine} + \text{sulfide} \rightarrow \text{cysteine} + \text{acetate} \qquad ...(10.14)$$

The *O*-acetylserine is an activated form of serine produced by a reaction with coenzyme A catalyzed by serine transacetylase:

$$\text{Serine} + \text{acetyl CoA} \rightarrow O\text{-acetylserine} + \text{CoA} \qquad ...(10.15)$$

Methionine can be formed by two possible pathways, transsulfuration or direct sulfhydration, involving the series of reactions. There is evidence that in higher plants transsulfuration is the primary pathway for methionine biosynthesis from sulfide. Following the synthesis of cysteine and methionine, sulfur can be incorporated into proteins and a variety of other compounds such as acetyl coenzyme A and S-adenosylmethionine. The latter compound is important in synthetic reactions involving the transfer of methyl groups, as in lignin synthesis.

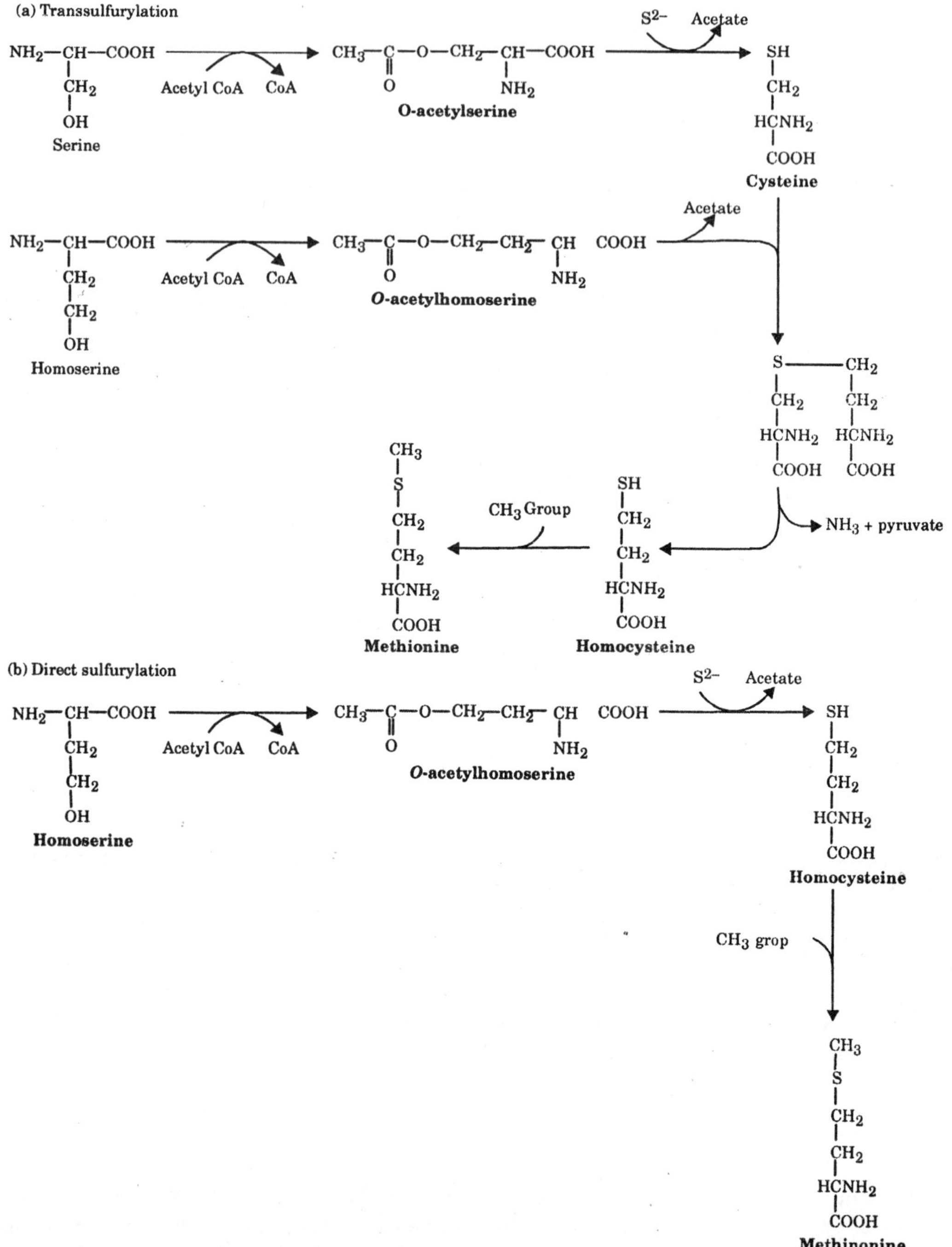


**Fig. 10.13. The two pathways of methionine bisoynthesis : (*a*) transsulfuration; (*b*) direct sulfurylation. The transsulfuration pathway is probably the major route of methionine production.**

## Phosphate Assimilation

Phosphate in the soil solution is readily taken up by plant roots and incorporated into a variety of organic compounds including sugar phosphates, phospholipids, and nucleotides. The main entry point of phosphate into assimilatory pathways occurs during the formation of ATP, the energy "currency" of the cell. The overall reaction for this process involves the addition of inorganic phosphate to the second phosphate group in adenosine diphosphate to form a phosphate ester bond. In mitochondria, the energy for ATP synthesis is derived from the oxidation of NADH produced during the breakdown of sugars in respiration, a process called oxidative phosphorylation.

The formation of ATP is also driven by light energy in the process of *photophosphorylation* that occurs in the chloroplasts. In addition to these reactions in mitochondria and chloroplasts, phosphate is assimilated by reactions in the cytosol. In glycolysis, inorganic phosphate is incorporated into 1, 3-bisphosphoglyceric acid, forming a high-energy acyl phosphate group. This phosphate can be donated to ADP to form ATP in a substrate-level phosphorylation reaction. Once incorporated in ATP, the phosphate group can be transferred in reactions to form the various phosphorylated compounds found in higher plant cells.

## Cation Assimilation

Assimilation of cations taken up by plant cells involves the formation of complexes with organic compounds without any chemical transformation of the nutrient. That is, the cation becomes bound to the carbon compound without the formation of covalent bonds. This occurs for macronutrient cations such as potassium, magnesium, and calcium as well as for micronutrient cations such as copper, iron, manganese, cobalt, sodium, and zinc. The assimilation of iron will be discussed in more detail in the next section.

## Cation Complexes

In the formation of a coordination complex, oxygen or nitrogen atoms of a carbon compound donate unshared electrons to form a bond with the cation nutrient. As a result, the positive charge on the cation is neutralized. Coordination complexes typically occur between polyvalent cations and organic molecules, such as copper and tartaric acid and magnesium and chlorophyll *a*, in which the magnesium is bound by coordination bonds with the nitrogen group atoms in the porphyrin ring.

The nutrients that are assimilated with the formation of coordinate complexes include copper, zinc, iron, and magnesium. Calcium can also form coordinate complexes with cell wall polygalacturonic acid. Electrostatic bonds are formed because of the attraction of a positively charged cation for a negatively charged group on an organic compound. In electrostatic, bonds, unlike coordinate bonds, the cation retains its positive charge. An important negatively charged group is the ionized form of a carboxylic acid. For example, monovalent cations such as potassium can form elctrostatic bonds with the carboxylic groups of many organic acids.

The incorporation of calcium into the pectic components of the cell wall involves coordinate bonds as well as electrostatic bonds with the carboxylic groups of polygalacturonic acid. In general, assimilation of magnesium and calcium that is not accounted for by the formation of coordination complexes takes place through the formation of electrostatic bonds with amino acids, phospholipids, and other negatively charged molecules. It should be noted that much of the potassium that is accumulated by plant cells remains as the free ion in the cytosol and the vacuole and functions in osmotic regulation and enzyme activation.

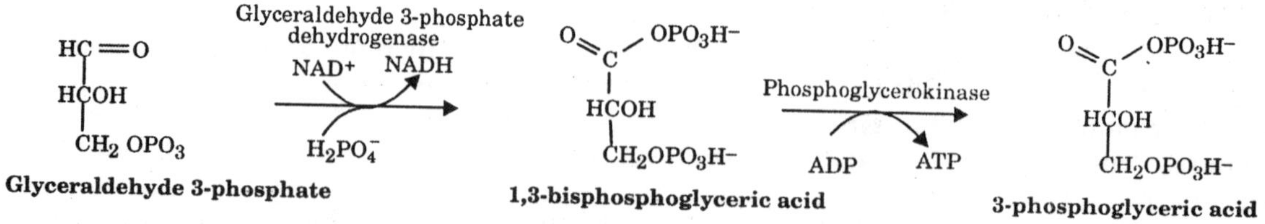

Glyceraldehyde 3-phosphate ($HC{=}O$, $HCOH$, $CH_2\ OPO_3$) $\xrightarrow[H_2PO_4^-]{\text{Glyceraldehyde 3-phosphate dehydrogenase; } NAD^+ \rightarrow NADH}$ 1,3-bisphosphoglyceric acid ($O{=}C{-}OPO_3H^-$, $HCOH$, $CH_2OPO_3H^-$) $\xrightarrow[ADP \rightarrow ATP]{\text{Phosphoglycerokinase}}$ 3-phosphoglyceric acid ($O{=}C{-}OPO_3H^-$, $HCOH$, $CH_2OPO_3H^-$)

**Fig. 10.14. Substrate-level phosphorylation with the production of ATP. This reaction, which occurs in glycolysis, involves the incorporation of inorganic phosphate into 1, 3-bisphosphoglyceric acid (first reaction) and the breakdown of this compound to produce ATP from ADP.**

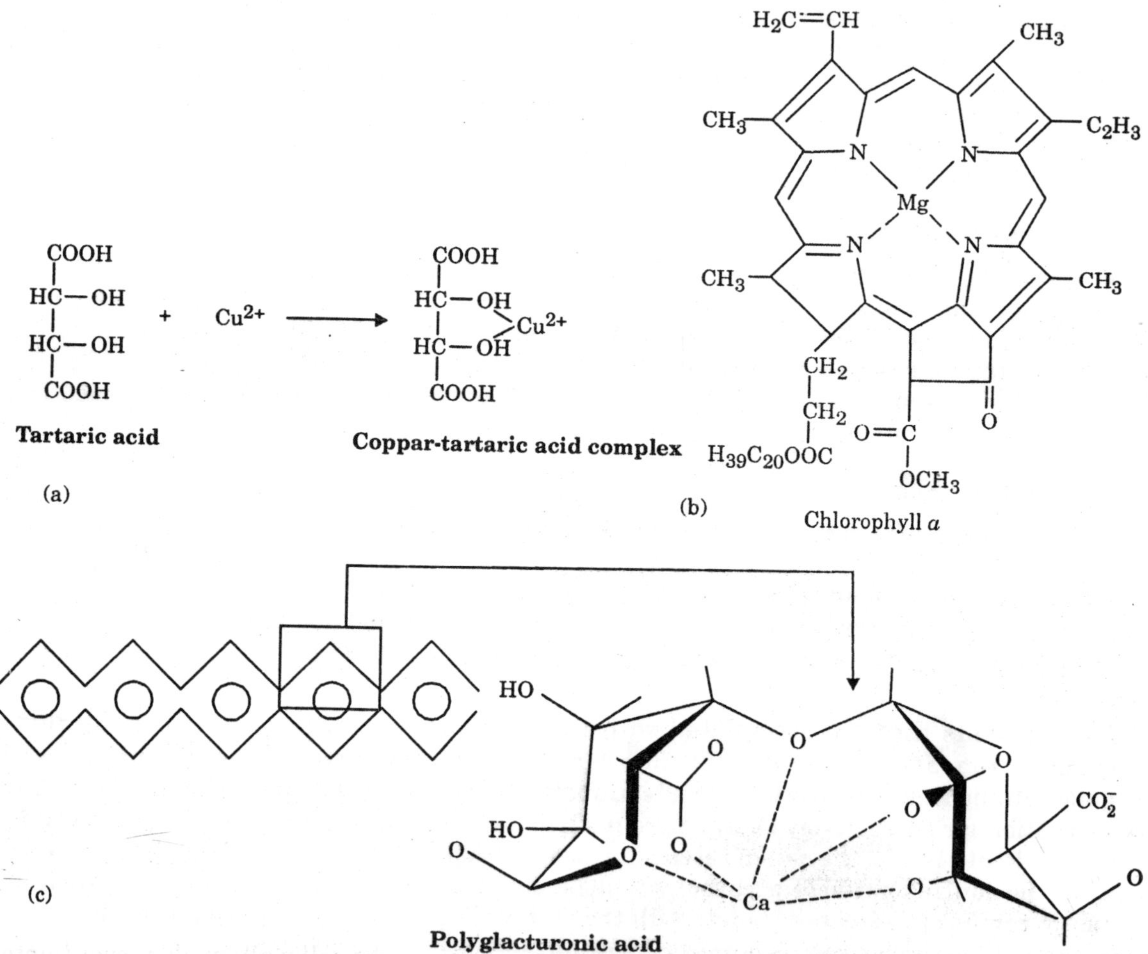


**Fig. 10.15. Examples of coordination complexes, (*a*) Copper ions share electrons with the hydroxyl oxygens of tartaric acid, (*b*) Magnesium ions share electrons with nitrogen atoms in chlorophyll a. (*c*) The "eggbox" model of the interaction of polygalacturonic acid and calcium ions. On the left, calcium ions are held in the spaces between two polygalacturonic acid chains, indicated by the kinked, horizontal lines. The arrow shows a blow-up of a single calcium ion forming a coordination complex with the hydroxyl oxygens of the galacturonic acid residues. Much of the calcium in the cell wall is thought to be bound in this fashion.**

## Iron Assimilation

Iron is important in plant cells as a catalyst in enzyme-mediated redox reactions and in chlorophyll biosynthesis. Iron is acquired by plant roots from the soil, where it is present primarily in the ferric form ($Fe^{3+}$) as oxides such as $Fe(OH)_2^+$, $Fe(OH)_3$, and $Fe(OH)_4^-$. This form of iron is extremely insoluble, as indicated by the $K_{sp}$ ($2 \times 10^{-39}$), which describes the equilibrium between the solid and dissolved species. It is therefore difficult for plant roots to obtain iron from the soil. The process by which plant roots can solubilize and acquire iron has been studied in detail for dicots. In response to iron stress, these plants: (1) acidify the medium surrounding the root, (2) chemically reduce soluble ferric chelates (Fe coordination complexes), (3) take up ferrous iron ($Fe^{2+}$), and (4) accumulate organic acids such as citrate.

Acidification and reduction are key events in the solubilization of iron, since acidification of the soil surrounding the roots greatly increases the solubility of the ferric form of iron and the reduced ferrous iron ($Fe^{2+}$) is much more soluble than the oxidized ferric form ($Fe^{3+}$). There has been a controversy as to whether iron reduction takes place as a result of the release of "reducing substances" such as caffeic acid into the soil from root cells or whether reduction occurs on the membrane surface of root cells by the action of a redox electron transport chain.

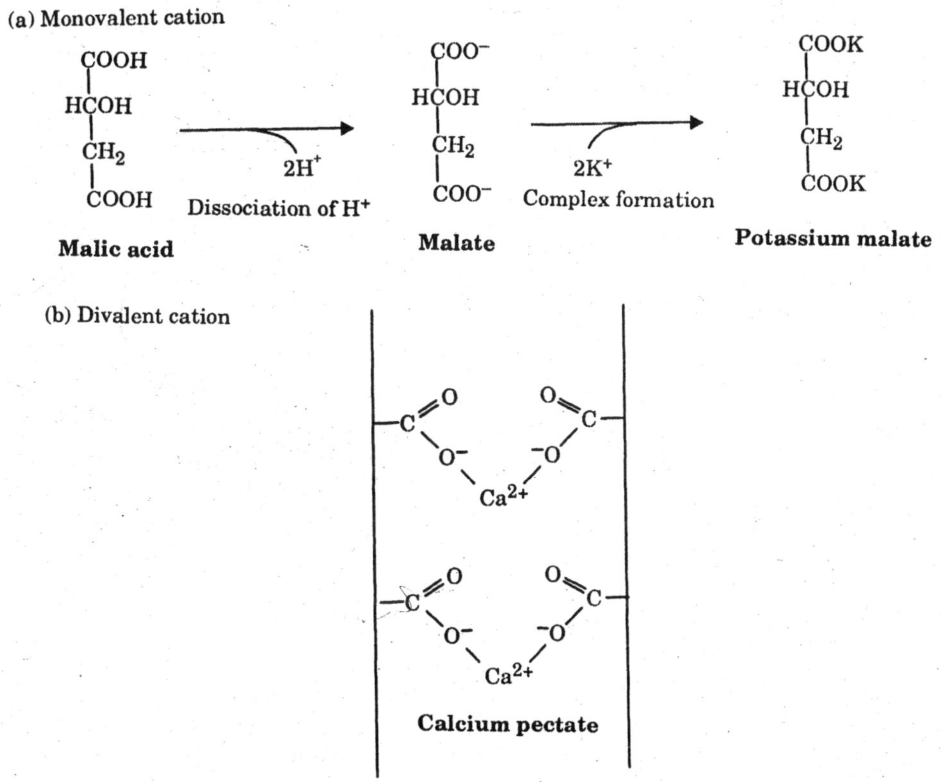

(a) Monovalent cation

COOH–HCOH–$CH_2$–COOH → (Dissociation of $H^+$, $2H^+$) → $COO^-$–HCOH–$CH_2$–$COO^-$ → (Complex formation, $2K^+$) → COOK–HCOH–$CH_2$–COOK

Malic acid · Malate · Potassium malate

(b) Divalent cation

$Ca^{2+}$ · $Ca^{2+}$

Calcium pectate

**Fig. 10.16. Examples of the formation of electrostatic (ionic) complexes, (*a*) Monovalent cation, potassium malate. (*b*) Divalent cation, calcium pectate. Divalent cations can form cross-links between parallel strands containing negatively charged carboxyl groups. Calcium cross-links may play a structural role in the cell walls of some algae and higher plants.**

Under conditions of iron deficiency, the activity of a trans-plasma membrane electron transport system greatly increases in bean roots and certain other species and may represent an adaptation to low $Fe^{3+}$ concentrations in the soil. In any case, reduction appears to be a prerequisite for the uptake of iron into plant cells. Most evidence suggests that the iron is then reoxidized in the cells of the root and transported to the leaves as an electrostatic complex with citrate. When iron-deficient plants receive iron, they may often take up massive amounts by an augmentation of the uptake system.

Large amounts of "free" iron in plant cells can pose problems because iron can interact with oxygen to form superoxide anions ($O_2^-$), which can damage membranes by degrading unsaturated lipid components. Plant cells can, however, provide safe storage for iron in the leaves through the use of an iron storage protein called phytoferritin. Phytoferritin is made up of 24 identical protein subunits that form a hollow sphere with a molecular mass of about 480,000 Da.

Iron incorporation into phytoferritin is thought to occur by oxidation of ferrous iron inside the phytoferritin protein and formation of an iron core. The core consists of about 5400 to 6200 iron atoms present as a ferric oxide-phosphate complex. How iron is released from phytoferritin is uncertain, but it is generally thought to involve a breakdown of the protein complex. It is also thought that the level of free iron in plant cells may regulate the de novo biosynthesis of phytoferritin. When iron acts as a redox component, it is often present as a complex within a porphyrin group in the active site of an enzyme. This is the case, for example, in the cytochromes that mediate electron transport in the mitochondrial inner membrane.

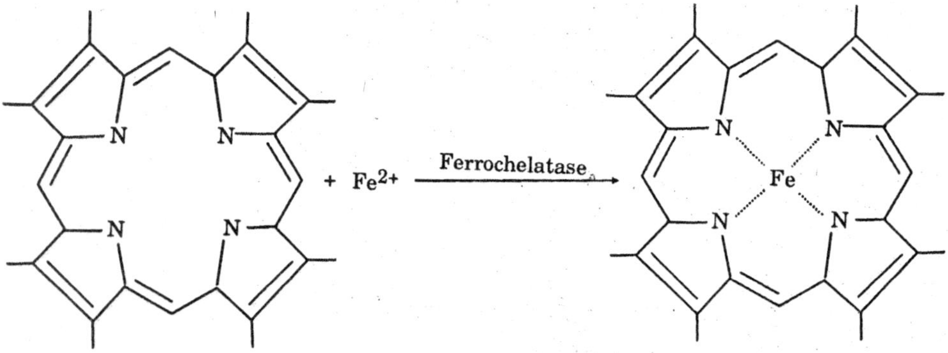

**Fig. 10.17. The ferrochelase reaction. The enzyme ferrochelase catalyzes the insertion of iron into the porphyrin ring to form a coordination complex.**

Therefore, an important assimilatory reaction for iron involves its insertion into the porphyrin system. This reaction is called the ferro-chelase reaction, and it is catalyzed by the enzyme ferrochelase. Ferrochelase is associated with chloroplasts and mitochondria and most likely functions in active metalloporphyrin synthesis for electron transport systems in these organelles. In addition, iron-sulfur proteins of the electron transport chain contain Fe covalently attached to the cysteine sulfur atoms of the apoprotein, as well as Fe atoms bound to other Fe atoms by means of sulfur bridges.

## Oxygen Assimilation

Although respiration accounts for the bulk of the oxygen ($O_2$) utilized by plant cells, a small proportion of oxygen may be directly assimilated into organic compounds in the process of *oxygen fixation*. Oxygen fixation involves direct addition of molecular oxygen to an organic compound in reactions carried out by enzymes known as loxygenases. Other than oxygen fixation, the major pathway for the assimilation of oxygen into organic compounds involves reactions in which the oxygen from water is incorporated.

For many years, it was assumed that only the latter pathway occurs in living organisms. Oxygenases are classified as dioxygenases or mono-oxygenases according to the number of atoms of oxygen that are transferred to an organic compound in the enzyme-catalyzed reaction. Examples of dioxygenases in plant cells are lipoxygenase, which catalyzes the addition of two atoms of oxygen to unsaturated fatty acids, and prolyl hydroxylase, the enzyme that converts proline to the rare amino acid hydroxyproline. Hydroxyproline is an important component of cell wall protein, *extehsin*.

Prolyl hydroxylase is considered a dioxygenase because one oxygen atome is incorporated into proline to form 1 hydroxyproline, while a second oxygen is used to convert α-ketoglutarate to succinate. Ferrous iron and ascorbate are also needed as cofactors but do not participate directly in the redox reaction. The synthesis of hydroxyproline from proline differs from the synthesis of all other amino acids in that the reaction occurs *after* the proline has been incorporated into protein and is therefore a post-translational modification reaction. Prolyl hydroxylase is localized in the endoplasmic reticulum, suggesting that most, if not all, proteins containing hydroxyproline are found in the secretory pathway. Monooxygenases add one of the atoms in molecular oxygen to an organic compound; the other oxygen atom is converted in.to water.

Monooxygenases are sometimes referred to as *mixed-function oxidases* because of their ability to catalyze simultaneously both the oxygenation reaction and the oxidase reaction (reduction of oxygen to water). The monooxygenase reaction also requires a reduced substrate (NADH or NADPH) as an electron donor, according to the following equation:

$$A + O_2 + BH_2 \rightarrow AO + H_2O + B$$

where A represents an organic compound and B represents the electron donor. An important monooxygenase in plant cells is the family of heme proteins collectively called cytochrome P-450, which catalyzes the hydroxylation of cinnamic acid to p-coumaric acid. In monooxygenases, the oxygen is first activated by combining with the iron atom of the heme group. NADPH serves as the electron donor.

The mixed-function oxidase system is localized on the endoplasmic reticulum and is capable of oxidizing a variety of substrates, including mono- and diterpenes and fatty acids. Another reaction in which oxygen is directly incorporated into an organic compound occurs during photo-respiration and involves the oxygenase activity of ribulose-l,5-bisphosphate carboxylase/oxy-genase (Rubisco), the enzyme of $CO_2$ fixation. The first stable product that contains oxygen originating from molecular oxygen is 2-phosphoglycolate.

## Concluding Remarks

Nutrient assimilation is the process by which nutrients acquired by plants are incorporated into the carbon constituents necessary for growth and development. For nitrogen, this is one step in a series of processes that constitute the nitrogen cycle, which encompasses the various states of nitrogen in the biosphere and their interconversions. The principal source of nitrogen available to most plants is nitrate ($NO_3^-$), which must be reduced to ammonia ($NH_3$) prior to incorporation into organic compounds. Under some conditions, ammonia is present in significant amounts in the soil and can be use directly by plants

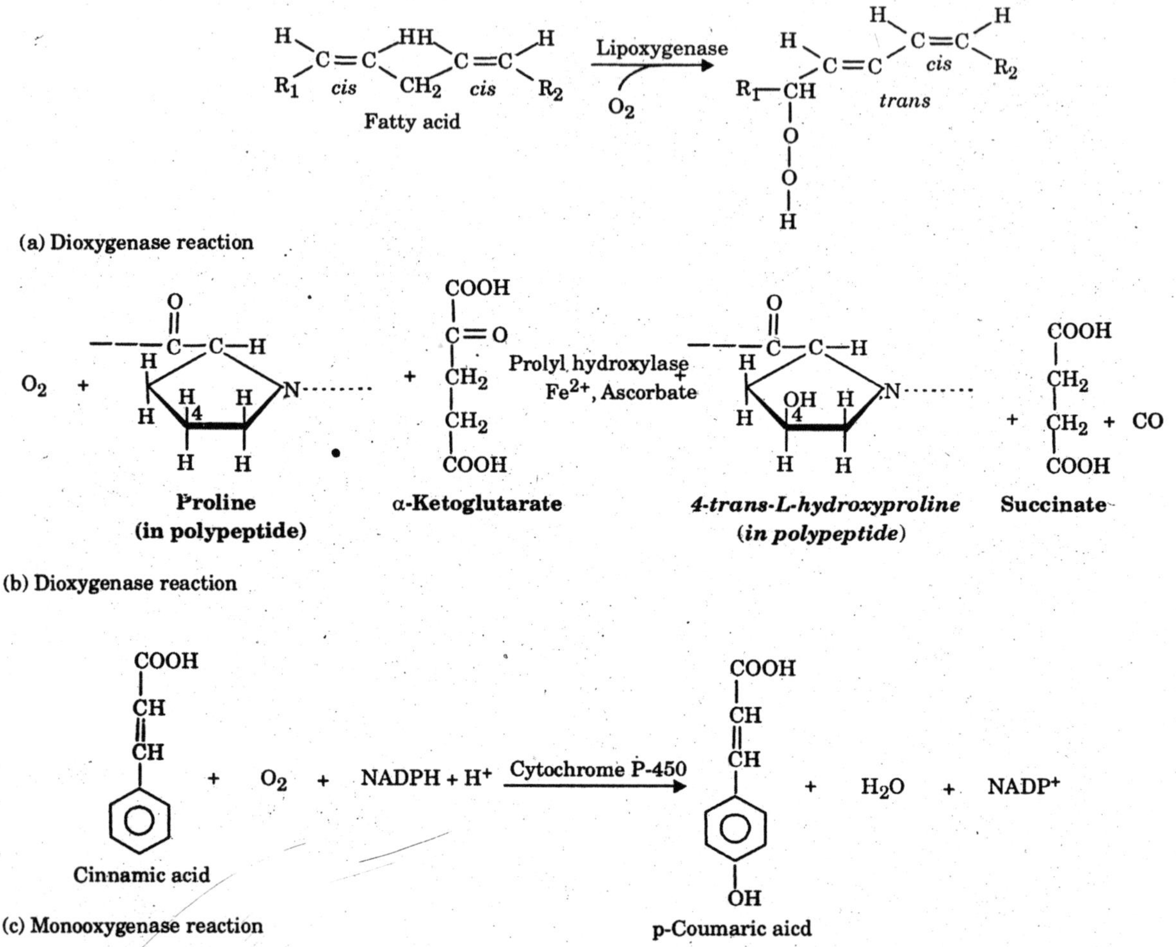

Fatty acid (*cis*, *cis*) + $O_2$ —Lipoxygenase→ hydroperoxide (*cis*, *trans*)

(a) Dioxygenase reaction

$O_2$ + Proline (in polypeptide) + α-Ketoglutarate —Prolyl hydroxylase, $Fe^{2+}$, Ascorbate→ *4-trans-L-hydroxyproline (in polypeptide)* + Succinate + CO

(b) Dioxygenase reaction

Cinnamic acid + $O_2$ + NADPH + $H^+$ —Cytochrome P-450→ p-Coumaric aicd + $H_2O$ + $NADP^+$

(c) Monooxygenase reaction

**Fig. 10.18. Examples of the two types of oxygenase reactions in cells of higher plants, (*a*) The dioxygenase lipoxygenase catalyzes the addition of two atoms of oxygen to the conjugated portion of an unsaturated fatty acid to form a hydroperoxide with a pair of *cis-trans* conjugated double bonds. The hydroxy peroxy fatty acid may then be enzymatically converted to hydroxy fatty acids and other metabolites. Such changes can contribute to membrane deterioration during senescence. (*b*) The dioxygenase prolyl hydroxylase catalyzes the addition of one oxygen to peptide proline and one oxygen to α-ketoglutarate. (*c*) The monooxygenase cytochrome P-450 uses one oxygen to hydroxylate cinnamic acid (and other substrates) and the other oxygen to produce water. NAD(P)H serves as the electron donor for monooxygenase reactions.**

A number of plants, primarily legumes, form asymbiotic relationship with nitrogen-fixing baceteria and utilize nitrogen ($N_2$) obtained from the atmophere. These bacteria contain an enzyme complex nitrogenase, that can reduce atmospheric nitrogen to ammonia. Nitrogen-fixing prokaryotic microorganism also exist that do not form symbiotic relationships with higher plants but benefit plants by enriching the soil nitrogen content. Like nitrate, sulfate ($SO_4^{2-}$) must be reduced in the process of assimilation.

Reduction of sulfate involves the formation of two activated forms of sulfate called adenosine 5'-phosphosulfate (APS) and 3'-phosphoadenosine 5'-phosphosulfate (PAPS). Sulfide ($H_2S$), the end product of sulfate reduction, does not accumulate in plant cells but is instead rapidly incorporated into the amino acids cysteine and methionine. Phosphate ($PO_4^{3-}$) is present in a variety of compounds found in plant cells, including sugar phosphates, ,lipids, nucleic acids, and free nucleotides. 'The initial product of its assimilation is ATP, which is produced by substrate-level phosphorylations in the cytosol, oxidative phosphorylation in the mitochondria, and photophos phorylation in the chloroplasts. Whereas nitrogen, sulfur, and phosphorus assimilation requires the formation of covalent bonds with carbon compounds, many macro and micronutrient cations simply form complexes. These complexes may be of the .electrostatic or coordination type.

Iron assimilation involves not only complex formation but also several steps of reduction and oxidation. To be taken up, $Fe^{3+}$ must be reduced to $Fe^{2+}$, most likely by a plasma membrane electron transport system. Iron is transported within the plant as $Fe^{3+}$ complexed with citrate. In order to store large amounts of iron, plant cells synthesize phytoferritin, an iron storage protein. An important function of iron in plant cells is to act as a redox component in the active site of enzymes, often as an iron-porphyrin complex. The insertion of iron into a porphyrin group occurs in the ferrochelase reaction. In addition to utilization in respiration, molecular en can be assimilated in the process of oxygen xation, the direct addition of oxygen to organic compounds. This process is catalyzed by enzymes known as oxygenases, which are classified as monooxygenases or dioxygenases.

## The Genes Involved in Symbiotic Nitrogen Fixation

The ability of rhizobia to become endosymbiotic nitrogen-fixing organelles in legume root cells supports the endosymbiotic origin of mitochondria and chloroplasts. Indeed, in the bacteroid we may be witnessing an intermediate stage in the evolution of a permanent nitrogen-fixing, self-replicating plant organelle. To carry out its dual existence as a free-living organism and an organelle, *Rhizobium* must to alter its genetic program.

Many of the genes necessary for nodulation and nitrogen fixation are present on a large circular plasmid, the Sym (for symbiosis) plasmid. "Curing" rhizobia of their Sym plasmid by incubation at high temperature renders them unable to infect roots or induce nodulation. Reintroduction of the plasmid restores the original infectious phenotype. The Sym plasmid genes involved in host-specific infection and nodule formation are referred to as *nod* genes. Our understanding of the *nod* genes is limited, and the biochemical functions of *nod* gene products have not yet been elucidated.

However, a number of *nod* gene mutants of *Rhizobium* have been described that are blocked at various stages of infection and nodulation. This has allowed mapping of *nod* genes to two distinct regions on the Sym plasmid corresponding to "common" genes, which are interchangeable with plasmid genes from bacteria covering a wide host range, and "host-specific" genes, which function only in a particular plant host. The genes involved in nitrogen fixation, which become active during the latter stages of nodule development, are called either *nif* genes or *fix* genes and are often clustered.

Mutations in either type result in bacteria that can induce nodulation but cannot carry out nitrogen fixation. Nif genes are present in both free-living and symbiotic nitrogen-fixers. They include the structural genes for nitrogenase as well as a number of regulatory genes. The *nifH* gene encodes the nitrogenase Fe protein subunit, while *nifD* and *nifK* encode the α and β subunits, respectively, of the MoFe protein of nitrogenase. *Fix* genes seem to be present only in symbiotic nitrogen-fixers and may occur either within or outside the *nif* gene cluster. One possible function of the *fix* genes is heme biosynthesis. The precursor for heme is δ-aminolevulinic acid (ALA).

Mutations in the *Rhizobium* gene for ALA synthase cause the production of white nodules that are unable to fix nitrogen. Apparently, heme production by the plant host cell is insufficient to support massive leghemo-globih synthesis. In addition to leghemoglobin, nodules contain a number of proteins of plant origin that are nodule-specific. Such proteins are called *nodulins,* and their "genes are termed Nod genes. It has been estimated that legume nodules contain 20-30 nodulins, many of which have been identified. Nodulins are expressed at specific stages of nodule development and include leghemoglobin and the enzymes involving in ureide metabolism. An important aspect of symbiotic nitrogen fixation is the interactions between the plant Nod genes and the bacterial *nod, nif,* and *fix* genes.

Just how the complex processes of infection and nodule development are coordinated remains a fascinating topic for future research. Such studies are not only of theoretical interest but also of practical importance, for they may one day enable plant scientists to broaden the range of *Rhizobium* to include nonleguminous crops such as cereals. This might greatly increase the productivity of nitrogen-poor soils as well as reduce the dependence of agriculture on energy-expensive nitrogen fertilizers.

# Index

❑❑❑